Developmental Biology

Springer

New York
Berlin
Heidelberg
Barcelona
Hong Kong
London
Milan
Paris
Singapore
Tokyo

Werner A. Müller

Developmental Biology

With 123 Figures

Springer

Werner A. Müller
Zoologisches Institut—Physiologie
University of Heidelberg
D–69120 Heidelberg
Germany

Cover: Foreground: a 13.5-day-old mouse embryo showing the expression of ptc mRNA by *in situ* hybridization. Courtesy of Ljiljana Milenkovic and Matthew P. Scott. Background: double immunofluorescence of wild type *Drosophila* eye imaginal disk cells stained for dTFIIA-S (red) and Elav (green). Image provided by Martin P. Zeidler and Mark Mlodzik.

This edition is published by arrangement with Gustav Fischer Verlag, Stuttgart. Original copyright © Gustav Fischer Verlag. Title of German original: Werner A Müller, Entwicklungsbiologie Einfuhrung in die klassische und molekulare Entwicklungsbiologie von Mensch und Tier

Library of Congress Cataloging-in-Publication Data
Müller, Werner A., 1937–
    Developmental biology / Werner A. Müller.
       p.   cm.
    Includes bibliographical references and index.
    ISBN 0-387-94718-3 (hard : alk. paper)
    1. Developmental biology.   I. Title.
QH491.M85   1996
574.3—dc20   96-11739

Printed on acid-free paper.

Production supervised by Natalie Johnson and managed by Impressions Book and Journal Services, Inc. Manufacturing supervised by Jeffrey Taub.
Typeset by Impressions Book and Journal Services, Inc., Madison, WI
Printed and bound by Walsworth Publishing Co., Marceline, MO
Printed in the United States of America.

9 8 7 6 5 4 3 2

ISBN 0-387-94718-3 Springer-Verlag New York Berlin Heidelberg   SPIN 10700741

Preface

⟞⟝

This book is addressed to students of biology and medicine whose interests encompass human development as well as the most important model organisms currently being investigated with methods from genetics, biochemistry, and molecular biology. *Developmental Biology* has been written for advanced undergraduates as well as graduate students and instructors who seek a succinct but thorough introduction to contemporary developmental biology.

A particular aim of this book is to bring together the rich heritage and pivotal ideas of the first "golden age" of developmental biology (1890–1940) with discoveries and hypotheses of recent research. Modern research in developmental biology is based mainly on biochemical and molecular methods, but increasingly incorporates computer modeling to cope with such intriguing and complex phenomena as biological pattern formation. The awarding of the 1995 Nobel prize in physiology or medicine for research on embryo development in *Drosophila* highlights a new "golden era" for developmental biology. This book takes full advantage of the molecular revolution that has swept through the biological sciences, but it also presents, in addition to data and observations, basic principles of development and points out the historical roots of our interpretations.

Many years of teaching experience have prompted me to follow an approach different from that used in other textbooks. After an introduction to terminology used in descriptive developmental biology, I introduce several "model" organisms of importance in research and teaching. I then introduce selected developmental processes and present them from a comparative perspective. Using this approach, students do not have to piece together the development of, for example, the frog (*Xenopus*) or the fruit fly (*Drosophila*), from fragmentary information scattered throughout many chapters (such as "Cleavage," "Gastrulation," and so forth). Nevertheless, the student will also find overviews of major themes, such as fertilization (Chapter 6), developmental genetics (Chapter 10), or sexual development (Chapter 20). It is my experience that any redundancy arising from this approach will serve to reinforce the concepts. An additional feature of the book is the presentation of seven boxed essays, each providing a brief overview of a single topic, such as the history of developmental biology (Box 1), signal transduction (Box 3), or current methods of research in developmental biology (Box 7).

Developmental biology is a discipline undergoing extremely rapid development: an explosion of papers appears daily in an ever-growing number of journals and monographs devoted to this field. It is my hope that reading this book will prepare a student to understand and appreciate information presented in more extensive (but perhaps less clearly structured) textbooks, and to follow the present state of knowledge by reading original research articles and current reviews.

This book is a revision and translation of a volume published in Germany in 1995. To illustrate the book, I drew most of the figures with my own hands; a minority were prepared on a computer using a drawing program. Approximately one-third of the figures were redrawn by Peter Adam, reproducing and modifying my drafts or the figures cited in the legends.

Following recent rules, the name of a gene is written in *italics*. To designate proteins derived from defined genes, this book uses capital letters. This enables the reader to easily distinguish, for example, the WINGLESS and EYELESS proteins from wingless and eyeless flies.

The following people have assisted in preparing this book for publication: the staff of Springer-Verlag, particularly the Editor, Robert C. Garber; Senior Production Editor, Natalie Johnson; and, in addition, the staff at Impressions Book and Journal Services, Inc. I also express my gratitude to all those who read (parts of) the original German edition and gave hints for improving the book. In particular, I would like to name Kurt Baumann, Günter Fachbach, Horst Grunz, Klaus Sander, and Einhard Schierenberg.

Heidelberg, June 1996

Contents

✳22.   **Life and Death: What Is the Major Mystery?**   323

       Bibliography   335

       Index   373

       Boxes

       Box 1   History: From the Soul to Information   3
       Box 2   Famous Experiments with Eggs and Embryos:
               Cloning, Chimeras, Teratomas, and Transgenic Mice   106
       Box 3   The PI Signal Transduction System   153
       Box 4   Models of Biological Pattern Formation   184
       Box 5   Signal Molecules Acting through Nuclear Receptors   200
       Box 6   How Cells Communicate and Interact   271
       Box 7   Contemporary Techniques in Developmental Biology   330

1

Development: Organisms Construct and Organize Themselves on the Basis of Inherited Information

—⟸⬥⟹—

Self-construction and **self-organization** are terms that convey the essential principles of development. The development of a multicellular organism starts, as a rule, with a single, seemingly unstructured cell—the fertilized egg cell. Similarly, when a creature such as the freshwater polyp *Hydra* clones itself by a process of asexual reproduction through buds, it is an association of only a few, poorly differentiated cells that constitutes the starting material for the development of the new organism. Whether egg or bud, the fully developed creature will have an internal complexity that surpasses our imagination. We are all familiar with the egg from our breakfasts—it is the amorphous yellow ball that constitutes the egg cell proper. But who is able to visualize the 10^{12} to 10^{13} cells of our brain and their 10^{14} to 10^{15} synaptic connections in their three-dimensional architecture?

All of the many and diverse **somatic** cells that originate from **generative** starting cells do not, however, attain autonomy as do the cells of a clone of single-celled protists. Somatic cells are viable only in the community of the cell association. For the sake of the supraindividual society—the organism— the cells assume different responsibilities and tasks. They **differentiate** morphologically and functionally, and together construct multicellular structures, tissues, and organs. In this process, cells form the same structures and patterns in the same temporal and spatial order, from generation to generation.

Thousands of years ago, this increase in diversity during the **ontogeny** of an organism and the autonomous shaping of form (**morphogenesis**) from simple, seemingly amorphous starting matter stimulated curiosity about the causal, organizing principles. Aristotle saw the ultimate form-determining principle in the soul (Box 1); nowadays the leading role is assigned to genetic information.

The ability to reproduce themselves is conferred upon organisms through genetic information (DNA in the nucleus and mitochondria of animal cells) and through further information-carrying structures in the egg cell (maternal information, cytoplasmic determinants); however, the finished organization is not encoded directly by this information.

It is commonly stated that the genome incorporates a *Bauplan*, an architectural plan or blueprint of the body. Actually, this is not the case; the genome is not a sketch or design of the finished body. The informational capacity of DNA is simply too low to store blueprints of the very complex final pattern of an organism. For example, a detailed design of the one hundred trillion to one quadrillion synaptic contacts in our brain alone would greatly exceed the capacity of the genomic memory.

What then does the genome really encode?

- It contains knowledge of how to make distinct proteins, rRNA and tRNA, and how to make replicas of the DNA itself.
- It apparently embodies some hierarchical organization: master genes (Chapter 10) dominate (via their products) whole sets of subordinate genes.
- The genome contains elements of a spatiotemporal program to control the order of gene expression (Chapter 10, Section 10.4).

We do not yet understand in detail how a developing organism is created on the basis of such minimal information, or how many organisms are able to regenerate lost structures.

Two partial explanations for this gain of complexity will be elaborated in the following chapters:

1. **Combinatorics at the level of genes.** In the various cell types, organs, and body regions, different combinations of genes become effective in time and space. Whether an organism's DNA contains 3,000 genes (the roundworm, *Caenorhabditis elegans*) or as many as 100,000 genes (human), the number of combinations of gene activities is practically unlimited.

2. **Cell sociology.** Cells interact; they are parts of a society in which the members influence each other. The cells distribute tasks and social roles. The pattern of their behavior is defined not only by internal determinants such as genes, but also by the flow of energy and information arriving from neighboring cells and, in later phases of the development, from the external environment of the living being.

Box 1

HISTORY: FROM THE SOUL TO INFORMATION

THE DAWN OF DEVELOPMENTAL BIOLOGY IN ANCIENT GREECE

Although embryos were described in ancient Sanskrit and Egyptian documents, the Macedonian **Aristotle** (384–322 B.C.) was the first to perform developmental studies in a systematic way, to interpret his observations in written words, and to coin lasting terms. An acknowledged philosopher and academic teacher (e.g., of the prince who was to become King Alexander the Great) as well as an enthusiastic naturalist, he wrote the first textbooks of zoology, and treatises on reproduction and development. Aristotle distinguished the following four possibilities as to how organisms might arise: (1) spontaneous generation from rotting substrate, where flies and worms were thought to originate; (2) budding; (3) hermaphroditism; and (4) bisexual reproduction. In his view, the egg was the instrument of reproduction in **oviparous** species. Mammals, human beings, and some other **viviparous** species lacked eggs. Females contributed to offspring by supplying unstructured material and males contributed by supplying semen, the purveyor or causative principle of form.

Aristotle described the development of the chick in the egg. Eggs were incubated for varying periods of time and then were opened. According to his observations, there is an unstructured material initially, which, in the course of **morphogenesis,** acquires a form. In the midst of the emerging figure, he observed a "jumping point," the beating heart.

He considered the aim of development to be the *ergon,* the finished work as it is the aim of the artisan. He thought of the modeling principle as *energeia* (energy), also called *entelecheia,* the principle bearing its aim, goal, and end in itself. Energy is both the efficient and final cause. To reach a particular species-specific end, the forming principle must have a "preexisting idea" of the final outcome. Hence the ultimate cause, the ultimate energy would be the soul (*psyche*).

To quote Aristotle (*De anima*, 416 b1–b27), "It (the soul) causes the production ... of another individual like it. Its essential nature already exists; ... it only maintains its existence.... The primary soul is that which is capable of reproducing the species."

Following Plato, Aristotle discriminated between the vegetative soul, which brings about life; the animal sensitive soul, which enables sensations; and the spiritual soul, which enables thinking.

The vegetative soul endows plants with the ability to regenerate. The vegetative soul also includes the formative power of animal development. In animals the mother (*mater*) supplies the matter (Latin: *materia*); in mammals the matter is supplied in the form of the menses. The semen was thought to coagulate the female material, and to trigger and govern its development. Aristotle's exposition on the residence and inheritance of the vegetative soul is not unequivocal: is it in the female matter or in the semen? In contrast, the second degree of soul, the animal soul, is inherent only in the semen and is transferred from the father to the future child in begetting. The animal soul then governs sensitivity and movement. The spiritual soul is eternal, immortal, painless, and sheer energy, and enters human beings from outside "through a door."

Aristotle's imprint on the Western educated world has endured for centuries. With all due respect for his great stature, what he said about begetting, fertilization, and determination of the female sex may be passed over courteously in silence.

THE RENAISSANCE OF DEVELOPMENTAL BIOLOGY

Embryology was revived in the 16th century. In the school of Padua (Vesalius, Fallopio, Fabricius de Aquapendente) the anatomy of ovaries and testes was studied. The idea of an egg arising in the ovary and the embryo arising in the egg was conceived by **Volcher Coiter** (1514–1576) upon completing his detailed study of chick embryo development, a study for which he finally has been recognized as the father of embryology.

The English anatomist and physician **William Harvey** (1578–1657), best known as the discoverer of the (greater) blood circulation in the vertebrate body, resumed the embryological studies of Aristotle, extending his research to insects and mammals (sheep and deer). Although Harvey was an admirer of "The Philosopher" (i.e., Aristotle), he maintained that spontaneous generation was restricted to lower organisms. In insects, however, development implies *metamorphosin* or **metamorphosis**—the transformation of already existing forms into other forms. Harvey considered the pupa as an egg, as Aristotle did before and several other investigators did after him.

In higher animals, however, Harvey regarded development to be not merely transformation but *epigenesin* or **epigenesis**—creative synthesis, incremental formation of a new entity out of nonstructured matter. Harvey wrote, "We, however, maintain ... that all animals whatsoever, even the viviparous, and man himself not excepted, are produced from ova; that the first conception, from which the foetus proceeds in all, is an ovum of one description or another, as well as the seeds of all kinds of plants."

Later literature shortened Harvey's phrase to *"omne vivum ex ovo"* ("all life from an egg"), probably inspired from the frontispiece of Harvey's embryological treatise, *Exercitationes de Generatione Animalium,* where an egg bears the inscription "*ex ovo omnia"* ("everything out of the egg"). However, Harvey's mammalian egg was not the same entity that we have in mind today, rather it was the blastocyst (young embryo) within the "shell" of the uterus. It was Carl Ernst von Baer (Chapter 4) who discovered the real mammalian egg.

PREFORMATION AND MECHANICISM

When the Swiss scholar **Konrad Gessner** (1516–1565), following the Roman naturalist Pliny, reported that the female bear gives birth to a lump of meat, thereafter licking it into shape, he probably was not yet influenced by the philosophy of mechanicism. In contrast, his later compatriot **Albrecht von Haller** (1708–1777) categorically maintained *"nulla est epigenesin"* ("there is no epigenesis"). With this notion, he followed the founders of microscopic anatomy. In 1683, **Antoni van Leeuwenhoek** (1632–1723) wrote "that the human fetus, though not bigger than a pea, yet is furnished with all its parts." Leeuwenhoek discovered **animalcules** or **zoa** within semen. (Later, von Baer renamed the zoa as the **spermatozoa**). Leeuwenhoek saw or conjectured that he would see **homunculi**—minute preformed human beings within the animalcules. Embryos were thought to result from the enlargement of homunculi.

Likewise, in the "eggs" (pupae) of insects, adult ants and butterflies were seen to be prefigured in miniature, as are leaves and blossoms in the buds of plants. The pre-existing beings only needed to be "evolved": unrolled and unwrapped. Such views advanced to meet the views of the mechanicists, who held that life merely obeys the laws of mechanics. Living beings were regarded as ingenious clockworks comparable to the marvelous astronomic clocks that contemporary artisans built. Whether these machines were viewed as soulless or animated entities depended on the religious and ideological position of the respective author.

The doctrine of preformation quickly led to some awkward problems such as the following:

- If ontogenetic development is only mechanical unwrapping of prefigured forms, must not all generations have been in existence from the very beginning of the world? *Emboitment* (encapsulation) was the answer: one generation lies within another generation like one Russian doll lies within another. According to computations of Vallisneri (1661–1730), the ovary of the primordial mother, Eve, contained 200 million humans packed into one another. This stock should suffice until the end of all days. The French-Genevan **Charles Bonnet** (1720–1793), who accurately described **parthenogenesis** in aphids, wrote the following in 1764: "Nature works as small as it wishes."

- The microscope showed cells and with the cells it showed a lower limit to the size of the preformed organisms. Microscopists showed not only egg cells but also spermatozoa. Now the prefigured *homunculus* was claimed to be prefigured and visible in the egg (**ovists**) or in the sperm (**animalculists, homunculists**). Among the ovists were the reknown anatomists **Marcello Malpighi** (1628–1694) and **Jan Swamerdam** (1637–1680).

- How can regeneration of body parts be explained if lost parts can only be made from preformed parts?

Lazzaro Spallanzani (1729–1799) was the first to perform **artificial insemination.** He reported that frog eggs degenerated in the absence of sperm. Working with dogs, Spallanzani finally laid preformation arguments to rest by proving that both the egg and the male semen are necessary to produce a new individual (although he erroneously believed that the animalcules swimming in semen were mere parasites).

EPIGENESIS AND VITALISM

Caspar Friedrich Wolff (1738–1794), who resumed the study of the chick embryo, again saw new formation—morphogenesis out of structureless yolk material. Wolff planted the seed, so to speak, of the germ-layer theory by describing "Keimblätter" (germ leaves) that later transform into adult structures. Like Aristotle before him, and all further vitalists after him, Wolff concluded that there are "immaterial" (noncorpuscular) virtues, a "vis essentialis" or "vis vitalis"—a force specific for life. The academic colleague of Immanuel Kant, Friedrich Blumenbach (1742–1840), postulated a particular physically acting Bildungstrieb (propensity, formative compulsion) that is inherited via the germ cells. Many important biologists were vitalists, among them Carl Ernst von Baer (1792–1876), who discovered eggs in several mammalian species and performed extensive comparative studies. Von Baer concluded that all vertebrates develop in a fundamentally similar way from germ layers. He established a rule, now known as von Baer's law, which states that all vertebrates go through a very similar embryonic stage and only thereafter do the developmental pathways diverge. Based on this rule, Ernst Haeckel (1834–1919) formulated his much-disputed "ontogenetic" or "biogenetic law" (Chapter 4). The theorem maintains that ontogeny is an abbreviated recapitulation of phylogeny.

Interest in human embryology was stimulated in 1880 by Wilhelm His with the publication of The Anatomy of Human Embryos.

Experimental embryology began in France in the tradition of morphology. Etienne Geoffry Saint-Hilaire (1772–1844), a zoologist and an opponent of the very influential Georges Cuvier, sought to elucidate the causes of developmental anomalies (terata) and disturbed, with crude methods, the development of the chick. In 1886 his compatriot Laurent Chabry began studying teratogenesis in the more readily accessible tunicate egg. Henceforth, invertebrates became preferred sources of eggs for studying very early animal development.

IMPETUS AND PROGRESS AT THE TURN OF THE CENTURY

From 1860 on numerous important discoveries were made, and the era of experimental embryology, cell biology, and genetics began. The first pioneer in developmental genetics, however, was an almost blind theoretician. With a presentiment of the role of genes, August Weismann (1834–1914), in his "Keimplasma Theorie" (germ plasm theory) (1892), ascribed his hypothetical, self-reproducing determinants to the chromosomes that had just been detected by investigators such as Eduard Strasburger and Walter Flemming. However, Weismann thought that the determinants become differentially distributed among the cells of the embryo, thus causing and directing cell differentiation.

Much of the credit for drawing attention to the nucleus as the seat of heredity goes to the brothers, Oscar Hertwig (1849–1922) and Richard Hertwig (1850–1937), who often worked together at the marine station in Roscoff (France). The Hertwigs supplemented the observations on fertilization of Otto Bütschli and recognized that the essential event is the fusion of the male and female gamete nuclei. Oscar Hertwig, working with sea urchin eggs, also identified the polar bodies and saw the nucleus within these small sister cells of the ovum. The sea urchin embryo came to be the most important subject of research in embryology in that period (O. and R. Hertwig, T. Boveri, H. Driesch, and T.H. Morgan).

With his careful observations and sagaciously interpreted experiments on eggs of the worm *Ascaris,* **Theodor Boveri** (1862–1915) advanced the **chromosomal theory.** He proposed that chromosomes were complex structures differing from each other even within the same nucleus and capable of producing qualitatively different effects in cells. For the first time Boveri experimentally demonstrated the significance of the chromosomes for development; he also realized that the cytoplasm and the nucleus interact, which was later verified by T.H. Morgan, H. Driesch, and Boveri for the sea urchin embryo. Boveri was the first to formulate the gradient hypothesis (Chapter 3, Section 3.1; Box 4).

The importance of cytoplasmic determinants was also verified by careful observations and subtle surgical experiments done by **Edmund Beecher Wilson** (1856–1939) and his students on the eggs and embryos of marine invertebrates, in particular, those with spiral cleavage such as the mollusk *Dentalium* (Section 3.5). **E.G. Conklin** performed similar studies on embryos of tunicates (Section 3.7). Wilson also discovered the sex chromosome in insects and wrote a very influential textbook (*The Cell in Development and Inheritance,* 1896).

Most of the pioneers of modern developmental biology, including Boveri, Wilson, Driesch, and Morgan, met at the Stazione Zoologica di Napoli, which Anton Dohrn founded in 1874. Besides studying eggs of sea urchins and other marine invertebrates there, they performed regeneration experiments on hydrozoans (*Tubularia*), following the example of the Swiss scholar **Abraham Trembley** (1710–1784), whose elegant and well-documented regeneration and grafting studies of *Hydra* were carried out as early as 1740. These regeneration studies marked the dawn of modern experimental biology. A compendium, still worth reading, of the regeneration studies of those days was written by T.H. Morgan (*Regeneration,* 1901).

FROM THE SOUL TO TODAY'S DEVELOPMENTAL INFORMATION

Of particular relevance are experiments on the egg of the frog by **Wilhelm Roux** (1850–1924) and on the sea urchin by **Hans Driesch** (1876–1941), and the establishment of *Drosophila* as the leading model organism of genetics by **Thomas Hunt Morgan** (1866–1945; the first biologist to win a Nobel prize, in 1933).

The classic experiment that Driesch performed, consisted of separating the first two daughter cells (blastomeres) that arise by division from the fertilized egg. The separated cells gave rise to whole sea urchin larvae. This experiment proved that living beings are not machines, as envisioned by the mechanicists, because no divided machine can restore itself.

Driesch revived the Aristotelian term "entelechy," but he did not assign a physical virtue to it (although entelechy was said to be able to "suspend" physical forces), instead attributing to it entities such as "knowledge" and "message." Thus, he anticipated the term **"information,"** which was introduced as a scientific concept by Norbert Wiener only in 1942. However, Driesch's entelechy was transcendental and was not dependent on a physical carrier, in contrast to the material, molecular genetic information of today.

Driesch also anticipated the idea of **positional information** by stating that "the prospective significance (fate) of a cell is a function of its position in the whole" and "each single elementary process or development not only has its specification but also its specific and typical place in the whole—its locality."

Fate-determining events and inductive interactions between the parts of an embryo were subsequently investigated with delicate and novel surgical methods, using amphibian embryos, by a student of Boveri, **Hans Spemann** (1869–1941, Nobel prize in 1935) and Spemann's student, Hilde Mangold. Their experiments culminated in the discovery of the **organizer** (Chapter 3, Section 3.8, and Chapter 9).

Classical, organismically oriented developmental biology was predominant until 1970 and had many important practitioners such as the following:

- **Gradient theory,** in regeneration: T.H. Morgan, C.M. Child; in sea urchin embryos: T. Boveri, Sven Hörstadius

- **Embryonic induction:** C.H. Waddington (chick), S. Toivonen, L. Saxen and P.D. Nieuwkoop (amphibians)

- **Cell interactions and cell cultures:** J. Holtfreter, V. Hamburger, Paul Weiss

- **Transdetermination and transdifferentiation:** Ernst Hadorn, Tuneo Yamada

- **Transplantation of cells and nuclei:** Beatrice Mintz (mouse, teratocarcinoma cells; Box 2), Robert Briggs, Thomas King, Jon B. Gurdon (frogs, Box 2)

- **Biochemical and molecular developmental biology:** Pioneers included W. Beermann and A. Ashburner (giant chromosomes in dipterans), Alfred Kühn (chains of enzyme and gene activities in insects), Jean Brachet (RNA in the amphibian egg), Heinz and Hildegard Tiedemann (inducing factors)

The role of **DNA** as the carrier of genetic information, mechanistically interpreted for the first time in 1953 by James D. Watson and Francis Crick; the interaction of cytoplasmic determinants with DNA; and the exchange of signals between cells are points of major effort in recent research.

- *Drosophila* has become the model of reference in genetic and molecular developmental biology, thanks to the pioneering works of the geneticists E.B. Lewis, Eric Wieschaus, Christiane Nüsslein-Volhard, David S. Hogness, Walter Gehring, Gary Struhl, and others. Sydney Brenner succeeded in establishing the nematode *Caenorhabditis elegans* as another model system.

- Many investigators are now resuming classical studies, extending them by molecular methods; among them are Eric Davidson (sea urchin), Marc Kirschner (*Xenopus,* cell cycle), John B. Gurdon (embryonic induction), and Lewis Wolpert (positional information, pattern formation in the avian limb bud).

- A pioneer of a different kind in developmental biology was the British mathematician Alan Turing. Based on mathematically formulated theoretical concepts, models for computer simulation of biological pattern formation are being developed (Box 4).

Many more scientists with comparable achievements deserve mention if space allowed.

Recent research in developmental biology can be followed in the current literature, notably by reading reviews that journals offer, such as those listed in the Bibliography.

2

Basic Stages, Principles, and Terms of Developmental Biology

2.1 Most Animals Pass through an Embryonic Phase, a Larval Stage, Metamorphosis, and an Adult Phase

Development, except in the case of vegetative reproduction (natural cloning), begins with **fertilization.** Fertilization is the fusion of two generative cells, or **gametes,** the sperm and the egg cell (ovum), to form a **zygote,** the fertilized egg. Sperm and egg cells are haploid; each contributes a single set of chromosomes, termed "paternal" in the sperm and "maternal" in the egg cell. Gamete fusion provides the embryo with two complete genomes (the diploid set of chromosomes), a source of information sufficient to construct the new organism.

Embryogenesis follows fertilization: within a protective envelope, and in viviparous organisms such as humans within the maternal body, numerous cells are produced from the zygote by continuous mitotic divisions. The

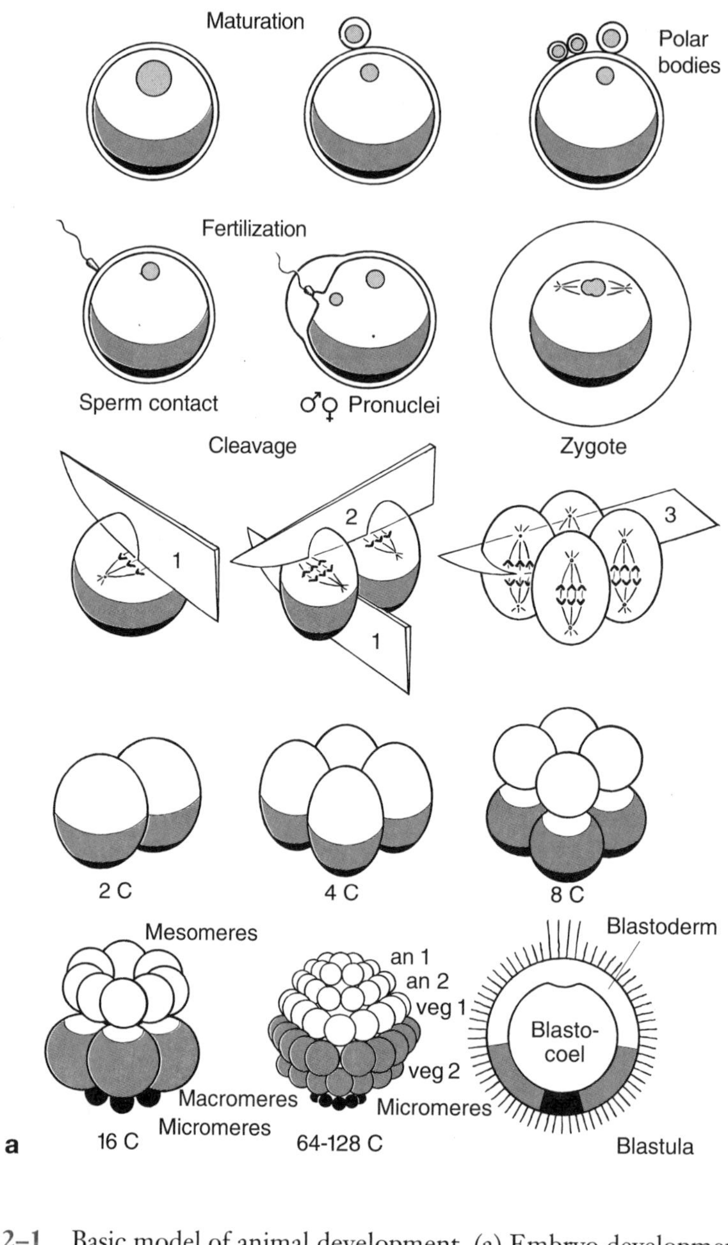

Figure 2–1 Basic model of animal development. (a) Embryo development in the sea urchin I: from fertilization through the blastula stage. The first two cleavages pass through the animal-vegetal axis and are called meridional because they divide the egg along meridians. The third cleavage is equatorial and perpendicular to the animal-vegetal axis. Further cleavages feature oblique spindle orientation, become asynchronous and unequal, and give rise to several tiers of cells. Finally, the cells adhere more strongly to each other, form an epithelial layer (the blastoderm), and give rise to a central cavity (the blastocoel). (b) Embryo development in the sea

GASTRULATION

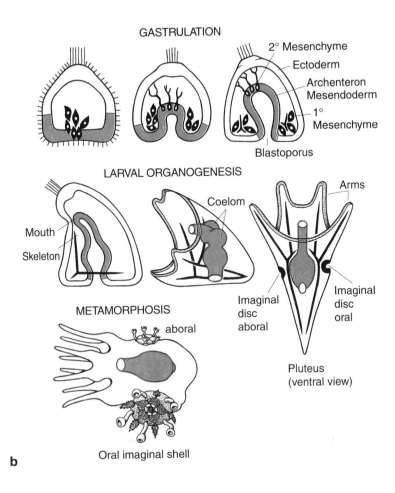

LARVAL ORGANOGENESIS

METAMORPHOSIS

b

Oral imaginal shell

urchin II: from gastrulation to the onset of metamorphosis. Gastrulation takes place in several phases and provides the interior cavity with cells from which the interior organs arise. The resulting larva is called the pluteus. From the metamorphosis of the bilaterally symmetrical larva to the pentameric adult sea urchin, only an initial stage is shown.

resulting cells stay together, collectively modeling the basic architecture of the new organism, and provide it with all the essential organs necessary to begin an autonomous life when the young animal hatches out of the envelope or the mother gives birth. In the majority of animals, embryogenesis results in a first phenotype, the **larva.** An example is the sea urchin, which can be taken as a prototype of animal development (Fig. 2–1). The larva itself undergoes development, which usually brings about only minor, unobtrusive changes, followed by a dramatic **metamorphosis** to a new phenotype, which is termed

imago or **adult,** and which typically settles in an ecological niche different from that in which the larva lived. Larva and adult exploit different resources in their environments.

After a **juvenile** phase, life culminates when the adult reaches **sexual maturity. Senescence** finishes the phase of sexual maturity, and **death** ends the life of the **somatic** cells of the individual. Before death, its **generative** cells should have passed life on to a new generation.

2.2   The Egg Cell Is Internally Asymmetrical, or Polar

Even when the egg is spherical rather than elongated and elliptically shaped (the eggs of insects are elliptic, for example), in its internal structure an egg cell is always **anisotropic,** or asymmetrical. In the terminology used in biology, an egg has a **polar** structure. At the very least, this polarity (anisotropy) is expressed in the location of the nucleus. Even in the **oocyte** (the diploid precursor cell) the nucleus is usually not located centrally but in the periphery near the surface of the cell. In the course of the meiotic divisions that give rise to the egg, the **polar bodies** are formed at this location (Fig. 2–1). Polar bodies are miniature sister cells of the egg cell (in the terminology of reproductive biology, polar bodies are "gones," as is the egg cell).

The location where the polar bodies are pinched off is usually shown as the "North Pole" of the egg and is called the **animal pole.** The opposite, "South Pole," is named the **vegetative pole** or **vegetal pole.** Material is generally deposited at the vegetative pole, which later in development is used in the formation of the primordial gut (**archenteron**), or is incorporated into the lumen of the gut.

In this context, the adjective "animal" refers to typical animal organs such as eyes or the central nervous system, which often are formed in the vicinity of the egg's animal pole. The adjective "vegetal" refers to the future "vegetative" organs that derive from the primordial gut and serve "lower" functions of life such as processing of food. The axis that extends from the North Pole to the South Pole and passes through the center of the globe is termed the **animal-vegetal egg axis.** However, when the site where the archenteron will be formed coincides with the location where the polar bodies are pinched off rather than being opposite it, as for example in coelenterates, the traditional terminology often causes confusion and gives rise to erroneously oriented and labeled illustrations of eggs and embryos.

Animal egg cells are surrounded by stabilizing and protective acellular envelopes. There are only a few exceptions; for instance, the eggs of coelenterates are covered only by an unobtrusive halo of glycoproteins. The innermost acellular sheet, which directly covers the surface of the egg cell, is made of glycoproteins and is termed the **vitelline membrane,** and in mammals, the **zona pellucida.** The term **chorion** is also used for envelopes, but can refer to different, nonhomologous structures. In insects chorion designates the acellular envelope of the egg; in reptiles, birds, and mammals chorion means a cellular extraembryonic epithelial structure that is made by the embryo itself.

Zoological textbooks like to list classifying terms, which inform readers familiar with ancient Greek about the amount and distribution of yolk (end syllable -*lecithal*) in the egg. Prefixes are as follows: *oligo* = few, *poly* = much, *iso* = uniform, *centro* = in the middle, *telo* = at the end, concentrated at one of the poles. Amount and distribution of the yolk affect the type and pattern of cleavage.

2.3   Cleavage Is a Series of Rapid Cell Divisions

Fertilization and activation of the egg are followed by cleavage. After fusion with the sperm, the fertilized egg is still unicellular. Its task is to give rise to a multicellular organism that may comprise many millions of cells. There now follows a series of rapid cell divisions. At high speed the zygote is divided, without increase in volume and mass, into more and more cells that therefore are smaller and smaller. This stage of development is called **cleavage.** It is indicated by the appearance of furrows on the surface of the egg.

The cleavage can take one of two forms:

- **Holoblastic** (= total), in which the egg is completely subdivided into individual cells; or
- **Meroblastic** (= partial), in which the egg is not completely divided, at least not at the beginning of its development. Whether or not holoblastic cleavage with regular cell divisions is possible depends on the egg's spatial dimensions and its content of yolk.

Where holoblastic cleavage occurs (examples are sea urchin, Fig. 2–1, and frog, Fig. 3–28), the first (still large) daughter cells are called **blastomeres** (Greek: *blastos* = germ-bud, seedling, embryo; *meros* = part). Depending on whether the first daughter cells are equal or unequal in size, we refer to **equal** or **unequal cleavage.** Where unequal cleavage occurs, a blastomere may give rise to a **macromere** and a **micromere.** Examples of meroblastic cleavage are the **superficial** cleavage in insects (Fig. 3–14) and the **discoidal** cleavage in fishes (Fig. 3–38), and reptiles and birds (Fig. 3–39).

The time course of holoblastic cleavage and the spatial arrangement of the daughter cells often follow strict rules: directed by the centrosomes of the mitotic spindle, in each embryo of those species the blastomeres come to lie in a distinctive geometrical configuration. As a rule, the first two **cleavage planes** cut the egg along the animal-vegetal axis, starting at the animal and ending at the vegetal pole, the second cleavage plane being perpendicular to the first (Fig. 2–1a). Contractile rings, consisting of actin and myosin filaments twisted together, bisect the egg along the meridians. Accordingly, the first two cleavages are said to proceed **meridionally.** By this time the four-cell stage has been reached. The third cleavage, which gives rise to the eight-cell stage, proceeds in the equatorial plane and therefore is called **equatorial.**

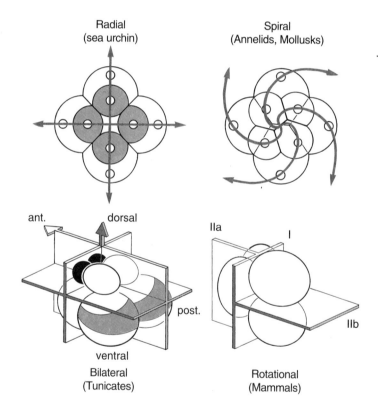

Figure 2–2   Cleavage patterns of holoblastic eggs. Holoblastic eggs usually contain little yolk and are entirely cleaved during cell division (cytokinesis). Radial and spiral cleavage patterns are most conveniently recognized looking down upon the top (animal pole) of the developing embryo. The bilateral and rotational cleavages are shown from a side view.

Further development is not uniform among the various taxonomic groups of animals, but there are frequently observed basic patterns. The three dominant patterns are the **radial type, spiral type,** and **bilaterial type** of cleavage (Fig. 2–2).

Even though cell division will continue, cleavage is considered terminated when the **blastula** stage is reached. This is a hollow ball of cells whose interior is filled with fluid or liquified yolk. The cellular, epithelial wall of the hollow ball is termed **blastoderm,** and the interior space, **blastocoel.**

Although development into a blastula is typical, there are several different routes embryogenesis can take. The term "blastula" must not be confused with the similar term **"blastocyst."** The mammalian blastocyst resembles the blastula of the sea urchin, but must not be equated with it, because the fate of the cellular wall of the cyst is completely different in sea urchins and mammals.

2.4   Gastrulation Prepares the Construction of Internal Organs

An animal needs internal tissues and organs. In the simplest animals, this may be a tubular space in which food can be enzymatically broken down and dissolved. Consequently, cells must be placed into the cavity of the blastula cyst, either singly or as a coherent sheet. The displacement into the interior of the blastula is termed **gastrulation,** and an embryo undergoing gastrulation is called a **gastrula.** The syllable *gastr* (Greek: *gaster* = belly, stomach) alludes to the future fate of the shifted cells: they give rise to the primordial intestine, the **archenteron (endoderm),** which will be largely responsible for forming the inner parts of the digestive tract and associated organs (such as the liver in vertebrates).

Gastrulation is a phase of dramatic events. Rapid RNA transcription retrieves large amounts of genetic information, marking the dawn of individuality at the molecular level. Gastrulation commonly is the point of no return for embryonic determination. From that point on, cells move irreversibly down their developmental pathways. Actual cell movements make the dynamics visible to the eye of the inquisitive researcher.

The means by which cells are placed in the interior is quite different among various animals. Commonly, the following five basic modes of active movement or passive displacement are distinguished (Fig. 2–3):

1. **Invagination:** infolding, inward buckling of the wall of the cyst

2. **Epiboly:** overgrowing, spreading of a sheet of cells over cells that thus come to be internalized
3. **Delamination:** detachment of cells
4. **Immigration** or **ingression:** ameboid movements of cells
5. **Polar proliferation:** cell divisions taking place at one of the poles; daughter cells are released into the cavity of the embryo

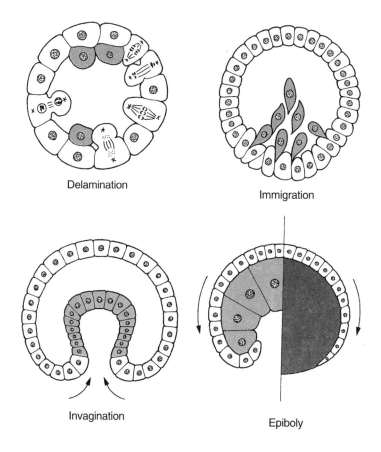

Delamination

Immigration

Invagination

Epiboly

Figure 2–3   Patterns of gastrulation, a process that brings cells into an interior cavity where they give rise to the archenteron and other internal parts of the embryo. In **delamination,** cells are displaced into the interior cavity directed by the orientation of the mitotic spindle. In **immigration** or **ingression,** cells move actively like amoebae (a case of polar immigration is illustrated). In **invagination,** a sheet of cells buckles inward by actively generating bending moments at curves and edges. In **epiboly** an epithelial superficial layer expands and spreads over other cells or the still uncleaved remainder of the egg cell, forming an envelope. Expansion of the outer layer may be brought about by crawling movements supported by cell division and changes in cell adhesiveness.

In most animal embryos, gastrulation occurs as a combination of several of these basic modes.

Germ-layer formation is another traditional expression for this and the following embryonic stage. **Gastrulation** can be defined as the process by which the embryo acquires two or more **germ layers,** thus becoming **diploblastic** or **triploblastic.** Diploblastic gastrulae, and diploblastic animals derived from them (sponges, coelenterates), have two germ layers: the **ectoderm,** forming the outer layer, and the **endoderm,** forming the inner layer. In all animal organisms above the organizational level of the coelenterates, gastrulation not only prepares the space for digestion and coats it with endoderm, but a further sheet is inserted between the gut rudiment (endoderm) and the outer wall (ectoderm) of the gastrula, the **mesoderm.** The gastrula is now **triploblastic.**

The mesoderm is an extremely versatile layer. As a rule, it will form muscles; connective tissue; blood vessels; the epithelial linings of the interior cavities, nephric, nephridial, and other organs engaged in excretion and osmoregulation; and skeletal elements, provided the genetic program of the respective species includes such faculties.

The expression "germ layers" is a relict from historic descriptions of the embryogenesis of the chick. In the avian embryo, all three "germ leaves," **ectoderm** (in the chick and mammalian embryo called epiblast), **mesoderm** (mesoblast), and **endoderm** (hypoblast), are temporarily arranged in stratified layers one upon another (Fig. 3–39).

How are three layers produced? In case of a multiphasic gastrulation, such as takes place in sea urchins (Fig. 2–1b), cellular material is brought into the blastula cavity in several batches. In another mode of gastrulation, seen in amphibians, first the blastula wall buckles and displaces into the interior cavity an archenteron consisting of mesendoderm, which subsequently divides into the endoderm of the stomach and gut tube, and a mesoderm.

The remaining epithelial outer wall of the gastrula, the ectoderm, is used to form the **epidermis,** the outer layer of the skin, but may also give rise to some sensory organs and parts of the nervous system. Consequently, in amphibians and insects the dorsal or ventral part, respectively, of the ectoderm is often termed **neuroectoderm,** because it gives rise to the nervous system and sensory organs (such as the inner ear in vertebrates).

It should be emphasized that germ layers are topographically defined structures. They do not necessarily have exactly the same fate or form the same tissue types in different species. The commendable attempt of linguistically educated authors of developmental treatises to replace the often improper term "derm" (skin) by the term "**blast**" (building material) has not yet been consistently adopted. Nonetheless, the term **"mesoblast"** is better than "mesoderm."

2.5 The Formation of Organs and the Differentiation of Tissues Give Rise to Autonomous Organisms

With gastrulation, the development of organs (**organogenesis**) is initiated. In vertebrates the first phase of organogenesis, which eventually leads to the formation of the central nervous system, is termed **neurulation.** Further stages cannot be subsumed under common, comprehensive terms because the results of organogenesis are species specific. To achieve the final organization of different species, varied pathways must be taken. Typically, embryonic development results in a first "edition" of the animal, the **larva,** which later in metamorphosis is remodeled into a second edition, the **imago** or **adult.**

Thus, embryonic development is completed with the hatching of the larva out of the egg envelope. The larva not only displays a form different than the adult, but also has acquired particular structural and physiological specialties that allow the larva to settle in another ecological niche—for instance, to make use of particular food supplies not accessible to the adult. Development that includes a larval stage is designated **"indirect."** Some animal groups, including the true land vertebrates, omit a larval stage; in the course of **direct development,** the adult form arises from the embryo gradually.

2.6 General Developmental Principles Suggest Possible General Mechanisms

In spite of the great diversity in animal development, there are some basic recurring events. The development of a multicellular organism consists of the following:

1. **Cell proliferation** (recurring cell divisions)
2. **Cell differentiation,** which must occur in a defined spatial order and therefore proceeds along with
3. **Pattern formation,** meaning that the various cell types are not chaotically arranged but occur in spatial patterns

In animal development, unlike that of plants, cell differentiation often gives rise to

4. **Cell movements and cell migration**
   A close look will reveal that in animal development
5. **Programmed cell death (apoptosis)** occurs.
   All these events together lead to
6. **Morphogenesis,** the modeling of form.

Cell divisions must be precisely controlled, as the organism's requirement for various cell types is different. Control of proliferation, therefore, is closely connected with control of differentiation.

Cell differentiation has a twofold meaning. The term may mean: (1) cells becoming different from each other, that is, the divergence of developmental pathways taken by the various cell types; or (2) cells becoming different in time: each cell takes a cell-type-specific pathway until it reaches maturity and becomes **terminally differentiated.** The development of each cell begins with a pluripotent founder cell—at the outset, the fertilized egg—which is capable of dividing and gives rise to several cell types. An act of decision, called **determination** or **commitment,** directs the developmental path of the descendants to different goals.

A cell that has become committed (programmed) to following a distinct pathway is designated by the suffix *-blast* (e.g., neuroblast, erythroblast). Such a committed cell may retain the ability to divide. If so, it will be the **founder cell** of a cell clone whose members are all destined to form one and the same cell type. After some rounds of cell division, the derivatives of the founder cell proceed to undergo terminal differentiation. They acquire their specific molecular equipment, specific form, and function. In so doing, the cells will, as a rule, lose their ability to divide. The suffix *-cyte* indicates that this has happened.

Determination proceeds along with **pattern formation.** Pattern is the ordered spatial configuration in which the various cell types and supracellular structures (tissues, organs) come to appear. Pattern formation in associations of cells is often based on position-dependent assignment of specific duties, that is, on position-dependent determination. However, patterns can also result from active migration of already committed and differentiated cells to defined locations.

In the experimental analysis of developmental processes, one starts from working hypotheses. Such hypotheses have to take into account the established fact that cell differentiation is based on **differential gene expression** ("differential gene activity," Chapter 10). Determination is viewed as a kind of programming by which it is decided which genetic information can be recalled in the future and which sets of genes will be blocked.

According to old hypotheses (Box 1), such programming or allotment of tasks may arise early in development from **cytoplasmic determinants,** which are deposited in the egg in a defined spatial pattern. The effect of such cytoplasmic determinants, stored in the egg cell, becomes clear when the Mendelian rule of reciprocity is invalidated, and the genetic constitution of the mother is of particular significance (e.g., hinny compared to mule, left-hand or right-hand coiling in snails, Fig. 3–10). **Maternal genes** provide maternal information, which in the course of **oogenesis** (development of the ovum in the ovary) becomes deposited in the egg cell as cytoplasmic determinants. Maternal effects make oogenesis an important factor in subsequent development.

Embryonic development that is heavily dependent on stored molecular determinants is called **autonomous development** or **mosaic-type development,** because a mosaic pattern of determinants leads to early allotment of duties and endows the cells with the capability of going their own independent way.

The programming of the cells can also be based on mutual consent. Cells communicate with each other via signals and make agreements. Such interactions make the development of various cell types dependent on their neighborhood, but also allow corrective regulation in case of disturbance. This aspect is expressed by the terms **"dependent"** and **"regulative development."**

3

Model Organisms in Developmental Biology

⸺◈⸺

The ability to understand developmental processes requires appropriate organisms. The field of genetics established the precedent of focusing research on a few reference or "model" organisms such as *Drosophila* or maize. Recent developmental biology has depended on a small number of organisms for much of its spectacular progress. This concentration of effort facilitates attempts to advance in the analysis of the basic processes down to the molecular level. On the other hand, there is no single organism that could be selected to study all fundamental events and aspects of development because each developmental pattern leads to a particular species and not to a generalized animal. From the egg of *Drosophila*, a fruit fly arises, not an insect in general, not a fish, and not a human being. General principles are only recognized when the events of development are studied in several diverse organisms, which develop differently yet display some common features. In addition, laboratory work quickly reveals that even the best model organism exhibits, in addition to its particular advantages, specific disadvantages. Thus the zebra fish has transparent embryos, but large-scale genetic studies require hundreds of aquariums and a staff of workers, and the fruit fly has a wealth of developmental mutants but cannot be conveniently frozen for long-term storage.

3.1   Sea Urchin Gametes and Embryos: Models for Studying Fertilization and Early Embryogenesis, and Subject of Historically Important Experiments

The sea urchin embryo is important in the history of developmental biology. The translucent embryos of sea urchins not only captivate us because of their beauty but, in addition, they trigger many seminal ideas and are favorable material to study very early development.

As early as 1890, pioneering work on fertilization was performed with eggs and sperm of sea urchins. Even today, sea urchin eggs remain the best investigated system for studies on fertilization, egg activation, and embryonic cell cycle. However, sea urchins are not good model organisms in all respects. Mutational analysis is not practical because of the long generation times and the difficulties in rearing larvae in the laboratory through metamorphosis. Mature sea urchins are not available throughout the year and must be removed from their natural marine habitat. This may pose problems with respect to species protection. These disadvantages are balanced by several particular advantages. Eggs and sperm can be harvested separately and in large amounts. Eggs are small (0.1 mm in diameter) and transparent, and are surrounded by a translucent and easily removed envelope (jelly coat). The eggs develop in water, even under the microscope, and, after artificial insemination, in perfect synchrony. Development up to the hatching larva takes 1 to 2 days.

3.1.1 Shedding of the Gametes

Eggs and sperm are released through genital pores on the dorsal side of mature sea urchins. In the laboratory, sea urchins are placed upside down onto beakers and are stimulated to shed gametes by intracoelomic injection of 0.5 M KCl. Eggs are released into seawater, while sperm is shed "dry" into dishes and can be used to inseminate eggs at a chosen time by mixing the milky sperm suspension into the egg suspension. The transparency of the envelope and the embryo facilitates in vivo observation of fertilization, cleavage, and gastrulation under the microscope.

The egg of sea urchins and the pattern of its cleavage have become prototypes in textbooks for animal eggs and development in general (Fig. 2–1). The egg is covered by an elastic, acellular coat, called **vitelline membrane,** and further wrapped up in a **jelly coat.** In its internal structure the egg is polarized along its **animal-vegetal axis.** By definition, the **animal pole** is that pole where the **polar bodies** are extruded during the course of the meiotic divisions, when the egg is still in the ovary. Polar bodies are miniature sister cells of the egg cells; they will decay. The haploid nucleus of the egg is found near the animal pole. Another marker of polarity is the subequatorial band of orange pigment found in some batches of the Mediterranean species *Paracentrotus lividus* (Fig. 2–1).

In sea urchin eggs, unlike mammalian eggs, all of the polar bodies are formed before fertilization. Insemination, fertilization, and activation of the sea urchin egg will be described in Chapter 6.

3.1.2. Start of Development, Cleavage

The sperm nucleus is gated and transported into the egg. Triggered by the contact of the sperm with the egg membrane, a **fertilization envelope,** serving as a barrier against the penetration of further sperm, is raised from the surface of the egg, and the egg becomes **activated.** Following activation, cleavage divisions dissect the egg into successively smaller cells, without appreciable changes in its volume. Sea urchins undergo synchronous, radial, holoblastic cleavages until the blastula stage. The first daughter cells that can be individually recognized in the microscope are called **blastomeres.** Doubling of their chromosomes takes place rapidly, and the cells divide every 20 to 30 minutes, driven by an internal molecular clockwork. The embryo thus passes through the 2-, 4-, 8-, 16-, 64-, and 128-cell stages. As development proceeds, the cell cycle lengthens and cleavage is no longer synchronous in the various regions of the embryo. A fluid-filled cavity now develops. Cells rearrange themselves to construct an outer epithelial wall enclosing the central cavity: the embryo has arrived at the **blastula** stage. The blastula's wall of cells is termed the **blastoderm** and its central cavity, the **blastocoel.** On the exterior surface of the blastoderm cilia are formed; the coordinated beat of the cilia causes rotational movements of the blastula within its envelope.

At the animal pole the first larval sensory organ is formed, recognizable as a bundle of long, nonmotile cilia: the **apical tuft.** At the vegetal pole, the blastoderm flattens and thickens to form the **vegetal plate.** The vegetal plate includes a group of cells, **micromeres,** which are filled with ribosomes. The micromeres begin to move and thereby open the phase of gastrulation.

Gastrulation begins shortly after the blastula, now comprising about 1,000 cells, hatches from its fertilization envelope. Gastrulation takes place in several phases.

1. The onset of gastrulation is marked by the immigration of the descendants of the micromeres into the central cavity. The cells become bottle shaped, release their attachment to the external hyalin layer and to their neighboring cells, and eventually detach to migrate individually into the blastocoel. They move in the fashion of amoebae. The immigrated cells represent the **primary mesenchyme.** The descendants of the smallest micromeres, which were previously located around the vegetal pole, are thought to undergo terminal differentiation only after metamorphosis. However, most of the cells of the primary mesenchyme soon become skeletogenic: clusters of reaggregated mesenchyme cells fuse into **syncytial cables.** These cables form the larval skeleton by secreting insoluble calcium carbonate that crystallizes into glittering spicules.

2. Micromeres are emitters of inducing signals. Transplanted to another site of the blastoderm, they induce their neighbors to invaginate. Invagination (Fig. 2–3) is the mode by which the archenteron is formed in the second phase of gastrulation. The vegetal plate buckles inward and elongates within the cavity of the blastocoel to form the tubelike larval intestine (Fig. 2–1). The invagination is radially symmetric around the animal-vegetal axis. When isolated surgically from the rest of the larva, the vegetal plate undergoes autonomous bending and involution.

The archenteron stretches along the length of the blastocoel. Lobelike pseudopods on the cells at the tip of the archenteron are transformed into filamentous pseudopods or **filopods.** Reaching out and moving around the inner surface of the blastoderm, the filopods explore the underside of the blastoderm cells. The filopods direct the tip of the elongating archenteron to a site where the definitive mouth will later break through. The primary mouth, the blastopore, will function as the larval anus. Sea urchins and all other members of the Echinodermata belong to the **deuterostomians,** as does our own chordate phylum.

3. Having fulfilled their function as pathfinders, the cells at the tip of the archenteron reduce their filopodia and detach. They represent the secondary mesenchyme, which gives rise to muscle cells surrounding the gut and some other cell types.

With the ingression and involution of cells into the blastocoel, germlayer formation still is incomplete. As the larva develops, **mesoderm** is formed by **coelomic pouches** that balloon out from the archenteron as evaginations (Fig. 3–3). Following the segregation of the mesodermal vesicles, the remainder of the archenteron constitutes the endoderm.

3.1.3 The Larva

Embryogenesis terminates with the hatching of the beautiful larva, which is transparent and fairylike. It is called **pluteus.** The larva floats through the water and uses its cilia to swirl microscopic food into its mouth.

3.1.4 Metamorphosis

The transformation of the free swimming (planktonic), bilaterally symmetrical larva into the pentameric sea urchin requires a fundamental reconstruction. The new construction starts from groups of cells that were set aside from participation in embryogenesis itself. The process resembles metamorphosis in "holometabolous" insects, such as butterflies, and the groups of cells that build up the new body are termed **imaginal discs** in sea urchins as well as in insects. The complicated "catastrophic" metamorphosis will not be described here. Experiments with embryos of sea urchins end with the observation of the pluteus larva.

3.1.5 Landmark Experiments 1: Embryos Are Not Machines and Are Capable of Regulation

The following experiment of historical significance was performed by Hans Driesch at the Stazione Zoologica in Naples. If the first two blastomeres are separated from each other at the two-cell stage, each blastomere gives rise to a complete larva. Although reduced in size to half of the volume of a normal larva, both larvae are harmoniously diminished (Fig. 3–1). They are identical, monozygotic twins. If the blastomeres are separated at the four-cell stage, quadruplets are formed. Even blastulae can be bisected, resulting in identical twin plutei, provided the embryo is bisected before the beginning of gastrulation and the cut is performed along the animal-vegetal axis (Fig. 3–1). Driesch concluded that living organisms are not mere machines, because pieces of machines cannot supplement themselves autonomously to restore the complete machine. This interpretation was in striking contrast to the prevailing view of the mechanicists (Box 1). Biologists now understand that regulation is possible because all cells are provided with the whole set of genetic information and because both halves receive animal and vegetal cytoplasmic components when the cut is made along the egg axis.

From the eight-cell stage on, it is possible technically to bisect the embryo in the equatorial plane at a right angle to the egg axis. Now the result is different: the animal half develops into a blastula-like hollow sphere, which, however, is incapable of forming an archenteron. Conversely, the vegetal half is able to form an archenteron, but the emerging pluteus displays considerable

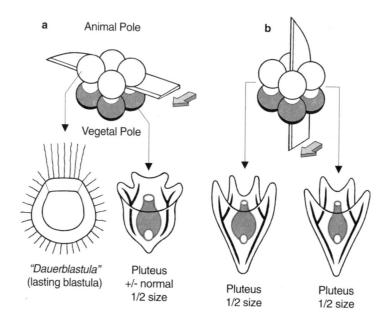

Figure 3–1   Experimental bisection of an early sea urchin embryo at the eight-cell stage. (a) Bisection perpendicular to the animal-vegetal egg axis in the equatorial plane results in two unequal embryos. (b) Bisection along the animal-vegetal axis results in two normal larvae of half the usual size.

deficiencies. For instance, it is mouthless and has short arms (Fig. 3–1a). By displacing cytoplasm through pressure or centrifugation, Theodor Boveri, Hans Driesch, and Thomas Hunt Morgan accumulated experimental evidence suggesting that **cytoplasmic components** are responsible for the different developmental potentials. Furthermore, Driesch concluded that the future fate of a cell, its "perspective significance," is "a function of its position in the whole." Boveri launched the idea of graded potentials along the animal-vegetal axis. The graded differences of the cells along the animal-vegetal axis were in the focal point of the following experiments.

3.1.6 Landmark Experiments 2: Interactions and Gradient Theory

Among the most remarkable experiments in the history of embryology were those performed by Sven Hörstadius investigating the developmental potentials of cells along the animal-vegetal axis of the early embryo. He separated tiers of cells from cleavage stages by transverse cuts, and juxtaposed tiers of cells taken from different locations along the animal-vegetal axis. To explain his results, Hörstadius advanced the gradient theory (first formulated by Boveri). The experiments are described here in some detail because they document the scientific approach of the classic developmental biology and prompted similar experiments in many animal systems.

If the animal cap *an1* is isolated by micromanipulation at the 64-cell stage (Fig. 3.2), the cap gives rise to a hollow sphere unable to gastrulate. A close look at the blastula-like sphere reveals a strange anomaly: the long cilia that mark the first larval sensory organ (the apical organ) are not concentrated in a small area, forming the apical tuft. Instead, the whole blastula bears these long cilia. Even the tier *an2*, whose descendants are normally equipped only with short cilia, develops, following isolation, into a blastula that is largely covered with long cilia. The ability to form an apical organ has spatially expanded. Such an exaggeration of structures characteristic of the animal sphere is called **animalization.** The tendency to form exaggerated animal structures diminishes with increased distance to the animal pole.

In the intact embryo the strong animal potency must be repelled or compensated by influences originating from the vegetal part of the embryo. These influences have been systematically investigated by Runnström, Hörstadius, and other Scandinavian researchers, who performed delicate transplantation studies. By juxtaposing micromeres onto isolated tiers of cells derived from the animal hemisphere, the exaggerated animal tendency can be compensated. Transplanted micromeres were by far the best **vegetalizers.** By determining the number of micromeres needed for compensation, one can titrate the strength of the animal potency (Fig. 3–2).

The tier *an1* is normalized by adding four micromeres; the conglomeration develops into a largely normal dwarf pluteus. To normalize *an2*, two micromeres suffice. In *veg1* one micromere suffices, but the embryo has such a weak animal component that the resulting ovoid larvae have short arms and show further defects. Apparently, an animal component is required for normal development. This requirement is even more apparent when the experiment is extended to *veg2*. The tier *veg2*, which ordinarily gives rise to a rudimentary, pluteus-like creature, is "**vegetalized**" by the addition of micromeres. It loses the ability to form arms and instead forms an intestine that is oversized relative to the rest of the body. As a result, the intestine cannot be invaginated into the interior cavity (**exogastrula**).

These results gave rise to the so-called **double-gradient model:** two physiological activities along the animal-vegetal axis occur as mirror-image gradients. The activities were assigned to **morphogenic substances,** nowadays called **morphogens** following a proposal by Turing (Box 4). The molecular identities of these hypothesized substances are still unknown.

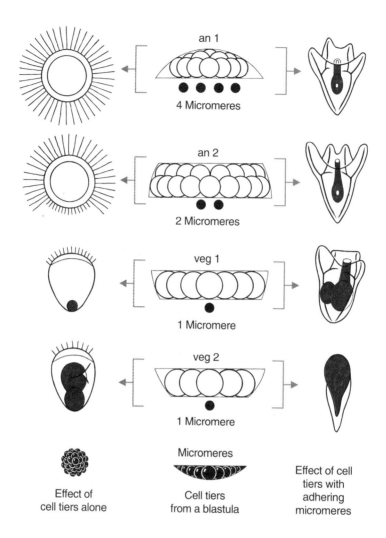

an 1

4 Micromeres

an 2

2 Micromeres

veg 1

1 Micromere

veg 2

1 Micromere

Micromeres

Effect of
cell tiers alone

Cell tiers
from a blastula

Effect of cell
tiers with
adhering
micromeres

Figure 3–2   Sea urchin. An experiment supporting the **gradient hypothesis** (conducted by Sven Hörstadius). From middle to the left column: Tiers of cells from a 64-cell stage blastula were isolated and their developmental potentials observed. For example, tier *an1* gives rise to an "**animalized**" blastula with an expanded apical tuft of long cilia covering the entire surface. Tier *veg2* gives rise to incomplete larvae lacking the oral arms and the mouth but with an enlarged hindgut; they show features of "**vegetalization**." Isolated micromeres are unable to continue development. From middle column to right column: Development of the same tiers following the addition of micromeres. If four micromeres are added to *an1*, the enhanced animal potential is compensated and a normal pluteus results. To normalize *an2*, it takes only two micromeres. The addition of one micromere to *veg2* leads to a highly vegetalized development: the "larva" lacks all oral structures and the enlarged archenteron cannot be incorporated into the interior. Such a malformation is called an "exogastrula."

An "animalizing" morphogen that pushes determination toward animal properties is thought to peak at the animal pole, while a vegetalizing morphogen pushes vegetal properties and peaks at the vegetal pole. Because the opposite morphogens neutralize one other, the strength of each factor decreases with the distance from its origin. Therefore, the local developmental pattern is determined by the ratio of two morphogens.

A popular hypothesis maintains that the morphogens can spread from cell to cell, freely crossing cell membranes. Alternatively, the relevant signals mediating interaction might spread only in the interstitial spaces between the cells or must be exposed on the cell surface and directly presented to neighboring cells. In either case, the signal molecules have to be picked up by receptors. (Such a view would connect the classic gradient hypothesis with the hypothesis of Davidson, outlined in Section 3.1.7).

Remarkably, animal hemispheres can be normalized not only by juxtaposed micromeres but also by lithium ions. High concentrations of Li^+ even bring about vegetalization up to exogastrulation. Lithium is known to block signal transduction from membrane-anchored receptors into the interior of the cell by a pathway known as the PI-PKC system (Chapter 5; Box 3).

3.1.7 Recent Views

Eric Davidson's work has aimed at detecting factors controlling development in sea urchin eggs. He has sought proteins possessing DNA-binding domains, and mRNA has been extracted encoding such proteins. The following picture emerged.

As in *Drosophila* (Section 3.6) and *Xenopus* (Section 3.8), the unfertilized egg contains maternal mRNA transcribed and stored during oogenesis that codes for at least 10 different transcription factors. These maternal components are deposited within the egg in a mosaic pattern and are distributed differentially among the blastomeres. They act as cytoplasmic determinants.

After the sixth cleavage the blastula is divided into the following five major territories: (1) the oral territory is directed by local determinants to form the ectodermal epithelium of the mouth region and the ciliated band that will extend along the arms; (2) a second territory forms the aboral ectoderm; (3) a third forms the vegetal plate and later, the archenteron; (4) the fourth territory constitutes the skeletogenic material; and (5) the fifth comprises the small micromeres that induce the invagination of the archenteron (Fig. 3–3).

However, the maternal transcription factors that have been allocated to the various territories do not enable autonomous development (except in the

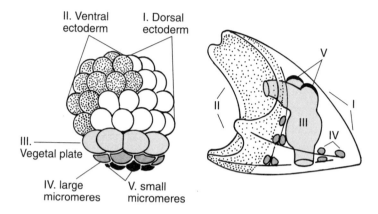

Regions in the distribution of maternal gene products

Figure 3–3 Sea urchin. Various regions of the egg cell and the early blastula are equipped with different sets of maternal gene products. The different areas give rise to different parts of the pluteus larva. (After investigations by Davidson.)

fifth territory). Rather, the cells of the various territories exchange signals, leading to a **sequence of inductive interactions.** For example, by presenting signal molecules the micromeres induce, in the neighboring tier *veg2*, enhanced expression of some genes and repression of others, leading to the suppression of the most vegetal developmental potency in *veg2*. In more general terms, by the mutual exchange of signals, the state of determination experiences fine tuning and stabilization. Lithium interferes with the signal exchange by blocking the PI signal transduction system (Box 3).

3.2 Pattern Formation Decoded: *Dictyostelium discoideum*

3.2.1 Life Cycle

Dictyostelium is a simple eukaryotic microorganism that displays the appearance of amoebae but is usually called a "cellular slime mold." However, its phylogenetic relationship is unclear. The cellular slime mold *Dictyostelium* is unrelated to multinucleated slime molds such as *Physarum*. It is also unrelated to yeasts or filamentous fungi. Some zoological textbooks classify *Dictyostelium* as member of the group of "social amoebae" (*Acrasiales*). Yet, a close relationship to the category of Rhizopoda is not evident. The organism is unusual in several aspects.

Under standard laboratory conditions, the life cycle does not start from fertilized eggs but rather from single haploid amoebae liberated from an envelope of **spores**. The life cycle of *Dictyostelium* is shown in Fig. 3–4. Note that it is entirely asexual.

> A curious type of sexual reproduction is known but does not occur under standard conditions. Two cells fuse and enlarge to a giant cell by cannibalistically engulfing neighboring amoebae. The giant cell encysts and later undergoes meiotic and mitotic divisions that give rise to new haploid amoebae.

During the vegetative life cycle, an unusual transition takes place from the unicellular to the multicellular state: numerous individual amoebae assemble to form a social community capable of making provisions for unfavorable environmental conditions. Thus, *Dictyostelium* is a "part-time multicellular

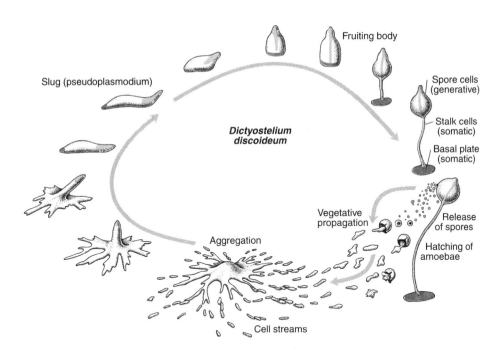

Figure 3–4 Developmental cycle of *Dictyostelium discoideum*. Haploid cells hatch from the spore envelope, live as soil amoebae, and reproduce asexually by mitotic cell division. As the food supply is exhausted, the amoebae present within a certain area assemble at a point into an aggregate. The size of the area is determined by the range of chemotactic signals emitted by the cells in the center of the aggregate. The aggregate takes the form of a slug, migrates to an appropriate place, and forms a fruiting body that releases new spore cells. (Redrawn after Gilbert.)

organism" (Kay et al. 1989) formed from previously independent unicellular amoebae.

Dictyostelium lives in rich organic soils. When moistened, spores seeded by fruiting bodies release haploid cells that exhibit the appearance and mode of living of amoebae. They live in films of water, eating bacteria and multiplying by binary fission (**vegetative phase**). Only when their food supply is exhausted, or the substrate is exposed to the risk of drying out, do hundreds to thousands of amoebae assemble, migrating singly or collectively like caravans, towards a common meeting place. The aggregate absorbs all of the converging streams of cells. Eventually, the aggregate forms a multicellular association that eventually acquires the form of a slug. Correspondingly, the association is colloquially called a "slug." In scientific terms the association is called **grex** or **pseudoplasmodium.** The slug is surrounded by a slimy acellular sheet and is able to move much like a true slug. It migrates to a bright location where it transforms into a **fruiting body,** consisting of a basal plate and a stalk that bears a spherical accumulation of new spores beneath its tip. The basal plate and the stalk consist of **somatic cells.** The somatic cells form walls made of cellulose and eventually die. By contrast, the spore cells survive; they are **generative cells** whose formation and release perform the functions of asexual reproduction. Thus, the fruiting body fulfills the criteria defining a true multicellular organism.

Although the life cycle of *Dictyostelium* is not at all typical for multicellular animals (in contrast to the life cycle of sea urchins), several features of *Dictyostelium* development can be taken as paradigms for similar events in higher eukaryotes. These include aggregation, cell differentiation, and pattern formation (the spatial order in which the various cell types occur).

3.2.2 Aggregation

Approximately 5 hours after removal of the food supply, hungry cells emit an attractant chemical serving to guide nearby starving cells to a central location. In *Dictyostelium discoideum* the chemical signal is **cyclic adenosine monophosphate (cAMP).** cAMP is emitted by the starving cells in synchronous pulses every 5 to 10 minutes, and spreads by radial diffusion in the water film. The signal is detected by a surface receptor protein having seven transmembrane domains like many receptor proteins in animal cells. The receptor is coupled to a signal transduction system of the PI type (Box 3).

To speed up the propagation of the signal and to increase the range of signaling, *D. discoideum* has evolved a courier system for relaying the signal. Neighboring amoebae that have received a signal at their surface receptors respond to it by releasing cAMP of their own. Thus, the local cAMP concentration is increased, the signal is amplified, and diffusion is accelerated. To ensure that the signal is transmitted from the pulsing central pacemaker to the periphery of the aggregation field, but not backward, each amoeba is unresponsive to a further cAMP pulse for about 3 minutes. The transitory "deafness" prevents the cells from being disturbed by their own signal, or that emitted by their immediate neighbors. Only when the signal has reached inaudible remoteness are the amoebae sensitive to a new signal arriving from the center.

Each pulse of cAMP induces a jerky movement toward the center. As each wave arrives, the amoebae take another step toward the central meeting point. Although each cell of the population can approach the meeting place independently, they usually migrate with a caravan. Cells join each other to form streams, the streams converge into larger rivers of cells, and eventually they all merge at the center. The final aggregate may comprise up to 100,000 cells.

3.2.3 Cell Differentiation and Pattern Formation

During the aggregation phase, which ends in the formation of the slug, the genetically uniform cell population segregates into several subpopulations: **prestalk cells,** representing about 20% of the population and positioned at the tip of the migrating slug; **prespore cells;** and the cells of the future **basal plate,** cells that resemble the stalk cells but form the rear of the slug.

The following two hypotheses have emerged regarding the division of the amoeboid cell population into prestalk, prespore, and basal plate cells:

1. The **positional information** hypothesis (Box 4): the location of the cell within the slug determines its fate.
2. The **sorting out** hypothesis: cells differentiate before or during aggregation and seek out their place within the slug according to their future role.

Both hypotheses must explain how a characteristic ratio is achieved in the numbers of the three cell types. The ratio is restored after experimental manipulation such as bisecting the slug. Surgical removal of either anterior or posterior cells causes the remaining cells to revise their developmental commitment. Whichever cells find themselves at the anterior end will become stalk cells, and whichever cells are in the posterior region will become spore cells or basal plate cells. Numerical regulation is possible until cell differentiation has become irreversible in the developing fruiting body. Much evidence favors hybrid hypotheses that combine features of both positional information and sorting out.

Initiation of aggregation appears to require, besides starvation, the accumulation of a secreted protein in the cell's environment. The concentration of protein may be an indirect measure of cell density, so that development only starts if sufficient cells are present to form a fruiting body. Experimental evidence also emphasizes a pivotal role for low molecular weight signal substances in the numerical control of cell differentiation and in positioning of the various cell types. Thus, cells are committed to become stalk cells in the presence of high concentrations of **cAMP** and **differentiation inducing factor (DIF)**, but low concentrations of **ammonium (NH$_3$)**, conditions that are thought to exist at the tip of the slug.

The choice between prestalk and prespore differentiation is critically dependent on the concentration of **DIF.** DIF is a chlorated aromate, the nucleus of which is a phenol moiety [1-(3,5-dichloro-2,6-dihydroxy-4-methoxy-phenyl)-hexan-1]. Like cAMP, DIF is liberated from starving and aggregating amoebae. NH$_3$ is formed when surplus protein is catabolized, for instance, in future stalk cells, which synthesize the N-free cellulose of their rigid and enduring cell wall before they die. The conversion of amino acids to cellulose releases NH$_3$.

There is evidence that additional low molecular weight substances are released to convey information in the transactions that are negotiated among the cells of the community. For instance, the nucleoside **adenosine** is produced from cAMP at the tip of the migrating slug and inhibits the formation of spore cells at the place of its origin. On the other hand, evaporation of NH$_3$ at the tip of the slug favors the formation of stalk cells there. Evaporation is facilitated at the elevated tip. Thus, both chemical and physical conditions determine the mode of cell differentiation and the location where a particular cell type will arise.

In summary, *Dictyostelium* has become a model to study periodic emission of signals, signal relay, chemotaxis, and the establishment of cell contacts through cell adhesion molecules. In recent studies aspects of cell differentiation and pattern formation prevail.

3.3   The Immortal *Hydra* and the Dawn of Modern Experimental Biology

Systematically planned regeneration and grafting studies on the small freshwater polyp *Hydra*, performed under a magnifying glass with forceps and scalpels by the Swiss scholar Abraham Trembley 20 years before Mozart was born, rang in the era of modern experimental biology. Trembley's observations are well documented in detailed descriptions and masterful engravings. Two and a half centuries later, developmental biologists still make use of the enormous capacity of reconstitution and regeneration that *Hydra* and its marine relatives display. *Hydra* readily replaces lost body parts—head, foot, or any other part of its tubelike body. Even more astonishingly, the *hydra* polyp can be **dissociated** experimentally into single cells, which sink to the bottom of the dish, creep around like amoebae, reestablish contact with one another, and form clumps of cells (**reaggregates**). Initially, reaggregates are disorderly and unrecognizable. However, within days or weeks the aggregates organize themselves to form new, viable polyps.

Hydra (Fig. 3–5) is a member of the phylum of the coelenterates: the simplest multicellular organisms possessing typical animal cells such as sensory, nerve, and epithelial muscle cells. Despite being equipped with mortal cells (such as nerve cells), *Hydra* as a whole is potentially immortal. Immortality is

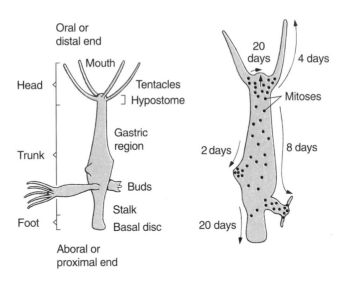

Figure 3–5 *Hydra vulgaris* with buds. Right: Dots along the animal indicate sites of cell division; the arrows indicate the direction in which the epithelial cells are displaced. The shift of the epithelial cells results from two events: (1) in the terminal body regions aged cells undergo cell death and are phagocytosed or sloughed off, while (2) new cells are born in the body regions between the terminal regions. A surplus of cells is exported in the form of buds. Once a bud has formed a head and a foot, it detaches to live as a new, independent individual. It is genetically identical with its parent.

achieved by an unlimited capacity for self-renewal: the polyp can replace any aging cell, or cells having fulfilled their tasks, by substitutes generated from immortal **stem cells.** In this process of perpetual renewal all cells are replaced, even the nerve cells. Cell proliferation, cell differentiation, and cell migration never come to a standstill. *Hydra* is a **perpetual embryo,** and although its terminally differentiated cells die, the cell community survives.

Recent *Hydra* research has emphasized patterning and the regulation of cell differentiation.

1. **Pattern formation and positional information** has been investigated in the context of regeneration and reorganization. Depending on where a group of cells is located, the cells can organize themselves to form a hypostome with tentacles (colloquially called the "head"), gastric region, stalk, or basal disc (the "foot"). Regulation of the body pattern is discussed in Chapter 9, Section 9.8.

2. **Renewal by stem cells, control of cell proliferation, and cell differentiation.** *Hydra* consists of two main categories of cells (Fig. 3–6).

1. **Epithelial cells** determine the basic architecture of the body. Two adjoined epithelial layers, the ectoderm and the endoderm, form the walls of the tubelike body.
2. **Interstitial cells** occur in spaces (*interstitia*) between the epithelial cells. Interstitial cells include sensory nerve cells, ganglionic nerve cells, four types of stinging cells, a type of gland cell, gametes, and the stem cells of all these cell types.

The epithelial cells, even those of the head and the foot, arise from epithelial stem cells in the middle of the body column. The cells born here are displaced toward the mouth; they become integrated into the growing tentacles or into the mouth cone. Other cells are shifted toward the foot. Having arrived in the terminal body regions (in the mouth cone, the tentacles, or the foot), they undergo terminal differentiation and eventually die. They are sloughed off or phagocytized by neighbors.

In well-fed polyps a surplus of cells generated in the gastric region is used to produce buds and thus the surplus is exported. Buds detach as self-reliant animals. By budding, *Hydra* clones itself.

Interstitial cells arise from interstitial stem cells (**I–cells**) that have retained the ability to differentiate into various pathways and therefore are classified as **pluripotent.** Both the number and the type of newly produced cells must be precisely regulated. *Hydra* in particular, and coelenterates in general, are animals in which tumors or other cancerous aberrations have not yet been found. This suggests a very efficient system of proliferation control.

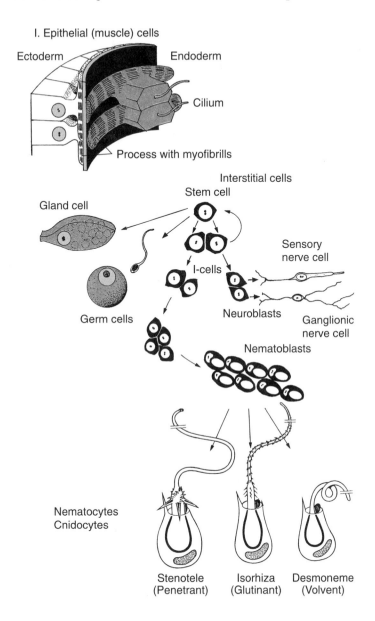

I. Epithelial (muscle) cells

Ectoderm

Endoderm

Cilium

Process with myofibrills

Interstitial cells
Stem cell

Gland cell

Sensory
nerve cell

I-cells

Germ cells

Neuroblasts

Ganglionic
nerve cell

Nematoblasts

Nematocytes
Cnidocytes

Stenotele
(Penetrant)

Isorhiza
(Glutinant)

Desmoneme
(Volvent)

Figure 3–6 Basic cell types of a hydra. Diagrammatic presentation; sizes are not to a consistent scale. In the ectoderm, muscle fibers run longitudinally and mediate the contraction of the body column; in the endoderm, muscle fibers run circumferentially and mediate the expansion of the body column. Derivatives of the interstitial stem cells reside in the spaces between the epithelial cells (except the gland cells that become integrated into the endodermal epithelium, and the mature nematocytes that are mounted within ectodermal "battery cells" of the tentacles).

However, *Hydra* is not always cooperative. It is not a useful model to study early embryogenesis from an egg, and a classical genetic approach is very tedious, if not impossible. As a rule, gametes are produced only under adverse and stressful environmental conditions, and only in small numbers. The eggs are enclosed in a rigid, nontransparent envelope. The investigator who is interested in embryogenesis (from the egg to the planula larva) or metamorphosis (from the planula to the polyp) should consider several marine relatives of *Hydra*, such as *Hydractinia*. This colonial hydroid grows on the shells of hermit crabs and is found in the North Sea of Europe and along the Atlantic coast of North America. The *Hydractinia* life cycle is illustrated in Fig. 3–7.

As with *Dictyostelium*, there are active efforts in *Hydractinia* to find and identify signal molecules involved in pattern formation and the control of proliferation

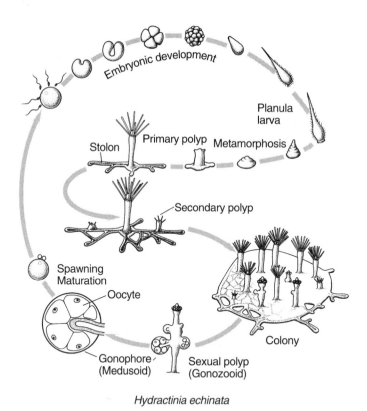

Hydractinia echinata

Figure 3–7   *Hydractinia echinata*, a colonial hydrozoan frequently found along the coast of Northern Europe on shells inhabited by hermit crabs. *Hydractinia symbiolongicarpus* is a closely related species with an identical life cycle that is found along the Atlantic coast of North America.

and differentiation. In *Hydractinia*, a factor **PAF** (proportion-altering factor, of unknown chemical nature) dramatically alters patterning in embryogenesis and metamorphosis. As metamorphosis begins, neuropeptides, such as Lys-Pro-Pro-Gly-Leu-TrpNH$_2$, act as an internal signal, triggering and synchronizing the transformation of the planula into a primary polyp. Several low molecular weight substances containing transferable methyl groups, such as **N-methylpicolinic acid** and **N-methyl-nicotinic acid,** have been discovered and are thought to participate in the control of metamorphosis and postmetamorphic pattern formation. The glycoprotein **SIF** (stolon-inducing factor) is liberated from stolons (a network of tubes similar to the vascular system); it induces stolon branching just like angiogenic factors induce branching of blood capillaries in vertebrates. In *Hydra* as well as in *Hydractinia*, **arachidonic acid** and other **eicosanoids** participate in the control of developmental processes. Arachidonic acid induces supernumerary head formation.

Although genetic crossing is not currently possible in *Hydra*, the use of reverse genetics and molecular screening techniques (Box 6) have resulted in the identification of many genes involved in the control of body pattern development, such as homeobox-containing genes.

The state of differentiation is not always irreversibly fixed in cells of coelenterates. Instead, cells can change their state and a muscle cell may transform into another cell type. **Processes of transdifferentiation** (Chapter 11, Section 11.1, and Chapter 21) are studied using isolated cross-striated muscle cells prepared from the medusae, for instance, of the species *Podocoryne carnea* (Atlantic Ocean, Mediterranean) (Fig. 21–1).

3.4   *Caenorhabditis Elegans:* Example of Invariant Cell Lineages

Thirty years ago the molecular biologist Sydney Brenner proposed intensive investigation of the small transparent roundworm, *Caenorhabditis elegans* (Fig. 3–8). *C. elegans* is now among the acknowledged reference models in developmental biology and has proven to be a superb subject for studying eukaryotic developmental genetics, cell biology, neurobiology, and genome structure.

In nematodes, embryogenesis proceeds in a precise, species-specific pattern that is faithfully repeated in every generation. Through careful observation the ontogenetic tree for every somatic cell of the body can be reconstructed. The most complete reconstruction of the cellular pedigree has been achieved in *C. elegans* (Fig. 3–9). This success rests on the worm's anatomical simplicity, its transparency, its genetic tractability, and especially on the high precision with which cell lineages are reproduced in each individual and each generation. Development terminates in each adult with an identical number of cells (**cell number constancy**).

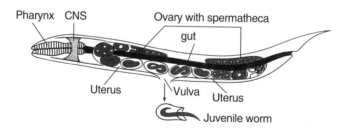

Figure 3–8   *Caenorhabditis elegans.* Morphology of the hermaphroditic, viviparous nematode (roundworm). The gonad has two arms; each arm contains not only oogonia but also sperm stored in the spermatheca. Embryogenesis takes place in the last part of the gonadal tubes, the uterus. CNS = central nervous system.

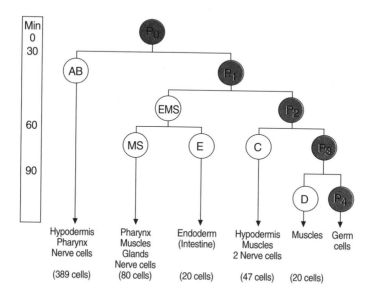

Figure 3–9   Cell genealogy of *Caenorhabditis elegans.* The P line (red circles) represents the germ line; the white circles represent somatic founder cells. The lineage and exact number of cells is nearly invariant.

The natural environment of *C. elegans* is the soil. Like *Dictyostelium, C. elegans* feeds on bacteria, and in the laboratory a similar technique is used to culture both organisms. Bacterial lawns are grown on agar plates and then the plates are inoculated with the organisms of interest: spores of *Dictyostelium* and in *C. elegans,* embryos enclosed in their transparent egg shell or hatched worms. The life cycle is short (3.5 days); embryogenesis lasts about 12 hours at 25°C and 18 hours at 16°C.

   C. elegans is usually hermaphroditic. The hermaphrodites have XX sex chromosomes and are female in appearance and anatomy, but they are capable of producing not only eggs but also sperm in their tubular gonads. Self-fertilization results in endogamy, and as a convenient consequence of repeated inbreeding, mutated genes (new alleles) are homozygous as early as in the F_2 generation. The occasional loss of an X chromosome by nondisjunction results in X0 males at a frequency of 0.2% (the corresponding XXX embryos are not viable). The X0 males mate with hermaphrodites, which now play the part of true females. Thus, in *C. elegans* cross-fertilization and self-fertilization are possible. New alleles can be imported by cross-fertilization.

   Embryonic development takes place in the proximal half of the gonadal tube, called the **uterus.** In addition to the transparency of the mother and the embryo, the embryo can be removed from the uterus and its envelope without being killed. A favorite method facilitating the mapping of cell fates is to inject permanent markers such as a fluorescent dye, labeled antibody, or reporter genes into the embryo (Box 7). Not only will the injected blastomeres be labeled, but their descendants will also be labeled. Some cell lines have natural differentiation markers. In the E-cell lineage leading to the gut, for example, rhabditin granules (of unknown function) can be detected with polarization optics. The original descriptive analyses of cellular ontogeny in *C. elegans* have been supplemented by the study of numerous mutants displaying disturbed cell lineages. For eliminating defined cells, a laser beam can be used. This powerful combination of tools from genetics and cell biology is unique to *C. elegans* and has justified Brenner's prescient enthusiasm on its behalf.

   The worm terminates its embryonic development and is born with 556 somatic cells and 2 primordial germ cells. It undergoes four larval stages separated by molts. The larval period lasts 3 days. At the end of its development, the mature worm has 959 somatic cells and about 2,000 germ cells if it is a hermaphrodite, or 1,031 somatic cells and about 1,000 germ cells if it is a male. The nervous system consists of 302 nerve cells; these arise from 407 precursors, of which 105 undergo a **programmed cell death (apoptosis).**

Cell divisions frequently are **asymmetric.** That is, the two daughter cells resulting from mitotic division of a progenitor cell are equivalent with respect to their inherited genetic information but not with respect to their cytoplasmic contents, and they have different destinies. The germ line is particularly impressive in nematodes. By definition, a **germ line** leads from the fertilized egg to the primordial germ cells and thus comprises the cells capable of forming gametes (Fig. 3–9). In *C. elegans* the cells lying on this line are termed P_0, P_1, P_2, P_3, and P_4. The last P cell, P_4, is the primordial germ cell. The cells P_0 to P_4 are characterized by a peculiar legacy: asymmetrical cell divisions allocate cytoplasmic **P granules** only to these generative cells but not to their sister cells destined to become somatic cells. In the roundworm *Ascaris* (synonym, *Parascaris*) the chromosomes remain complete only in the germ line, whereas the somatic cells generated by asymmetric divisions lose some chromosomal material (**chromatin diminution**). (Theodor Boveri, who described and analyzed this phenomenon in 1910, suggested the abandoned chromatin might be of significance in the development of the germ line. Today, the function of the eliminated DNA is still a matter of speculation.)

We return to *C. elegans* where chromatin diminution is not known. Analysis of the cell pedigree in *C. elegans* revealed that a few founder cells produce only one tissue: blastomere E gives rise to all of (and only) the gut. (The term "stem cell" is avoided in the recent literature on *C. elegans* because no daughter cell with the capacity for self-renewal is left in the hatched worm; instead, the term "founder cell" is used.) However, this specialization is not the rule. Normally, equivalent lineages give rise to cells of more than one type along the length of the embryo. Conversely, most tissues are derived from several founder cells, which in turn give rise to other tissues. Such tissues generated by several embryonic founder cells are of **polyclonal origin.** For instance, nerve cells originate from cells whose descendants can also participate in the generation of muscles. On the other hand, muscle cells originate from three pluripotent founder cells. Thus, besides muscles, blastomere C produces neuronal and hypodermal cells.

The careful analysis of numerous mutants as well as surgical deletions of founder cells have led to the view that the fate of each cell is determined not only by cytoplasmic constituents allocated to it in early embryogenesis (such as RNA), but also to a great extent by early and precise interactions between adjacent cells.

The genetic control of development in *C. elegans* is thought to be accomplished by about 1,600 genes, many of which are so-called "selector genes" or "master genes" (see Section 3.6, or Chapter 10). The total worm genome is estimated to comprise only 3,000 genes, embodied in a total genome of about

8×10^7 bp. The small size of the genome (approximately 3% the size of the human genome) has led to a project to sequence it. The project is scheduled for completion by 1997.

3.5   Spiralians: A Recurring Cleavage Pattern

In the animal kingdom spiral or oblique cleavage (Fig. 2–2, 3–10) occurs in several phyla: among the plathelminthes, in the acoelous turbellarians, and in nemertines, annelids, and mollusks (except cuttle fish). These phyla are grouped under the term **spiralians.** As a rule, the embryonic development of spiralians passes through a classic blastula and gastrula, and terminates in a **trochophore** (Fig. 3–11) or a trochophore-like larva, such as the veliger.

As in *C. elegans,* an invariant pattern of cleavage and cell genealogy makes it possible to trace the origin of various parts of the body back to distinct founder cells. Particular attention has long been paid to blastomere D of the four-cell stage and to its descendant **4d.** The cell 4d is the **primordial mesoblast,** the founder cell of the mesodermal inner organs. In several mollusks a conspicuous part of the egg's vegetal cytoplasm is sequestered in the

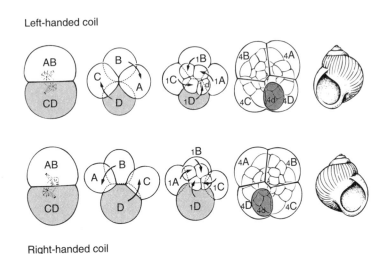

Figure 3–10   Spiral cleavage in right-handed and left-handed coiling snails. We are looking down upon the animal pole. The left-handed and right-handed snails display mirror-image patterns of cleavage (after Morgan 1927). The red-labeled cell 4d represents the founder cell of the mesoderm.

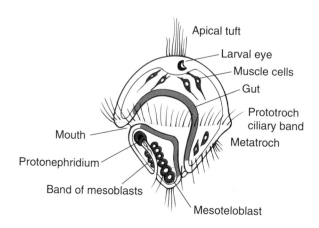

Figure 3–11   Trochophore larva of an annelid. The mesoteloblast is a descendant of the 4d cell (compare Fig. 3–10) and gives rise to mesodermal tissues.

form of a **polar lobe.** If the vegetal cytoplasm is removed with a micropipette prior to its translocation into the lobe, the embryo fails to develop a dorsoventral body axis.

The anucleate lobe is extruded periodically in the process of cleavage. During cell division, the lobe remains attached to one of the two daughter cells, into which it is absorbed after cytokinesis. The lobe extrudes again prior to the next cleavage. The cells marked by the pulsating bulb are the cells of the D line. The lobe contains components of unknown identity that are essential for future mesoderm formation. When the lobe is cut off and fused with, for example, cell A, cell D loses the ability to give rise to mesoderm. Instead, cell A can now generate the mesoblast.

Among the spiralians, no single species has gained the position of a dominant reference model. Of the mollusks, development has been examined in the pond snail *Lymnea stagnalis*, the land slug *Bithynia*, the marine snails *Littorina* and *Ilyanassa*, the scaphopode *Dentalium* (tusk shell), and the marine mussel *Spisula* (trough shell). In *Spisula* the activation of the egg has been particularly interesting.

Among the annelids, the marine polychaete worms *Chaetopterus* and *Platynereis*, or the tropical leeches *Helobdella* and *Haementeria*, are often brought into the laboratory. In the leech embryo microinjecting tracers enabled reconstruction of the pedigree of each neuron back to its founder cell.

Leeches and other annelids are superbly suited models for investigating how animal diversity arose by developmental changes during evolution. The segmental body plans of arthropods and annelids are presumed to have arisen from common segmented ancestors, yet the type of segmentation can differ extensively among the various members of these groups. For instance, in annelids and many arthropods segments are added in postembryonic development near the rear of the growing animal. By contrast, in *Drosophila* all segments are formed simultaneously (Section 3.6). The "reverse genetics" approach (Box 7) using probes derived from *Drosophila* genes, such as *engrailed*, is being applied successfully in annelids, and certainly will yield exciting results.

3.6 *Drosophila:* Still the King of Genetic and Molecular Developmental Biology

Nearly 100 years have passed since Thomas Hunt Morgan made *Drosophila* the acknowledged leading model of classical genetics. The rapid life cycle, ease of breeding, and giant polytene chromosomes that allow gene localization made the fruit fly superbly suited to analyze heredity. However, *Drosophila* came to be the most important model subject in developmental biology only in the past years, after E.B. Lewis introduced the complex of homeotic genes as early as 1978. Two complementary lines of research were particularly successful: the search, by mutagenesis, for genes that specifically control development, carried out by Christiane Nüsslein-Volhard and Eric Wieschaus; and the use of new tools from molecular biology by researchers such as David S. Hogness, Gary Struhl, Walter J. Gehring, and many others.

3.6.1 A Short Curriculum Vitae

The life cycle of the fruit fly from hatching to hatching takes 2 to 3 weeks (Fig. 3–12). Embryogenesis takes only 1 day; the larva passes the three larval stages, separated by molts, in 4 days. The larva then encloses itself in a new cuticle, called the **puparium,** in which it undergoes metamorphosis in 5 days. The adult fly lives for about 9 days. Many peculiarities of development in *Drosophila* reflect paths taken during the evolution of its unusually rapid life cycle. *Drosophila* has also developed a particularly efficient means of producing mature eggs at high speed.

3.6.2 Oogenesis: Providing for the Future

In *Drosophila* production of the egg cell already anticipates rapid embryonic development. Many features of fly developmental genetics make eminent sense when seen from this perspective.

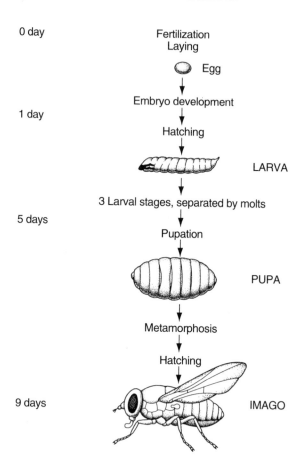

Figure 3–12 *Drosophila melanogaster.* A short curriculum vitae. (Redrawn after Alberts et al.)

The eggs are formed within tubes called **ovarioles,** which are divided into chambers by transverse walls (Fig. 3–13). The cells constituting the wall of an ovariole and surrounding the future eggs are termed **follicle cells** by analogy to the follicle cells in the ovary of vertebrates. In each chamber there is one female primordial germ cell, the **oogonium.** The oogonium undergoes four rounds of mitotic divisions, resulting in 16 cells. These remain interconnected by cytoplasmic bridges called **fusomes.** In the center of the cluster, 2 of the 16 cells are connected with 4 sister cells; 1 of these 2 cells will become the **oocyte,** the future egg cell. The remaining 15 sister cells are fated to become **nurse cells.** While the oocyte remains diploid, and later will become haploid in the course of two meiotic divisions, the nurse cells become polyploid by

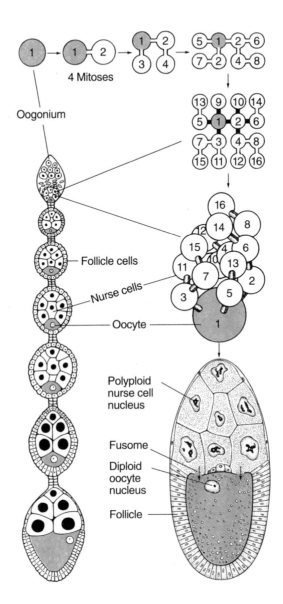

Figure 3–13 *Drosophila* oogenesis. Schematic drawing of one tubular ovariole, containing oogonia close to its distal tip. Oocytes of progressive size and maturity are seen along the tube. The oocytes are accompanied by nurse cells and are surrounded by follicle cells. Enlarged details are shown outside the tube of the ovariole.

replicating their DNA repeatedly; this amplification of the genome enables high transcriptional activity. The nurse cell will provide the oocyte with huge amounts of ribosomes and mRNA enclosed in **ribonucleoprotein particles (RNP).** Ribosomes and RNP are exported and shipped by the nurse cells through the fusomes into the oocyte and are thereby placed at the disposal of the developing egg, enabling it to pass the dangerous phase of embryogenesis rapidly, when escape from predators is impossible.

In the late stages of oogenesis, the follicle cells help nourish the oocyte. They mediate the supply of the yolk—the major constituent of the large egg. The yolk is synthesized outside the egg cell in the so-called "fat body," in the form of **vitellogenins** and **phosphovitins.** These lipoproteins and phospho-proteins are liberated into the fluid of the central body cavity (the hemolymph), gathered by the follicle cells, and passed over to the oocyte. The oocyte takes over the materials, storing them in yolk platelets and granules to build up a stock of amino acids, phosphate, and energy.

The egg cell is merely a consumer; its own nucleus is transcriptionally inactive. The maternal nurse cells, follicle cells, and the cells of the fat body use their own genes and cellular resources to manufacture all of the materials that are exported into the oocyte. Therefore, the gene products are maternal and carry **maternal information.** When such genes affect the embryonic development of the egg, they are termed **maternal effect genes.**

Finally, the follicle cells secrete the multilayered **chorion,** a rigid enve-lope encasing the egg. At the anterior end of the egg shell a canal called the **micropyle** is left open, allowing the entry of a sperm before the egg leaves the mother fly. Insemination occurs, as in terrestrial vertebrates, in the last sec-tion of the ovary that serves as an oviduct. Sperm, stored during copulation in the *receptaculum seminis,* is used throughout the life of a female fly.

3.6.3 Embryogenesis at High Speed

3.6.3.1 Cleavage   Embryonic development in *Drosophila* (Fig. 3–14) starts immediately following egg deposition and within only 1 day leads to a larva able to hatch. This is a remarkably rapid process. Cleavage, called **superficial,** is unusual. Nuclei are duplicated at a high frequency with intervals of only 9 minutes, until after 13 rounds of replication about 6,000 nuclei are present. In this phase the egg represents a **syncytium.** When the stage of 256 nuclei is surpassed, nuclei begin to move to the periphery of the egg and settle in the cortical layer. The first nuclei to reach the cortex are those moving to the pos-terior end of the egg. Previously, under the organizing influence of the mater-nal gene *oscar,* **pole granules** containing several species of RNA (including mitochondrial RNA) accumulated in this region. Transplanted posterior-pole cytoplasm from oocytes and early embryos can instruct early embryonic nuclei at any position within the syncytium to adopt the germ cell fate. (Such transplanted pole plasm can also direct the formation of a second mirror-image abdomen. See Section 3.6.4.)

At the posterior pole of the normal egg, the immigrant nuclei and pole granules are encased by cell membranes and extruded as **pole cells** by a process of budding (Fig. 3–14). These first embryonic cells are committed to become the primordial germ cells from which the next generation will arise. During gastrulation, the pole cells migrate through the midgut epithelium to arrive in the embryonic gonad (Fig. 3–15).

The other nuclei are arrayed in the cortex of the egg as a monolayer and the egg arrives at the stage of the **syncytial blastoderm.** After some more divisions, the final 6,000 nuclei are produced. Next, the plasma membrane is pulled down between the nuclei. Membrane invaginations continue to deepen past the nuclei and close off a compartment of cytoplasm around each nucleus, thus creating cells. This is the stage of the **cellular blastoderm.** The ventral part of the cellular blastoderm constitutes the **germ band,** which gives rise to the embryo proper.

Some nuclei are left in the central yolk mass; they represent the **vitellophages** (a misnomer, as nuclei do not eat yolk). Later in development these nuclei are enclosed in cells, too. In several groups of insects these central vitellophages are thought to participate in the formation of the midgut, but recent literature does not assign such a role to these cells in *Drosophila.*

3.6.3.2 Gastrulation and Early Organogenesis

The formation of the body begins at the ventral midline of the germ band, and proceeds to the dorsal side of the egg. Gastrulation comprises two main, independent events.

1. **Endoderm formation** (Fig. 3–15): the endoderm is formed by anterior and posterior midgut invaginations. Both invaginations deepen into the interior, approaching each other. Eventually, they fuse, forming the midgut. As the indentations deepen, the surrounding blastoderm is drawn in, forming the **stomodeum,** which will give rise to the foregut, and the **proctodeum,** which will give rise to the hindgut.

2. **Formation of the mesoderm and the ventral nerve cord** (Fig. 3–14, 3–20): along the ventral side a wide band of cells undergoes coherent morphogenetic movements. The cells change shape to form an indentation in the epithelium, creating the **ventral furrow (primitive groove),** which deepens and invaginates. The invaginated groove is incorporated into the interior of

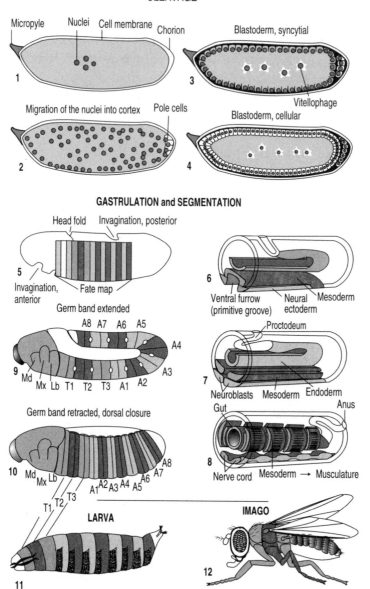

CLEAVAGE

Micropyle Nuclei Cell membrane Chorion

Blastoderm, syncytial

1

3

Vitellophage

Migration of the nuclei into cortex Pole cells

Blastoderm, cellular

2

4

GASTRULATION and SEGMENTATION

Head fold Invagination, posterior

5

6

Invagination, anterior Fate map

Ventral furrow (primitive groove) Neural ectoderm Mesoderm

Germ band extended

A8 A7 A6 A5

A4

A3

9 Md

A2

Mx Lb T1 T2 T3 A1

7

Proctodeum

Neuroblasts Mesoderm Endoderm

Anus

Germ band retracted, dorsal closure

Gut

10 Md

A8

Mx Lb

A7

A6

A1 A2 A3 A4 A5

8

T3

Nerve cord Mesoderm → Musculature

T1 T2

LARVA

IMAGO

12

11

Figure 3–14 *Drosophila*. Summary of embryo development. In the "superficial" cleavage nuclei multiply to establish a preliminary syncytial blastoderm, and only later are complete cells formed constituting the cellular blastoderm. Before and during gastrulation a longitudinal "germ band" is formed that expands longitudinally so that the tail comes to lie close to the head. Later, the germ band is retracted. Gastrulation takes place as invagination at three sites. Along the ventral midline a long furrow, known as the "primitive groove," appears. The mesoderm (red) and the cells of the future nerve cord invaginate in this groove.

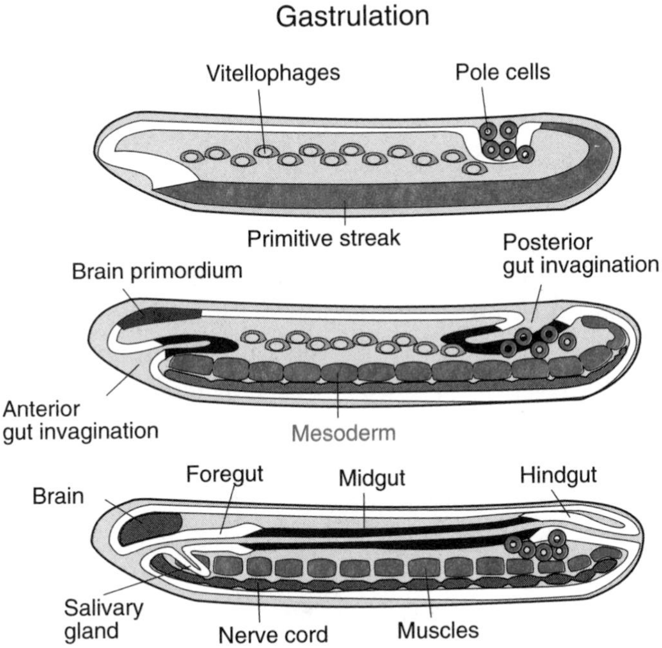

Gastrulation

Figure 3–15 *Drosophila*. Gastrulation and early steps of organ formation in an idealized fly. Schematic longitudinal section. Ahead and behind the primitive groove, local invaginations give rise to the foregut and the hindgut. Both extend into the interior and approach each other. Transitory changes associated with germ band extension and retraction are omitted.

the egg and forms the bandlike **mesoderm.** Soon thereafter, the band becomes dispersed into groups of cells that will form the larval muscles. Two further bands of cells, which formerly accompanied the mesoderm band of the ventral midline on each side, contain the **neurogenic cells.** Neurogenic cells are singled out (Fig. 9–3), segregated from the remaining epidermoblast cells, delaminated into the interior, and come to lie between the mesoderm and the ectoderm. Incorporated into the interior, the neurogenic cells separate and give rise to groups of **neuroblasts** from which derives the **ventral nerve cord.**

3. **Brain:** The primordium of the optic lobe appears as a placode (local thickening of the blastoderm) and is invaginated. Delaminating neuroblasts provide further building material to construct the brain.

4. **Dorsal closure:** The dorsal edges of the ectoderm as well as the edges of the internal organs grow over the central yolk dorsally until they meet each other and fuse along the dorsal midline.

Events of minor importance, such as the transitory expansion of the germ band and the formation of an internal envelope called the amnionserosa, will not be described in this book.

3.6.3.3 Segmentation and Basic Body Pattern

In the meantime, **segmentation** has begun. The body comes to be divided into periodically repeated units. Segmentation begins at the level of gene expression at the stage of the syncytial blastoderm but is morphologically visible only when the mesoderm is divided into packets and furrows appear in the outer ectodermal epithelium. In the outward body pattern 14 segments can be distinguished; these are initially quite uniform (homonomous) but become dissimilar (heteronomous).

The germ band subdivides itself into three main groups (**thagmata**) of segments. In examining the fly larva the various segments and groups of segments can be recognized only by identifying the numerous visible specializations of the larval cuticle, including dorsal hairs, ventral denticles, tracheal spiracles, and sense organs. After metamorphosis, the differences are conspicuous in the imaginal fly. The fusion of the terminal **acron** (bow) with probably seven **cephalic segments** results in the **head.** The number of seven fused segments has been derived from the expression pattern of genes such as *engrailed* and *wingless*, which normally are expressed along the borders of segments (see as follows, segment polarity genes). All seven cephalic segments (three pregnathal and four gnathal) supply neuroblasts for forming the central nervous system (CNS). The CNS consists of the supraesophageal ganglion (brain) and the subesophageal ganglion. The three posterior (gnathal) segments, termed *mandibular, maxillar,* and *labial,* respectively (*Mb, Mx, Lb*), manufacture the tools for eating around the mouth. In the larva, however, the head is retracted into the interior of the body. In its outward appearance, the body of the larva begins with the following three **thoracic segments:** *T1 = prothorax, T2 = mesothorax,* and *T3 = metathorax.* In the finished fly, each thoracic segment bears a pair of legs; the mesothorax, a pair of wings; and the metathorax, a second pair of structures that once were wings but now are evolutionarily reduced to oscillating bodies called **halteres,** structures that are equipped with sensory organs to control wind-induced torsions during the flight.

The **abdomen** of the fly or its larva consists of eight segments (*A1* through *A8*); the terminal **telson** (after-deck) at the rear is not classed with true or complete segments, nor is the terminal acron at the anterior end.

3.6.4 Genes Controlling the Body Pattern

The early embryonic development of *Drosophila* is governed by genes that affect, via their protein products, the state of activity of other subordinate

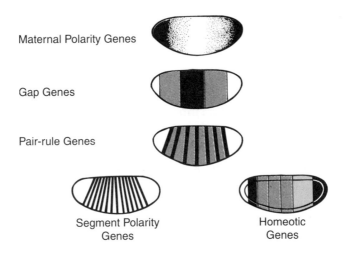

Maternal Polarity Genes

Gap Genes

Pair-rule Genes

Segment Polarity
Genes

Homeotic
Genes

Figure 3–16 *Drosophila.* Temporal sequence of expression of master genes controlling embryo pattern formation. The stripes show the distribution of proteins encoded by such genes. The pattern is established (1) by maternal effect genes that determine the main body axes and induce the expression of (2) zygotic gap genes; these define broad territories and turn on (3) pair-rule genes that are expressed in alternating stripes and forecast the location of future segments. (4) The segment polarity genes initiate the actual segmentation and their subdivision into smaller units. (5) The homeotic genes ultimately define the individual identities of the segments. (Redrawn after Gilbert.)

genes. These regulatory gene products contain DNA-binding domains and act by controlling transcription. At present, three classes of genes controlling development are distinguished (Fig. 3–16).

3.6.4.1 Maternal Genes Affecting the Establishment of the Body's Coordinates (Axis Determination)   The primary axes or coordinates predetermine the basic bilaterally symmetrical architecture of the fly body. To reach such an organization, two axes must be set up: the anterior-posterior axis and, perpendicular to it, the dorsoventral axis. In *Drosophila* axis determination takes place under the influence of genes, but it is the genes of the mother and not of the embryo itself whose products set up the coordinates. The existence of such maternal genes is demonstrated when eggs, produced by mutant females, give rise to embryos that fail to form distinct body regions, or to form them in correct locations, although the eggs may appear normal in their shape. Genetic analysis reveals that the mutant genotype of the mother, and not that of the embryo, causes this failure.

Axis determination genes are active in the ovary tubes of the maternal organisms—in the nurse cells or in the follicle cells. The products encoded by these genes are channeled into the oocyte, as a rule, in the form of mRNA enclosed in RNP particles, and translated into protein after the egg is fertilized and laid. The proteins produced with maternal information do not directly participate in the construction of the embryo but form gradients of morphogens, thus mediating the subdivision of the whole egg space into subspaces with different fates. (For the definition of the term "morphogen," see Box 4.) About 30 different gene products have been identified that bring about the subdivision of the egg space and determine the basic architecture of the embryo.

These **maternal effect genes** are classified into the following groups (Fig. 3–16, 3–17):

1. **Genes effecting anterior-posterior polarity**
   * Anterior group with *bicoid* (*bcd*)
   * Posterior group with *nanos* (*nos*)
   * Terminal group with *torso* and *caudal*
2. **Genes effecting dorsoventral polarity**
   * *dorsal* (*dl*) and *toll*

1. **Genes effecting anterior-posterior polarity.** The product of the *bicoid* gene has attracted particular attention. *bicoid* mRNA is directed into the egg by means of its untranslated 3′-region and is anchored at the egg's anterior pole. The *bicoid* message enables the anterior end to become an organizing center. In the first minutes of development the message is translated into protein. The BICOID protein has some restricted capacity to diffuse; thus, a gradient in BICOID protein is established having its highest concentration at the anterior pole and extending over half the length of the egg. In the *bicoid* mutant the product is defective. If both alleles are defective (a homozygous null mutation), the larva lacks a head and a thorax, and the acron is replaced by an inverted telson. The morphological counterpart is the mutation *bicephalic*: the larva has a head at either end, and the two heads are mirror-symmetrically arranged. Such Janus-headed larvae arise from oocytes bordering on nurse cells at either end (Fig. 3–18); the eggs were supplied with *bicoid*⁺ mRNA from both ends.

Biheaded larvae have also been produced experimentally by injection of cloned wild-type *bicoid*⁺ message into normal eggs near the posterior pole. The resulting biheaded larvae are **phenocopies** of the true mutant *bicephalic*.

The BICOID protein is incorporated into the embryonic nuclei in the anterior region of the egg. The protein contains a **helix-turn-helix domain**

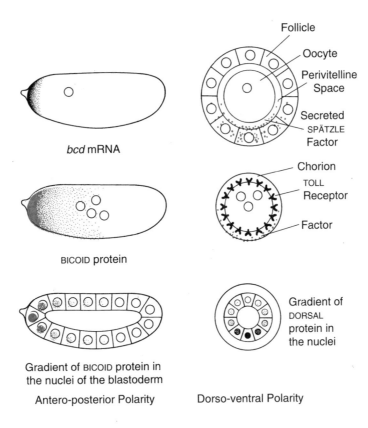

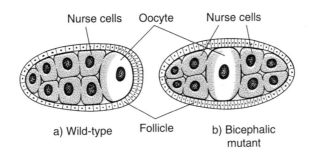

Figure 3–17 *Drosophila.* Establishment of the anteroposterior body axis (left) and the dorsoventral polarity (right). Along the longitudinal axis, the distribution of the products of the maternal effect gene *bicoid*⁺ are shown. The cross section shows the secreted SPÄTZLE factor, the corresponding TOLL receptor, and the distribution of the transcription factor coded by the *dorsal*⁺ gene.

Figure 3–18 *Drosophila.* Chamber of an ovariole from (a) wild-type fly; (b) the maternal effect mutant *bicephalic*. (Redrawn after Alberts et al.)

derived from the **homeobox** of the gene. With this domain the BICOID protein attaches onto the promoters of other genes, thus bringing them under its control. The activated subordinate genes are genes of the embryo; they are termed **zygotic.**

The gradient of the BICOID protein is said to provide **positional information.** When the BICOID concentration is increased in the anterior part of the egg (by genetically manipulating the mother), the borders of the head and the thoracic region are shifted posteriorly, and the relative size of the head and the thorax is increased (Fig. 3–19). A large amount of BICOID protein switches on head-specific combinations of genes, low concentrations of tho-

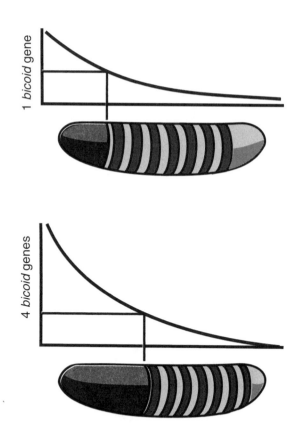

Figure 3–19   *Drosophila.* Gradient in the distribution of the *BICOID*⁺ protein. The gradient specifies the position and dimensions of the head-thorax region. If more *bicoid*⁺ genes have been active in oogenesis due to genetic manipulation of the parental flies, the height and the range of the gradient are enlarged. As a consequence, the dimension of the head-thorax region also becomes enlarged.

rax-specific combinations, and the pattern-forming system uses threshold concentrations to define the borderlines of a body region (Box 4). Among the zygotic (embryonic) genes turned on by the BICOID transcription factor, *hunchback* is one of the first to be expressed. Initially the *hunchback* message is widely distributed in the syncytial egg chamber, but its spatial expression domain in the body becomes restricted to the anterior two-thirds of the egg by the suppressing influence of the NANOS protein.

The posterior organizing center is dominated by the products of the maternal effect genes *nanos* and *oskar*. The message of the genes is localized at the posterior pole. Transplantation of posterior pole plasm to the anterior pole can result in larvae with a second mirror-image abdomen instead of a head. Such a phenotype is also known from the mutation *bicaudal*, in which a complete second mirror-symmetrical abdomen is formed instead of an anterior body.

In wild-type embryos NANOS protein diffuses away from the posterior tip of the egg, forming a concentration gradient running in an opposite direction to the BICOID gradient. However, unlike BICOID, NANOS is not a transcription factor and does not bind to DNA. Instead, NANOS suppresses the translation by ribosomes of the *hunchback* message in the posterior region of the egg.

The *torso* gene codes for a transmembrane protein serving as the receptor for an extracellular signal molecule. The signal molecule or its precursor has been deposited by follicle cells at either end of the egg into the perivitelline space between the cell membrane of the egg and the vitelline membrane covering the egg. Occupied by the ligand, the TORSO receptor (a tyrosine kinase) mediates the formation of the terminal body structures, the acron and the telson. The homeobox gene *caudal* is also involved in specifying the telson.

2. **Genes effecting dorsoventral polarity.** The effect of *torso*, described previously, also holds for *toll*. The maternal *toll* message is used to provide the egg cell with transmembrane receptors to sense an external cue that tells the embryo where to make the ventral side (Fig. 3–17). As in the case of the terminal signal, the external cue specifying the ventral side has been laid down and anchored in the perivitelline space surrounding the egg cell. The signal itself appears to be a product of the maternal effect gene *spätzle* (named after a German dumpling) and is liberated from the anchoring complex by means of a protease. Because the SPÄTZLE precursor is present on the ventral side but not on the dorsal side of the egg, only the receptors on the ventral side will find a ligand. Receptors occupied by ligands then will organize and direct an internal mechanism effecting the redistribution of the DORSAL protein. This redistribution is enabled by receptor-mediated phosphorylation of the DORSAL protein in the ventral region of the egg.

The newly translated, native DORSAL protein is initially found throughout the egg, but upon phosphorylation, it is translocated into the nuclei of the embryonic cells. As in the case of BICOID, we eventually see a gradient of DORSAL accumulated in the nuclei, but for now the highest concentration is in the nuclei on the ventral side of the embryo (Fig. 3–17, 3–20). For those unfamiliar with the genetics terminology, the designation "*dorsal*" may be misleading, as the protein defines the ventral side in the normal embryo. However, when the gene *dorsal* (and, consequently, the protein DORSAL) is defective, the embryo is unable to make ventral structures. It will have a dorsal appearance all around; the embryo is said to be **dorsalized**—hence the name of the gene and its product.

Cells in the blastoderm of normal embryos that lack DORSAL protein will in turn liberate a factor into the dorsal perivitelline space. It is the DECAPENTAPLEGIC factor encoded by the *ddp* gene and belonging to the TGFβ family of "growth" factors (as does the inducing factor **activin** in *Xenopus*). The extracellular DPP gradient and the intracellular DORSAL gradient help to subdivide the cellular blastoderm into several territories of different fates: along the ventral midline is the territory of the future

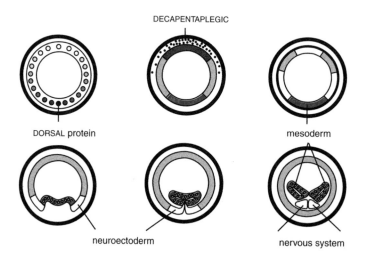

Figure 3–20 *Drosophila.* Dorsoventral patterning and gastrulation. Cross sections showing (upper row, left to right) the distribution of the DORSAL protein, the expression domain and secretion of the DECAPENTAPLEGIC factor, and the fate map along the dorsoventral body axis. The lower row shows gastrulation: formation of the primitive groove, invagination of the mesoderm (red), and of the future nervous system (white).

mesoderm, flanked on both sides by bands of **neural ectoderm** that will give rise to the nervous system. Adjoining dorsolaterally are the bands of the **dorsal ectoderm** that will form the epidermis of the larva (Fig. 3–20).

3.6.4.2 Genes Effecting Segmentation of the Body   Along its main axis, the body of arthropods, including insects, is composed of repetitive **modules.** These are organizational and structural units, which eventually become visible as **segments.** The zygotic (embryonic) genes turned on by the transcription factors encoded by maternal effect genes, such as *bicoid,* are involved in **pattern formation,** eventually resulting in the establishment of those segments.

However, before segmental furrows actually form, bands of gene expression do not exactly coincide with the future visible segments but are shifted anteriorly against the visible segments by half a segment. Drosophilists speak of **parasegments.** One parasegment includes the posterior part of a future segment and the anterior part of the posteriorly adjoining segment.

Segmentation occurs in steps. In the syncytial blastoderm the embryonic cells begin to produce mRNA and protein of their own. Many of the new proteins are gene regulatory factors. These are not uniformly expressed along the body but in spatially restricted **expression zones.** First, these zones are broad; later, expressed products appear in smaller but more numerous bands (Fig. 3–16, 3–21). About 25 genes have been identified that participate in the elaboration of future segments. In the morphologically uniform blastoderm the sequence of expression is as follows.

1. **Gap genes** emerge in broad, overlapping zones. These include the expression zones of the genes *hunchback* (*hb*), *Krüppel* (*kr;* German for cripple), and *Knirps* (*kni;* German for doll or manikin). Defective products made by the eponymous mutants cause groups of consecutive parasegments to be deleted in the larva. In *hb* null mutants a zone comprising several parasegments is missing in the middle of the body, and another zone in the posterior third. The expression of the gap genes is followed by the expression of the pair-rule genes.

2. **Pair-rule genes:** the embryo is not yet cellularized; nevertheless, the future segmentation is heralded by repetitive bands of gene expression in the embryonic nuclei. The transcription pattern is unexpected and striking. Pair-rule genes are expressed in patterns of seven stripes in alternating segments: one vertical stripe of nuclei expresses a gene, the next stripe does not express it, the next adjoining segment expresses it, and so forth.

The pair-rule genes comprise remarkable genes, such as *hairy, even skipped,* and *fushi tarazu* (*ftz;* Japanese for "too few segments"). *Fushi*

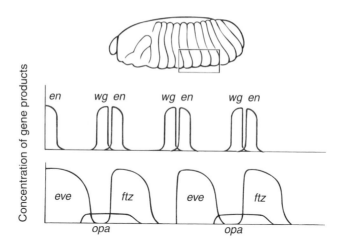

Figure 3–21 *Drosophila.* Expression patterns of pair-rule genes (*eve* = *even-skipped*, *ftz* = *fushi tarazu*) and of segment polarity genes (*en* = *engrailed*, *wg* = *wingless*). The diagram refers to a detail in the larva. Note the alternating expression of *eve* and *ftz*, and the bordering on each other of *en* and *wg*. The border between *wg* and *en* marks the boundary between parasegments. The boundaries of the final segments are localized posterior to the stripes in which *engrailed* is transcribed. Besides showing the expression patterns of genes, the figure also suggests interactions. The *engrailed* gene is expressed when the cells contain high amounts of either EVEN-SKIPPED or FUSHI TARAZU proteins. The *wingless* gene is transcribed when neither of these two genes is active. (Redrawn after Gilbert.)

tarazu is expressed in the odd-numbered parasegments and *even skipped* is expressed in the even-numbered parasegments. As the embryo is going to make 14 visible segments, it expresses 7 stripes of *fushi tarazu* and 7 of *even skipped*. In fly embryos mutant for *even skipped*, 7 stripes are of no use and will be skipped, and the larva will be left with the remaining 7 *fushi tarazu* segments. Conversely, embryos with mutant *fushi tarazu* genes give rise to larvae with 7 stripes that correspond to the expression bands of the intact *even skipped* gene.

3. Once the stage of the cellular blastoderm is reached, **segment polarity genes** subdivide the various segments into smaller stripes. Several genes are of particular significance in the demarcation of the final, visible segment boundaries in the middle of the former parasegments. These genes are *engrailed* (*en*), *wingless* (*wn*), and *hedgehog* (*hh*), supplemented by the gene *patched* (*ptc*) (Fig. 3–21, 9–4). The genes are found in other arthropods as well,

apparently fulfilling similar tasks, but genes related to *engrailed, wingless,* and *hedgehog* are also found in vertebrates (Chapter 9).

The ENGRAILED protein appears in 14 narrow stripes, only a few cells wide. In the absence of the proper ENGRAILED protein, a stripe in the posterior part of each segment is replaced by a duplicated and inversely oriented anterior stripe. Hence, the designation segment *polarity* genes. Homozygous *engrailed* mutants display segments with mirror-image duplicated anterior stripes.

In normal embryos the stripes expressing *engrailed* are adjacent to stripes expressing the WINGLESS protein. The border between these adjacent stripes marks the future boundary between two visible segments. WINGLESS is not incorporated into the nuclei. Instead, it is spit out from the producing cells as a signal molecule (Fig. 9–4).

However, signals are not emitted unidirectionally. There is a back-and-forth pattern. For example, the cells producing ENGRAILED also produce HEDGEHOG. The HEDGEHOG protein is exposed on the cell surface and is presented to the anterior neighbor expressing WINGLESS. Stimulated by the presented signal, the anterior cells in turn continue to emit WINGLESS. Thus, the expression zones of *wingless* and *hedgehog* stabilize and hold each other in check.

In other parts of the body, such as the **wing imaginal discs,** another variant of HEDGEHOG is made. This second version of HEDGEHOG is cut off at the cell surface and spreads by diffusion in the interstitial spaces between the epithelial cells. For example, it spreads in imaginal discs that later, in metamorphosis, are used to construct the adult fly. In some imaginal discs the row of cells adjacent to the row of cells producing HEDGEHOG respond to the HEDGEHOG signal by expressing WINGLESS, as is the case in the body segments. In other discs cells adjoining HEDGEHOG cells respond by secreting DECAPENTAPLEGIC protein.

In vertebrates freely diffusible proteins related to HEDGEHOG (Chapter 9) or DECAPENTAPLEGIC fulfill the function of signal molecules controlling and promoting embryonic development (Fig. 9–4).

In *Drosophila* mutual interactions between *engrailed* and *wingless* expression is also implicated in the segregation of the mesoderm from the ectoderm. The mesoderm, newly formed in gastrulation, emits WINGLESS; under its influence, the ectoderm continues to express *engrailed.*

3.6.4.3 Genes Effecting Segment Identity: The Homeotic Genes   Homeotic genes ultimately determine which particular type of body segment will appear. For instance, will a given segment become a wingless prothorax or a winged mesothorax, a metathorax with halteres, or an abdominal segment? The genes responsible are the **homeotic selector genes** or **homeotic master genes;** they define the individual identity of each segment. In *Drosophila* most of the

homeotic genes are on the third chromosome, arranged in two clusters. One cluster is called the *Antennapedia complex* (*Antp-C*) and the other is called the *bithorax complex* (*BX-C*). Together, both constitute the *HOM complex* (Fig. 3–22, 10–3).

The five genes of *Antp-C* are expressed in segments destined to form the head and the thorax; the three genes of *BX-C* are expressed in the thorax and the abdomen (Fig. 3–22). Defects in these genes may lead to spectacular **homeotic transformations:** a morphologically correct structure is made at the wrong place. For instance, the *Antennapedia* gene is needed to specify the wing-bearing mesothorax. In a dominant *Antennapedia* mutant, the gene is expressed in the head as well as in the thorax. Accordingly, parts of the head are transformed into thorax segments, but thorax segments are committed to bear legs and not antennae: therefore, a pair of legs, rather than a pair of antennae, arises from the head (Fig. 3–23). A similar effect is brought about by the *nasobemia* mutation of the *Antennapedia* locus. On the other hand, in homozygous recessive *Antennapedia* mutants, no functional ANTENNAPEDIA protein is present in the thorax, and antennae sprout out of the leg positions.

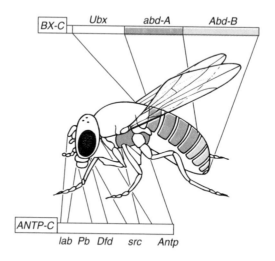

Figure 3–22 *Drosophila.* Homeotic genes of the *Antennapedia complex* (*Antp-C*) and of the *bithorax complex* (*BX-C*). Note: The sequential positions of the genes along the chromosome correspond generally to the sequence along the body of locations where the genes are expressed (see also Fig. 10–3). However, the spatial expression domains in the body are not always sharply demarcated and some domains overlap with others. For example, *Antp* is mainly expressed in the thorax segment T2 but, with decreasing intensity, in T1 and the abdomen as well. (Redrawn after Gilbert.)

In mutant flies in which several homeotic mutations were collected by inbreeding, the halteres are transformed into wings (Fig. 3–23). Thus, the evolutionarily original state of four-winged insects is reestablished (**atavism**).

The genes of *Antp-C* and *BX-C* are arranged on the third chromosome, one after another. The order of their alignment in the chromosome roughly reflects the temporal and spatial order of their expression. Moving along the *Drosophila* body from the head to the rear, one first will see ANTP-C proteins, and these continue to be present. In the posterior thorax, the first BX-C proteins are met. In the last abdominal segments, A8, proteins encoded by several genes of the *HOM complex* are present, most abundantly those encoded by *Abd-B*, the last of the *BX-C* genes.

The increasing abundancy of *BX-C* transcripts in the posterior body is attributed to a decreasing concentration of suppressing gene products, in particular, of the products of the genes *polycomb* (*Pc*) and *extra sex comb* (*esc*). The proteins encoded by these proteins are more abundant in the thorax than in the posterior abdomen.

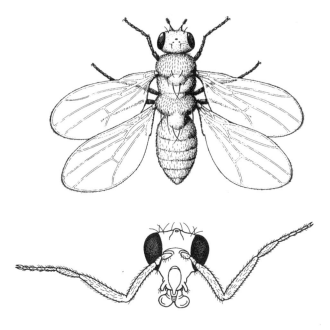

Figure 3–23 *Drosophila*. Homeotic transformations caused by inbreeding mutated homeotic genes of the *Antennapedia complex* (bottom, after a photograph by Lawrence) or the *bithorax complex* (top, after a photograph by Lewis). In the four-winged fly the halteres are transformed back into wings.

Like the BICOID protein, most of the proteins encoded by the segmentation genes and homeotic selector genes include a DNA-binding domain—for instance, a **zinc finger domain** (*hunchback*) or a **helix-loop-helix domain**—derived from the *homeobox* sequence of the respective gene. Thanks to their DNA-binding domains, these proteins in turn function as **transcription factors** controlling other subordinate genes. This is a remarkable result of recent research: in early development a hierarchical cascade of gene activation is initiated, whereby early expressed genes turn on, or off, batteries of subordinate genes to be expressed later.

3.6.4.4 Genes Governing the Formation of a Complete Organ: The Master Gene, *eyeless*

In *Drosophila* a daring experiment succeeded. For the first time in the history of biology, artificial induction of organ formation was achieved, not by transplantation of tissue, but by directed ectopic expression of a gene. The experiment was based on the detection of an eyeless mutant, and the identification of the corresponding gene, *eyeless*. When the functional wild-type allele was expressed in the imaginal discs committed to develop antennae, legs, or wings, flies arose with extra eyes on their antennae, legs, or wings (Fig. 10–2). More about this spectacular experiment can be found in Chapter 10. The term "imaginal disc" is explained in more detail in the following section.

3.6.5 The Metamorphosis of *Drosophila*

As early as the first larval stage, a fraction of cells remain diploid instead of becoming polyploid or polytene, as do most of the larval somatic cells. The particular task that is assigned to groups of such cells is to construct, beneath the pupal envelope, the **imago**, the adult fly. In the interior cavity of the larva flat, rounded bodies can be found wrapped in thin epithelial bags, the **imaginal discs** (Fig. 3–24, 16–2). In addition, single **imaginal cells** are dispersed within the inner organs, such as the intestine or the Malpighian tubules. The imaginal discs are unwrapped in the course of metamorphosis; they evaginate and expand, and piece together the exterior imago. The fly is a mosaic composed of expanded imaginal discs.

The numerous events of metamorphosis are triggered and synchronized by **hormones,** whose release is controlled by the brain. Hormonal control is outlined in Chapter 19. To understand the experiments described in the following paragraphs, it suffices to know that in the adult fly the production of **juvenile hormone** is resumed. It now functions as a gonad-controlling (**gonadotropic**) hormone, but exerts the same effect on imaginal discs as it did in the larva: the hormone permits the growth of the discs but prevents their prema-

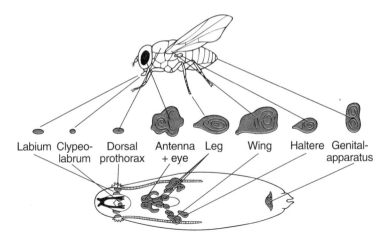

Figure 3–24 *Drosophila.* Imaginal discs. (Redrawn after Alberts et al.)

ture metamorphosis into adult structures. The experiments in question make use of this effect to multiply discs (Fig. 3–25). Imaginal discs are surgically removed from larvae, cut into pieces, and implanted into the body cavity of adult flies. Under the influence of juvenile hormone, the pieces regenerate and grow to normal size. Thus, from one single leg disc a clone of hundreds of leg discs can be raised.

If one wants to know whether a leg disc is indeed still a leg disc, that is, whether the state of determination has been maintained and transferred to the disc's offspring, the cloned discs can be placed back into larvae that are ready to metamorphose. Here they undergo metamorphosis together with their host, and one can find a leg, a wing, or some other fragment of an imago within the body cavity of the hatched fly. It has been shown by such experiments that normally the *state of determination is inherited and maintained over many cell generations.* One speaks of **cell heredity.**

Occasionally, however, the discs forget their imprinted tasks, and one finds a wing instead of the expected leg, or a leg instead of the expected wing. The phenomenon became known as **homeotic transdetermination.** Apparently, homeotic selector genes are involved, but how? How is the state of determination programmed and how is the program replicated in cell division? *Drosophila* will continue to provide answers that might be more difficult to obtain in other systems.

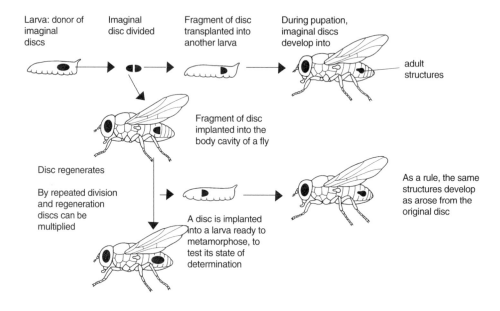

Larva: donor of imaginal discs

Imaginal disc divided

Fragment of disc transplanted into another larva

During pupation, imaginal discs develop into

adult structures

Fragment of disc implanted into the body cavity of a fly

Disc regenerates

By repeated division and regeneration discs can be multiplied

A disc is implanted into a larva ready to metamorphose, to test its state of determination

As a rule, the same structures develop as arose from the original disc

Figure 3–25 *Drosophila.* Transplantation and multiplication of imaginal discs, and assay of the state of determination. Experiments by Hadorn (explained in the text).

3.7 Tunicates: Often Quoted as an Example of "Mosaic Development" in the Phylum of Chordates

The expression "mosaic development" refers to the traditional theory that the egg is subdivided into regions of different qualities (Box 1). In the course of cleavage, these qualities would be differentially allocated into the daughter cells. In this way, the blastomeres would be able to continue development autonomously, independently of their neighborhood.

This view rested on old experiments, done with fine needles and aimed at deleting certain cells or separating them from others. In the 19th century, French researchers were interested in **teratology**—the causes of embryonic malformations. Seeking an experimentally accessible system, in 1886 Laurent Chabry set out to produce malformations by puncturing blastomeres in the tunicate embryo. Because the defects were permanent and were not corrected by the remaining blastomeres, Chabry concluded that each blastomere is responsible for generating a particular part of the body. In 1905 the American E.G. Conklin described how colored plasma is allocated into various blastomeres. By following the fate of these blastomeres, Conklin concluded that each of the colored regions of cytoplasm contains specific "organ-forming substances."

As is often the case in science, the experiments were continued by an international community of researchers that included G. Reverberi (Italy, 1947); J.R. Whittacker (England, between 1970 and 1990); and eventually H. Nishida and other Japanese researchers. The experiments were made using eggs of the ascidians *Styela picta* and *Halocynthia roretzi*.

The ascidian egg contains **cytoplasmic morphogenetic determinants** responsible for programming cell fates. These determinants are almost homogeneously distributed in the oocyte, but they become arranged in a distinct spatial pattern after fertilization but before cleavage begins, through a process of sorting out termed **ooplasmic segregation** (Fig. 3–26). In particular, the yellow plasm contains a component initiating muscle-specific development. Presumably, the component is a member of the MyoD1/myogenin family of transcription factors (Chapter 10, Section 10.2) because it shares with these factors the ability to release the program of muscle cell dif-

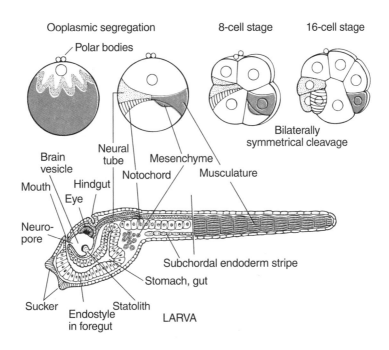

Figure 3–26 Tunicate: an ascidian, a simple chordate, showing development through the larval stage.

ferentiation. The ascidian factor comes to be segregated into 8 cells in the ventral-posterior part of the 64-cell embryo. These cells acquire molecular components early that are indicative of muscle cells (e.g., acetylcholinesterase, F-actin, myosin) and they give rise to the cross-striated muscle tissue in the tail of the larva. Upon isolation from other cells, cells containing myoplasm autonomously continue to develop into muscle cells. However, one also finds secondary muscle cells that only arise if the founder cells have contact with neighboring cells.

The situation is similar to that in sea urchins, in *Caenorhabditis elegans*, and in the spiralian embryo. The work of maternally inherited cytoplasmatic determinants, even if it is as elaborate as in tunicates, must be supplemented by cell-cell interactions between the cells of the embryo. Such interactions are needed for a more detailed allocation of individual cell fates, and are dominant in the vertebrate embryo.

3.8   *Xenopus:* Excellent Eggs and Exemplary Embryos for Vertebrate Development

Amphibians represent the archetype of vertebrate development. The modified development of reptiles, birds, and mammals can be deduced readily from the amphibian model exemplified by frogs and newts. The amphibian egg is rich in yolk and large (often 1 to 2 mm in diameter), yet the egg is able to undergo holoblastic cleavage, and cleavage converts a typical blastula to a gastrula displaying textbook features. In addition, amphibian embryos develop outside the mother and are therefore accessible to experimentation at all stages. The transparent jelly coat can easily be removed, surgically or chemically. Pieces cut out from an embryo are able to continue development in sterile salt solutions without added nutrients, thanks to a supply of yolk in each cell. The eggs are suitable for microsurgical operations, which can be performed by hand with glass needles prepared in the laboratory; no complicated and expensive apparatus is needed.

In the past, most studies on amphibian development were done using eggs of newts, but nowadays the African clawed toad, *Xenopus laevis*, is preferred. *Xenopus*, which always stays in the water, is easily maintained, and because *Xenopus* is bred in the laboratory, there is no need to take animals from natural populations. In particular, *Xenopus* can be induced to spawn at any time by injecting gonadotropic hormones. Females receive about 600 IU of human gonadotropin the day before the intended date of spawning; males receive 300 IU 2 days before, and an additional 300 IU 1 day before.

3.8.1 The Study of Oogenesis in *Xenopus* Yielded Basic Knowledge, Applicable to Oogenesis in Humans

One scarcely would understand oogenesis in humans (Fig. 3–45) if oogenesis in *Xenopus* had not been studied thoroughly already. In 1958 researchers at the University of Oxford discovered a mutant whose cells, although diploid, contained only one nucleolus in their nuclei instead of two nucleoli present in the wild type. Nucleoli are the factories where ribosomes are manufactured. The heterozygous mutant, called *1-nu*, was viable because one set of ribosomal genes (which are clustered in the nucleolus organizer region of a chromosome) suffices to make enough ribosomes. Crossing *1-nu* × *1-nu* resulted in homozygous *2-nu*, heterozygous *1-nu*, and homozygous *0-nu/0-nu* in the Mendelian ratios 1:2:1. The *0-nu/0-nu* offspring who lacked both nucleoli were not viable in the long term. However, surprisingly, embryonic development proceeded rather normally and the embryos even reached the stage of free-living tadpoles. Then they died.

This surprising finding prompted the working hypothesis that oocytes might be provided with ribosomal RNA and the message for ribosomal protein even before they undergo meiosis. This dowry would enable embryos arising from *0-nu* eggs and *0-nu* sperm to manufacture ribosomes without making use of their own genome, which lacks the corresponding genes. Perhaps, in the oocytes from which the *0-nu/0-nu* embryos arose, the message for the production of ribosomes was transcribed before meiosis led to the haploid stage. Before completion of the meiosis in the oocytes of the *0-nu/1-nu* mother, a normal chromosome was still present besides the mutant chromosome that was left after meiosis in the oocyte (while the normal chromosome was allocated to a pole cell). The analysis of the phenomenon has led to a deeper understanding of oogenesis. In fact, amphibian oocytes get a dowry enabling the embryo to pass through embryogenesis even without any contribution of its own genome. However, normal transcriptional activity commences in the stage called the **"midblastula transition"** shortly before gastrulation begins (see as follows).

In amphibian oogenesis meiosis starts in the ovary of a young female soon after metamorphosis and takes several months. The meiotic prophase proceeds up to the diplotene stage. Then the prophase is interrupted for a long time during which the oocyte undergoes extensive growth. In the nuclei of the oocytes high transcriptional activity is initiated. As an expression of this high activity, **lampbrush chromosomes, rDNA amplification,** and **multiple nucleoli** occur in the nuclei. Furthermore, the liver of the female supplies maternal **vitellogenins** (protein) that are carried to the ovary by the blood stream, taken up by the oocyte, and stored in the form of yolk granules. Oogenesis in vertebrates is described in more detail in Section 3.12 (human) and Chapter 5.

3.8.2 Amphibian Embryogenesis Is the Textbook Prototype for Understanding Vertebrate Development

3.8.2.1 Determination of the Body Axes Even before cleavage starts, processes are initiated that lead to the establishment of the egg's spatial coordinates—that is, to the establishment of bilateral symmetry. Unlike the egg of *Drosophila*, the amphibian egg does not receive a maternal dowry that completely predetermines the future bilateral body organization. In its outward appearance the *Xenopus* egg presents a black animal hemisphere and a white vegetal hemisphere. Besides this rotationally symmetrical pigmentation pattern, no other polarity axis is visible. Yet, being a bilaterally symmetrical organism with right and left sides, a belly and a back, the vertebrate animal needs two axes of asymmetry: an anteroposterior axis and a dorsoventral axis. How are they established? What is the relationship between the animal-vegetal axis and the two final coordinate systems of the body?

If we view the egg as globe of the earth, and project the future embryo onto this globe, the head of the embryo will come to lie near the animal "North Pole" (in other amphibians it will lie on the Tropic of Cancer) and the head-tail line will extend along a longitudinal line over the Northern Hemisphere, cross the equator, and terminate in the Southern Hemisphere near the Tropic of Capricorn (Fig. 3–27). The general directive that specifies that the head should be situated in the Northern Hemisphere and the tail in the Southern Hemisphere is laid down in the internal organization of the egg (animal-vegetal asymmetry). However, the embryo needs further instruction to determine along which longitude the line should extend. The specification of this head-tail line is called **dorsalization** and is accomplished by the interaction of several events. Besides the given animal-vegetal egg axis, there is also the point where the sperm enters the egg and the gravitational force of the (real) earth.

The sperm can only attach on the animal hemisphere, but the exact point is determined by chance. After its entry, the **"fertilization membrane"** is elevated, just as in the sea urchin egg. This elevation enables the egg to rotate so that the yolk-rich, heavy vegetal hemisphere points downward in response to gravity.

For dorsalization to occur, a uniform rotation of the entire, unaltered egg would not be much help. However, not all components of the egg move to the

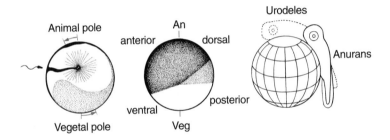

Figure 3–27 Amphibian. Determination of bilateral symmetry. The future dorsal-posterior site (tail) is opposite the entry point of the sperm. The projection of the embryos onto the egg sphere does not take into account cell movements during gastrulation and therefore does not represent a true fate map. The projections merely superimpose the initial egg sphere and the finished tail bud larva.

same extent and into the same direction. Induced by the sperm and influenced by gravity, active movements within the egg lead to an asymmetrical rearrangement of egg components. While the inner cytoplasm, heavy with dense yolk, sinks down and remains stabilized by gravity, the egg membrane and outer cortical layer of the cytoplasm shift relative to the inner cytoplasmic mass about 30° towards the point of sperm entry. This rotation of the cortical layer relative to the inner mass also causes a redistribution of pigment granules in the animal hemisphere. In eggs of the frog *Rana*, the redistribution sometimes leads to a partial depigmentation on the future dorsal side opposite the point of sperm entry. The region of diminished pigment is referred to as the **gray crescent.** The gray crescent or its spatial equivalent (in *Xenopus*, a gray crescent is not visible) marks the region where gastrulation will be initiated and the blastopore will form. The egg now has a bilaterally symmetrical organization. A line drawn from the gray crescent over the animal pole to the sperm entry point coincides with the line extending from the tail bud over the back to the head.

Such projections, however, must not be misunderstood as a definitive fate map. The material located between the animal pole and the gray crescent will be displaced largely into the interior during gastrulation. Cellularized egg material located around the gray crescent will shift into the head region and material near the poles will come to lie in the tail region.

The cellular and molecular means by which sperm and gravity effect the molecular organization of the egg is largely unknown. Presumably, the cytoplasmic rotation that occurs during fertilization leads to a segregation and patterning of maternal cytoplasmic determinants. In particular, the redistribution of **maternal RNA** coding for "induction factors," such as ACTIVIN, WNT, and NOGGIN, or transcription factors, such as GOOSECOID and DOR3, may lead to an enrichment of those components in the area of the gray crescent and, thus, in the region where the Spemann organizer (see Section 3.8.3) will appear.

3.8.2.2 Cleavage and Gastrulation (Fig. 3–28) Radial, holoblastic cleavage leads to a blastula. Gastrulation combines processes of invagination and epiboly. It starts in the center of the area corresponding to the gray crescent in frogs. A group of cells sinks into the embryo, forming a groove and pulling along adjoining cells: the **blastopore** is formed (Fig. 3–29). The crescent-shaped groove of the blastopore becomes elongated and eventually forms a circle. The dorsal sector of the circle is referred to as the **upper blastopore lip.** The cells constituting the lip are constantly changing because they are travelers merely passing through.

Attracted by signals emanating from the region around the blastopore, cells of the blastoderm stream together toward the blastopore (**convergence movement**); arriving at the blastopore, they roll over the lip and squeeze into the interior. Once inside the embryo, the cells move in an anterior direction as a sheet and spread along the entire inside surface of the blastoderm, thereby forming the **archenteron.** Involution and gliding are partly the result of changes in cell adhesiveness and are supported by the formation of lobopodia by the cells located at the leading edge (Fig. 12–1). By such active migration, cells of the animal hemisphere and of a belt of cells around the equator, termed the **marginal zone,** invaginate. The marginal cells passing over the lip pull along yolk-rich cells of the ventral blastoderm. All of these cells participate in forming the roof and floor of the archenteron. The archenteron becomes the principal cavity at the expense of the blastocoel.

Almost half of the wall of the blastula is removed from the surface and moves to the interior. Nevertheless, the circumference of the gastrulating ball remains unchanged because cells of the remaining outer wall spread out by dividing and flattening (epiboly), thus compensating for the loss.

Germ-Layer Formation. The archenteron sometimes is termed "mesendoderm" because it gives rise to both the endoderm and the mesoderm. The roof of the archenteron detaches from the floor; the roof comes to be the **mesoderm** (also called the **chordamesoderm,** as it will give rise to the notochord and the mesoderm). After the detachment of the roof, the walls of the remaining archenteron expand and converge to replace the lost roof. The archenteron has become the **endoderm.** The outer wall of the gastrula is des-

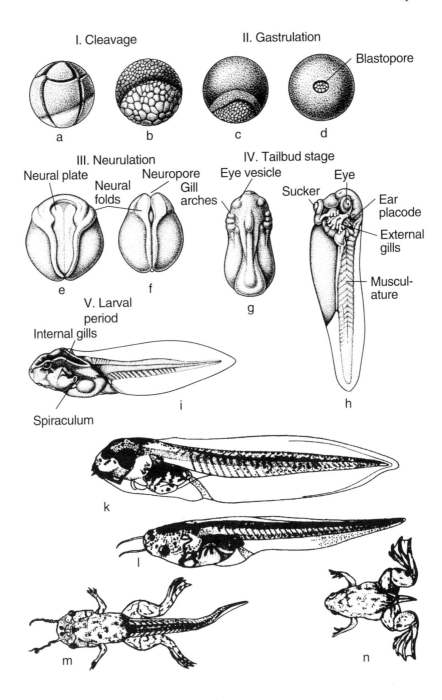

Figure 3–28 Amphibian. Development of the frog (*Rana*) and of the clawed toad (*Xenopus laevis*). (a–i) *Rana* (after Houillon); early stages of *Xenopus* look quite similar. (k–n) *Xenopus* larvae and metamorphosis (redrawn after Wischnitzer).

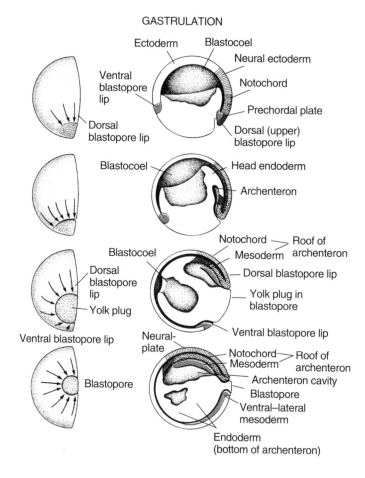

Figure 3–29 Amphibian. Gastrulation. Left column: Half of the embryo, seen from the outside. Arrows indicate the movement of the superficial cells toward the blastopore. Right column: Section along the anteroposterior body axis. (Redrawn after Saunders.)

ignated as the **ectoderm;** in the animal region of the gastrula it is also known as the neural ectoderm because the animal region will give rise to the nervous system.

3.8.2.3 Neurulation and Organogenesis

The Nervous System. **Neurulation** is a process by which cellular material is sequestered to form the brain and spinal cord (Fig. 3–30). The keyhole-shaped **neural plate** is formed above the detaching roof of the archenteron. The plate is delimited by **neural folds** that rise like surging waves and con-

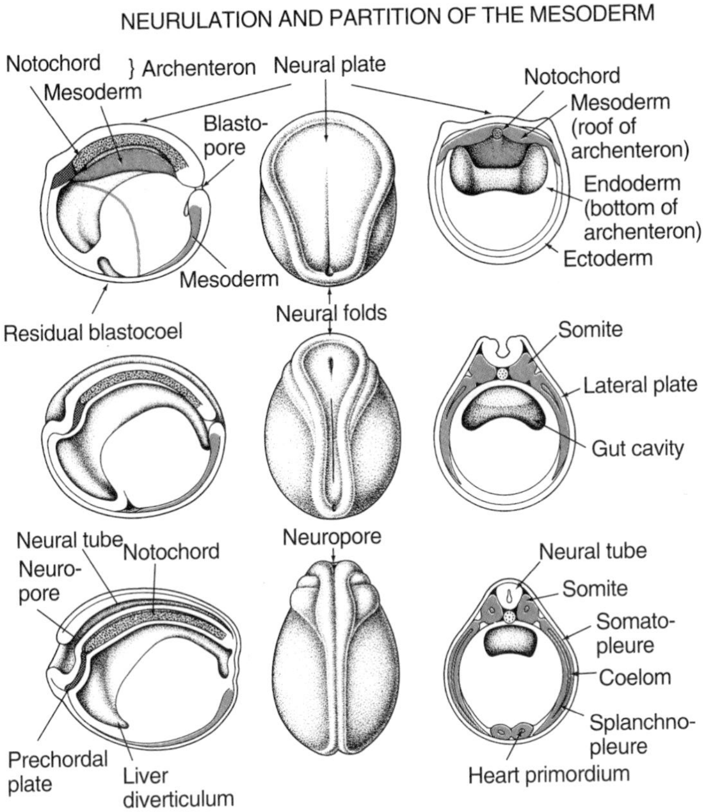

NEURULATION AND PARTITION OF THE MESODERM

Figure 3-30 Amphibian. Neurulation. Left column: Section along the anteroposterior body axis. Right column: Transverse section along the dorsoventral axis. Middle column: Looking down upon the neural plate. (Redrawn after Balinsky.)

verge along the dorsal midline of the embryo. Here the folds adhere to each other and fuse, forming the hollow **neural tube.** An embryo undergoing these processes is called a **neurula.** While the tube is being formed, it sinks into the interior of the embryo and detaches from the surface. For some time closure of the neural tube is incomplete: an anterior and a posterior **neuropore** is left open.

Above the neural tube, the ectoderm recloses the outer wall. The anterior, expanding half of the neural tube will form the **brain,** while the posterior half will elongate to form the **spinal cord.** Residual cell groups along the neural tube, termed "**neural crest cells,**" will supplement the central nervous system by forming the spinal ganglia and the **autonomous nervous system** (Fig.

4–3, 13–1). Further developmental potentials of the fascinating neural crest cells will be described in Chapter 13, Section 13.4, and Chapter 14, Section 14.1.

Mesodermal Organs (Fig. 3–30, 3–31, 3–32). The detached chordamesoderm subdivides itself in a process of self-organization into several parts.

1. The anterior part of the mesoderm, which was the first mesoderm to involute, gives rise to the **prechordal head mesoderm** that forms the muscles of the head region.
2. The head mesoderm is followed by the more posterior **chordamesoderm.** Along its midline, the round rod of the **notochord** detaches. The notochord is a transient structure needed to organize the development of the CNS and the vertebral column.

3 and 4. The dorsal axial mesoderm that is involved in notochord formation is flanked by two lateral bands, the **paraxial** or the **somitic mesoderm.** These bands separate into blocks of cells called **somites.** Somites are also transient structures, but they are extremely important in organizing the basic segmental body pattern, and they give rise to many tissues: vertebrae; muscles; and the inner layer of the skin, called the dermis.

5 and 6. The adjoining residual cells derived from the roof of the archenteron become flat, expanding sheets called **lateral plates.** Both plates

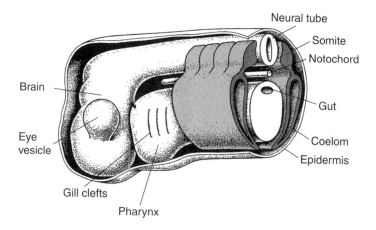

Figure 3–31 Amphibian, generalized vertebrate. Basic body plan or phylotypic stage. By this stage, the basic traits of a chordate are elaborated: dorsal neural tube, notochord, foregut with gill slits, ventral heart (not shown). In addition, eye vesicles and somites are formed in all vertebrates. (Redrawn after Waddington.)

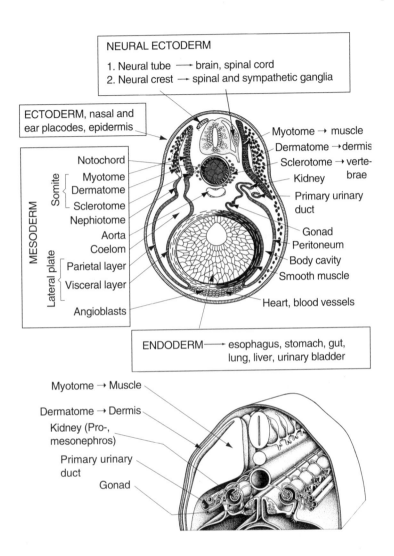

Figure 3–32 Amphibian, generalized vertebrate. Development following neurulation. Fate of the "germ layers." (Redrawn after Portmann, top, and Huettner, bottom.)

spread laterally and ventrally into the space between the ectoderm and the endoderm. While doing so, a cleft within the plates transforms them into flattened bags. The bags enclose a new cavity, called the **coelom** or secondary body cavity. The coelom will subdivide into the pericardial cavity, enclosing the heart, and the large body cavity. In mammals this cavity is subdivided into the pleural cavity, enclosing the lungs, and the peritoneal or visceral spaces around the intestine.

The wall of the lateral plate snuggling up and clinging to the ectoderm (**somatopleure,** or **parietal mesoderm**) will form the peritoneum and pleural lining, whereas the wall clinging to the endoderm (**splanchnopleure**) will provide the circular musculature of the stomach and the intestine, as well as the mesenteries needed to suspend the digestive tract in the coelom.

Somites and lateral plates are connected by intermediate mesoderm that forms a longitudinal series of thickenings, the **pronephrotic rudiments.** The solid rudiments hollow out and form funnels opening into the coelom. The pronephrotic tubules will later be transformed and supplemented by further mesodermal derivatives to yield the **urogenital system.**

In the anterior body beneath the archenteron, where the two lateral plates approach each other, migratory mesodermal cells accumulate. They are myogenic and organize themselves to form the **heart** and the adjoining major **blood vessels** (Chapter 15).

The most extensive developmental potential belongs to the **somites.** Somites are transient but the cells constituting them do not vanish. They separate into two main groups: the **sclerotome** and the **myodermatome.** The myodermatome further subdivides into the **myotome** and the **dermatome.**

The cells of the **sclerotome** lose contact to each other, emigrate, and accumulate at the notochord, thereby surrounding it. They become **chondrocytes** committed to construct the **vertebral bodies.** Most of the enclosed notochordal cells die, but some are left to form the nucleus of the **intervertebral discs.** These are the discs that slip in painful back injuries.

The packets of the **myotomes** expand and give rise to the **cross-striated musculature** of the body that occupies the largest spaces in the trunk and tail region.

The (dividing) cells derived from the **dermatome** migrate and spread extensively, clinging to the inner surface of the ectoderm. They will give rise to some additional muscle tissues but will mainly form the **dermis,** while the epidermis will be formed by the outer ectoderm covering the dermis.

The complete derivation of all connective tissue masses and **skeletal elements** is complex. Most precursors of the chondrocytes and osteocytes that construct the cartilage and bones of the body derive from the somites, but those osteocytes that form dermal bones appear to arise in the dermis. The skeletal elements in the ventral head and pharynx region, summed up as **visceral skeleton,** do not derive from the mesoderm but from the **neural crest cells** (Fig. 4–3, 13–1; Chapter 13).

Endodermal Organs.   The archenteron forms the **pharynx** with **gills** and **lungs,** and the digestive tube with adhering organs, such as the **pancreas,** the **liver,** and the **gallbladder.** The roof of the pharynx participates in forming the pituitary gland by contributing the anterior lobe, the **adenohypophysis** (Fig. 4–1, 14–2).

3.8.3 The Most Famous Embryological Experiments Were Done on Amphibians

3.8.3.1 Nuclear Transplantations
Experiments that led to the production of **cloned frogs** (Box 2) are justly famous. Originally, the experiments were planned (by R. Briggs, T.J. King, J.B. Gurdon, and others) not to clone animals but to examine whether somatic nuclei remain totipotent in the course of development, or if genetic information is irreversibly lost or inactivated (Chapters 10 and 11).

3.8.3.2 Identical Twins and Chimaeras
As early as 1920, Hans Spemann separated early amphibian embryos (from the two-cell stage up to beginning gastrulation) into halves using lassos made from baby hairs to make a ligature (Box 2). If the embryos were divided along the animal-vegetal axis so that each half was provided with material of the gray crescent, **monozygotic** (one egg) **identical twins** arose. Partial constriction resulted in "Siamese" **conjoined twins.** By pressing together two cleaving embryos from the newts *Triturus cristatus* and *Triturus vulgaris,* **chimeras** composed of a mosaic of genetically different cells were obtained.

3.8.3.3 Transplantations Ask How Determination Proceeds and Whether Cells Receive Positional Information
The most exciting experiments in the history of developmental biology were carried out by Hans Spemann and his students between 1920 and 1940, using amphibians. The experiments yielded the first information on the significance of signals sent out by parts of the embryo to instruct other parts (**embryonic induction**). By microsurgery, pieces were taken from donor embryos and were inserted into host embryos at various places (Fig. 3–33, 3–34). The experiments were aimed at finding out whether tissue pieces would behave **according to their (new) location,** meaning that they still could be reprogrammed by positional cues, or whether they would behave **according to their descent,** meaning that they already were irreversibly committed.

In a famous experiment that addressed the question of **positional information** (Spemann: "**development according to location**"), the nondetermined, prospective epidermis of the belly region was transplanted to the area fated to become the mouth region of a newt. The transplant formed the mouth and the teeth according to its new location (but horny teeth according to the genetic potency of the donor; tadpoles of frogs have horn teeth and not teeth of dentin as do salamanders!) (Fig. 3–33). In the course of such studies the phenomenon of induction was observed.

3.8.3.4 Embryonic Induction: One Part Gives Orders and the Other Part Responds   Systematically planned transplantation studies, which began in 1918, led to the discovery of extensive **embryonic induction** (published in 1924; for this discovery Spemann won a Nobel prize in 1935). In a classic example, the roof of the archenteron induces the overlaying animal ectoderm to form the neural plate and, hence, the central nervous system (this is so-called "primary" or neural induction). However, induction begins earlier and has a spectacular culmination when the early gastrula forms the blastopore. At this time and place the **organizer** becomes active.

Prompted by Spemann, Hilde Mangold transplanted the upper blastopore lip of the darkly pigmented *Triturus taeniatus* into the ventral ectoderm or (accidentally) into the blastocoel of the white blastulae of *Triturus cristatus* (Fig. 3–34). The result was exciting; when properly inserted into the ventral ectoderm, the donor tissue did not adapt to its new surroundings. Instead, it started invaginating; soon axial organs (notochord, somites) and a head emerged, and eventually a complete second embryo formed, joined to the other, a Siamese twin. However, even when the inserted lip dropped into the interior cavity, it initiated gastrulation in the surrounding host tissue. When the host embryo developed further, it was found that an additional system of axial organs appeared on the side where the graft came to lie. Partial or complete supernumerary embryos arose. Because black donor and white host embryos had been used, the supernumerary embryos were determined readily to be composed of host as well as donor tissue (Fig. 3–34). In fact, the largest part of the supernumerary embryo was made by the host; host cells were induced to participate in the formation of the second body.

Spemann referred to the upper blastopore lip as the **organizer** because it induces host cells to change their fate and is able to release an entire, coordinated developmental program leading to a well-organized, complete embryo. A general term designating the source of an inducing signal is **inductor** or **inducer.**

3.8.4 Induction Is Based on Signal Cascades; Signal Molecules Have Been Identified   **Organizing** is a complex event. A whole bundle of signal molecules appears to be involved, either simultaneously or consecutively effective.

Recent knowledge dates the starting point of the story back to fertilization or even oogenesis. During fertilization, dorsoventral specification is accomplished by rotation of the egg cortex and displacement of inner cytoplasmic

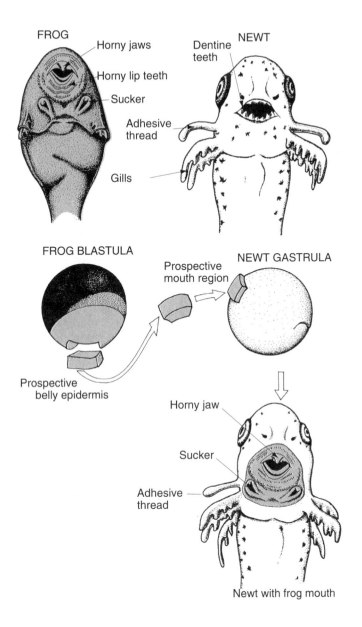

Figure 3–33 Amphibian. Classic transplantation experiment of Spemann. A future ("presumptive" or "prospective") piece of the epidermis of the belly, taken from a frog blastula, is transplanted into the future mouth region of a newt. The frog piece becomes committed in the mouth region of the host to form mouth tissue. However, it is a frog mouth that is formed, according to the genetic potential of the frog cells. Positional information is not species specific, but the positional information received by a cell can only be interpreted according to the cell's genetic program.

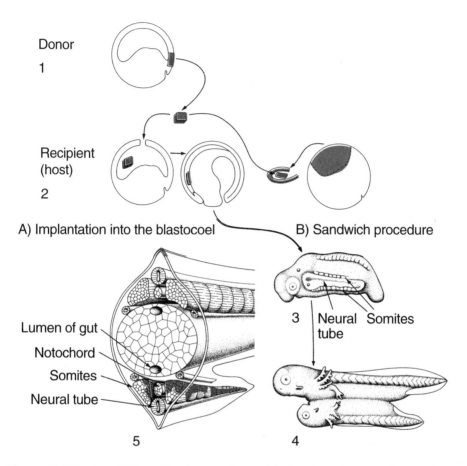

Figure 3–34   Amphibian. Classic experimental induction of a secondary embryo
by transplantation of an upper blastopore lip ("organizer"). The transplanted tissue
is marked in red, including derivatives that are found in the secondary embryo
(5, cross section). However, the secondary embryo is mainly formed by the host.
Experiment proposed and interpreted by Hans Spemann, and carried out by Hilde
Mangold.

constituents. Maternal mRNA coding for signal molecules and receptors
(among many other proteins) becomes localized in a distinct pattern. Incorporated into the cells of the blastula, the differentially disseminated mRNA
species enable the cells in the various regions of the embryo to generate and
to receive different sets of signal molecules. Important maternal information
is concentrated in and around the area exposed as the gray crescent.

Inductive interactions between the cells begin in the early blastula. They
are studied at several levels with methods that reflect the history of experimental developmental biology.

1. **Surgical methods such as explantations.** Different parts in the mid- and late-blastula stages are isolated, rejoined in various combinations, and cultured, and the result is analyzed microscopically. Antibodies recognizing defined differentiated cells or cells being about to differentiate facilitate the evaluation of the outcome. A historically important example of this type of experiment was the combination of isolated, uninduced animal caps with pieces comprising yolk-laden vegetal cells. While neither animal caps nor vegetal pieces by themselves gave rise to mesodermal structures, the combination did. The vegetal partner caused descendants of the animal cap to develop into mesodermal cells, such as notochord cells, muscle cells, and blood cells. By combining different parts of the embryos at different times, consecutive inductive interactions were disclosed. The observations were brought into a coherent picture (Fig. 3–35).

2. **Bioassays of possible inducers using animal caps.** Uninduced animal caps are exposed to extracts or fractions thereof, and the result is again examined microscopically (Fig. 3–36).

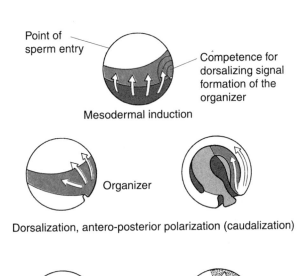

Figure 3–35 Amphibian. Sequence of inductive events in the blastula and the gastrula, suggested by the outcome of experiments in which different parts of early embryos were combined. Arrows indicate the assumed spread of inductive signals.

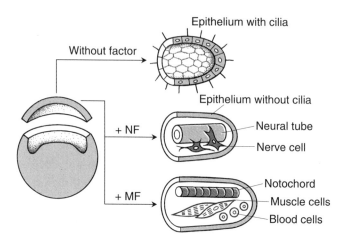

Figure 3–36 Amphibian. Assay to test the inducing potential of extracts or puri-fied factors. Uncommitted animal caps are removed from blastulae and exposed to solutions containing putative factors. The caps form hollow spheres with ciliated walls in the absence of inducing factors (top). If inducing factors are present in the solutions, the exterior layer forms an epidermal epithelium enclosing neuronal or mesodermal cells. Neuralizing factors (NF) or mesodermalizing factors (MF) have been extracted from heterologous sources (i.e., from other species) or from autolo-gous sources (i.e., from the same species from which the animal cap was taken).

3. **Injection of putative inducers, or of mRNA coding for such induc-ers.** Putative inducers, either isolated from *Xenopus* embryos (autologous source) or from alien donors (heterologous sources such as chick embryos, fetal calf serum, or a supernatant of cell cultures) is injected into a site that still contains uncommitted cells and normally does not develop the expected structures. If the putative inducer is a polypeptide, mRNA can be injected instead of the protein. Injection of antisense RNA, on the other hand, can be used to weaken or eliminate an inducing interaction.

From a variety of such experiments, a coherent picture must be recon-structed. The picture must be dynamic like a film. Different researchers tell stories differing in detail, but in general the play is divided into three acts.

3.8.4.1 **The Mesodermalizing Induction** Signals spread from the vegetal cells of the blastula and cause a broad, ringlike zone (marginal zone) around the equator to become the mesoderm in the future. The cells respond by expressing, for example, the gene *brachyury*, a marker of mesodermal progeni-tor cells. During gastrulation, this marginal zone is shifted into the interior, forming the roof of the archenteron that will give rise to the notochord, the somites, and the lateral plates. Similar results can be obtained by bathing omnipotent ectoderm, taken from the animal hemisphere (animal cap), in purified extracts from chick embryos or in solutions of various "growth factors" produced by cultured cells of *Xenopus* or mammalian embryos.

Employing classic biochemical techniques and bioassays, a mesoderm-inducing protein that turned out to be a member of the TGFβ family of growth factors was enriched and purified from many gallons of chick extract. Soon, additional mesoderm-inducing factors were found. All factors identified so far are proteins or glycoproteins. They belong to the following families:

1. The **FGF family** comprises basic and acid **fibroblast growth factors, bFGF and aFGF. bFGF** induces "**ventral mesoderm,**" that is, blood cells, mesenchyme, and some muscle cells.

2. The **TGFβ** family with the transforming growth factor TGFβ and the **activins.** Activins are heterodimers consisting of the moieties α and βA (activin A), or α and βB (activin B). With heterologous gene probes (Box 7), maternal mRNA has been identified in the blastula, encoding several activin-like proteins. The mesoderm-inducing factor XTC-MIF, secreted by a cell line derived from *Xenopus* embryos, has been identified as a member of the TGFβ family and finally as an activin. Historically, this identification was one of the breakthroughs in the chemical characterization of inducers.

   Another member of the TGFβ family, the **bone morphogenetic factor 4 (BMP4),** which is expressed in the anterior-ventral region of the embryo, mediates epidermal rather than mesodermal differentiation. However, besides evoking epidermal development, BMP4 cooperates with mesodermalizing factors to confer ventral properties to mesodermal structures. Ventral mesoderm forms, for example, the heart.

3. The **WNT family,** including WNT-1 and WNT-8. The members of this family display nonrandom sequence correspondence with the *wingless* gene of *Drosophila*.

In the uncleaved egg, maternal mRNA can be found that codes for such inducing factors. Once the egg is cellularized, the mRNAs are thought to be allocated to different blastomeres. The cells translate the message and secrete the produced factors into the interstitial spaces, where the factors spread by diffusion to reach the neighboring cells.

Diffusion results in the establishment of concentration gradients because the available space increases with distance from the source and released factors may be bound to receptors and thus may be removed from the interstitial fluid. It is thought that gradients contribute to the regionalization of the embryo. This idea is supported by findings from the laboratory. **Activin** is known to exert dose-dependent effects in animal caps cultured in vitro (Fig. 3–34): very low concentrations cause the cap to differentiate into the epidermis, but as the activin concentration increases, muscle cells, beating heart-specific muscle cells, and the notochord are evoked.

Remarkably, high concentrations of activin B, or of some strong mesodermalizing factors of the WNT family, can cause explanted animal caps of blastulae to form the **"dorsal mesoderm,"** that is, the prechordal head mesoderm (the progenitor of head muscles), the notochord, and the cross-striated muscle cells (derivatives of the myotome moiety of the somites). Moreover, when mRNA of *activin B* or some types of WNT is injected into early embryos at a ventral site that normally forms the epidermis of the belly, and is expressed there in high enough concentration, it causes the surrounding cells to form a second organizer. As a consequence, a (partial) second body axis may emerge at this ectopic place. Thus, activins and WNT proteins exert effects, directly or indirectly, that also must be attributed to the following, second category of embryonic induction.

3.8.4.2 Dorsalizing and Cephalocaudalizing Induction

Cell communities receiving dorsalizing signal molecules in high doses will form structures characteristic of the head-back-tail region, including the dorsal axial organs such as the spinal cord and the notochord. In addition, the signal molecules appear to mediate or contribute to the longitudinal subdivision of the dorsal axial organs into the head, the trunk, and the tail. Altogether, inductive events appear to occur in several steps:

1. In the early 32- to 64-cell blastula, a small group of blastomeres (referred to as the Nieuwkoop center) beneath the area of the former gray crescent has incorporated some maternal information enabling the cells to release a first set of signals. These signals, possibly factors such as **Vg1** and **activin,** help organize the foundation of an important broadcasting center in the adjacent area.

2. This broadcasting center is known as the **"Spemann organizer."** (Some researchers consider the Nieuwkoop and Spemann centers to be consecutive stages of the same area.) The establishment of the Spemann center may be demonstrated by the presence of a transcription factor encoded by the gene *goosecoid* in the nuclei of cells in this area. (*Goosecoid* has a homeobox sequence resembling, in part, the *bicoid* gene, and, in part, the *gooseberry* gene from *Drosophila*.) The transcription factor GOOSECOID

enables the cells to undergo morphogenetic movements: they begin to form the dorsal blastopore lip. Gastrulation is initiated.

3. However, GOOSECOID is not a secreted molecule and therefore cannot convey information to neighboring cells without the intervention of signal molecules that are secreted or exposed on the cell surface. Several potential signal molecules have been identified. Among them are NOGGIN protein, encoded by the *Xenopus* gene *noggin* and CHORDIN protein encoded by the *chordin* gene.

   The expression of the *noggin* and *chordin* genes starts in Spemann's organizer subsequent to that of *goosecoid*. Both NOGGIN and CHORDIN are expressed at the right place and at the right time to regulate the onset of gastrulation. Microinjection of their mRNA at an anteroventral place causes twinned axes and hence a more-or-less perfect supernumerary embryo (part of the head and the trunk), just as if an early blastopore lip had been implanted.

4. From the blastopore lip, signals cause the cells destined to become the mesoderm to stream toward the blastopore (**convergence movements**) in order to submerge there into the cavity of the blastocoel.

5. While the archenteron invaginates, the rolling blastopore lip continues to be the source of signals that spread anteriorly in the plane of the neuroectoderm and of the roof of the archenteron, forming gradients. Also, those cells that formerly constituted the lip, but now move anteriorly into the interior cavity, may continue to produce signals. The overall effect of these factors is to induce, in the emerging neural plate and in the roof of the archenteron, an **anteroposterior** and **dorsoventral asymmetry.** As a consequence of this polarization, the mesoderm, for example, acquires the competence to anteriorly form head muscles (dorsal) and the heart (ventral), and to posteriorly form tail muscles (dorsal) and the kidney (ventral).

CHORDIN and NOGGIN expression is not confined to the blastopore lip but continues in the notochord. This pattern suggests a role in the elaboration of the CNS, because classic transplantation studies have assigned the capacity to induce neuronal tissue to both these structures. This conjecture will be substantiated experimentally in the following section.

3.8.4.3 Neuralizing Induction   The induction of the CNS with the brain and the spinal cord is also a multistep process.

   1. **Acquisition of competence.** Once committed to become the mesoderm, the marginal zone along the equator radiates signals into the animal

hemisphere. These enable the animal cap to deviate from the pathway towards the epidermis. Instead, the animal cap acquires the competence to form neuronal tissue as the default state. This subdivision of the neuroectoderm into a territory that will form the epidermis and a territory around the animal pole that will form the neural plate implies a change in the expression pattern of the BMP4. This secreted factor, a member of the TGFβ family, mediates epidermal development. In the future epidermis expression of BMP4 continues, while in the future neural plate its expression is suppressed.

When the contacts between the cells are untied by partial dissociation in animal caps taken from *Xenopus*, the caps develop neural cells (neurons and glia) instead of forming a ciliated epithelium. This "autoneuralization" is thought to occur because dissociation is likely to wash away BMP4. In the embryo, FOLLISTATIN and CHORDIN appear to inhibit BMP4 function.

2. **Regionalization according to the "two-signal model."** When the mesoderm is displaced into the blastocoel and moves over the inner surface of the animal cap during gastrulation, it continues to radiate neuralizing signals toward the overlying animal ectoderm. These **"vertical signals"** eventually result in an irreversible neural determination.

Simultaneously, while moving along the ectoderm, the mesoderm confers gradual regional differences upon the neuroectoderm. This time-dependent regionalization is thought to be supported by **"planar signals"** transmitted from the dorsal blastopore lip through the plane of the neural plate. Instructed by the various neuralizing signals, the emerging CNS will subdivide into the forebrain, the hindbrain, and the spinal cord. To explain this subdivision, pioneers in the field of embryonic induction (namely Tiedemann, Nieuwkoop, and Saxen and Toivonen), proposed similar models known as two-step or two-signal models. These models assume that the regional type of neuronal tissue is specified by the combined action of two sets of signals produced by the dorsal mesoderm, that is, by the roof of the archenteron.

The first signal, or set of signals, is referred to as the **"activator"** (by Nieuwkoop) or the **"neuralizing inducer"** (by Tiedemann, and by Saxen and Toivonen), and initiates neural development, inducing competent tissue to form anterior neural tissue (the forebrain and the midbrain). The inducers are proposed to be produced by both the prechordal head mesoderm and the chordamesoderm. Candidates for this type of inducers are NOGGIN and CHORDIN.

The second set of signals, the **"transformer"** (Nieuwkoop) or **"mesodermalizing inducer"** (Saxen and Toivonen), converts the neural tissue induced by the first group into progressively more posterior types of neural tissue, such as the hindbrain and the spinal cord. These signals are thought to require induced anterior neural tissue as substrate, rather than being able to act directly as neural inducers of uninduced ectoderm. The second type of signals is thought to act in a graded manner, with increasing concentrations specifying progressively more posterior neural patterns. Candidate molecules for this second type of signals are some types of FGF, in particular, bFGF.

Beginners in this field of research may be confused when they read that the same molecules that have been said to be used for dorsalization, NOGGIN and CHORDIN, are now said to be used for neural induction. However, not only beginners are confused, as the results obtained from the different laboratories, or their interpretation, are not always in agreement. The reasons for conflicting results or interpretations are the following: (1) We have to do with chains of events: a factor may first induce dorsal mesoderm that in turn induces neural tissue. (2) Cells respond to the same factors differently, depending on the concentration of the factor, and as a result of the cell's previous history. When NOGGIN or CHORDIN are expressed ectopically at the ventral side of a blastula, they may induce the development of a complete secondary axis, including mesodermal and neuronal structures. However, when NOGGIN or CHORDIN are added to competent animal caps in high enough concentration, the cells express forebrain-specific markers without the mediation by mesodermal cells. (3) Further signal molecules that may be present in different amounts modify the response.

Such modifying signal molecules are the members of the HEDGEHOG family. For example, the notochord not only secretes NOGGIN and CHORDIN but also SONIC HEDGEHOG. The secreted factor is thought to spread toward the ventral part of the neural tube, and the **floor plate** of the spinal cord is induced (Fig. 9–5).

Additional molecules whose biological effects and distribution pattern suggest their involvement in regional patterning are the protein FOLLISTATIN and the carotinoid lipid **retinoic acid (RA)**. RA is a versatile derivative of vitamin A and will be introduced in another context (chick limb morphogenesis, Chapter 9, Fig. 9–8).

3.8.5 Homeobox Genes Demonstrate the Success of Induction

The establishment of the Spemann organizer can be seen by the appearance of transcripts and protein coded by the homeobox gene *goosecoid*. During gastrulation, many other homeobox-containing genes, collectively called *Xhox* in *Xenopus*, are expressed in temporal and spatial patterns that outline and define future structures, such as the notochord, the somites, and the neural plate. We learn more about this topic in Chapter 10.

3.8.6 Secondary and Tertiary Induction: An Eye View

When Spemann coined the term "primary induction," he was not yet aware that the cascade of inductive interactions leading to the formation of the central nervous system starts in the uncleaved egg. However, he was aware that further inductive events, called "secondary" and "tertiary," follow primary induction, for it was Spemann who detected the classic example of such a secondary induction: the optic vesicle induces in the epidermis of the **lens** to help complete the eye (Fig. 3–37). (However, in *Xenopus,* unlike other amphibians, poor lens formation would start at the correct place even in the absence of the optic vesicle.) After the lens placode has been formed, the story continues with a tertiary induction: the lens induces the overlying epidermis to supplement the optical apparatus of the eye by forming the transparent cornea.

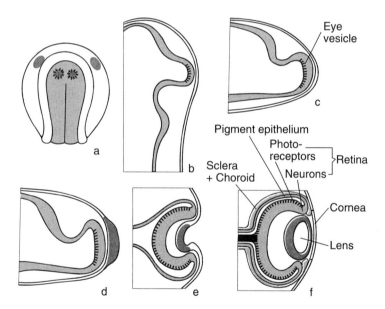

Figure 3–37 Amphibian. Induction of an eye lens. Illustration (a) looks down upon a neurula. Red spots indicate areas from which the future lenses originate (a,d) and where the first inductive signals arrive. In some species, early induction is sufficient; in others, the emerging eye vesicle (b, c, d) must continue to emit inductive signals toward the overlying ectoderm for a lens to be irreversibly determined. The lens (red part in d,e,f) in turn induces the formation of the cornea (f).

3.8.7 Twisted Eyes

Among many experiments done with later stages of amphibian embryos, we have space to mention only one. By removing, rotating, and retransplanting optic vesicles, we gain insight on how the eye is neurally connected with the brain (**retinotectal projection,** Chapter 14; Fig. 14–5). Recent interest in neuroscience prompted similar investigations in fish and the avian embryo.

3.8.8 Metamorphosis and Hormones

The larva of *Xenopus* is transparent. It feeds by sweeping in planktonic algae and other microorganisms. As in other amphibians, the fundamental reorganization into the adult clawed toad is controlled by the hormones prolactin and thyroxine (Chapter 19).

3.9 An Upstart: The Zebra Fish, *Danio rerio*

The zebra fish, *Danio rerio* (formerly *Brachydanio rerio)*, is a cyprinid found in the rivers of India and Pakistan. The adult fish is 4 to 5 cm long, displays a pleasant striped pattern, and lives socially in schools. In industrial laboratories all over the world, zebra fish are widely used for standard toxicological assays. However, increasingly the zebra fish is used for embryological and genetic studies.

Those who promote *Danio* list the following advantages:

- The fish is easy to breed in the laboratory. Mature animals may spawn every morning and a female releases up to 200 eggs. Typically, one female and two to three male fishes are put in a prepared container in the evening. The bottom of the container is replaced by a grid, through which the eggs can drop but which prevents the voracious males from eating the eggs. After spawning, an exhausted female needs 1 to 2 weeks for recovery.
- The eggs are transparent and have a diameter of 0.6 mm; embryo development is fast and is completed in 2 to 4 days, depending on the water temperature. Transparency is one feature that is an advantage in *Danio* compared with *Xenopus*. Another feature is its suitability for extensive genetic analyses.
- For genetic studies, recessive mutations can be produced using mutagenic agents: males are allowed to swim in a solution of the point mutation-inducing mutagen, ethyl-nitroso-urea. For large-scale studies, mutagenized males are crossed with normal females; the F_1 and F_2 generations are inter-crossed and homozygous mutants appear in the F_3 generation. For small-

scale studies, homozygous offspring can be obtained from mutagenized females by parthenogenetically activating their eggs with UV-irradiated sperm. Methods have been developed to double the haploid set of 25 chromosomes to the diploid set of 50 by applying heat shock to suppress the first mitotic division. Haploid eggs would develop until the hatching stage.

The zebra fish is now used extensively to study gene expression and wonderful pictures have been made showing the products of genes containing homeoboxes, zinc fingers, paired boxes, and so on. Our introductory treatment will provide a brief description of the embryo development in the zebra fish.

Teleosts (bony fishes) have taken their own evolutionary pathway. Compared with amphibians and land vertebrates they are an outgroup, as reflected by peculiarities in their ontogeny.

3.9.1 Determination of the Body Axes

After fertilization, cytoplasm streams to the animal pole of the egg and forms a cap of clear material in which the zygotic nucleus is incorporated. Beneath, yolk material accumulates in the vegetal part of the egg. The egg rotates and usually the animal-vegetal axis of the egg becomes horizontally oriented. The head will be positioned in the region of the animal pole; the midline of the animal's back will extend along the uppermost meridian.

3.9.2 Early Embryogenesis (Fig. 3–38)

Cleavage starts 40 minutes after fertilization; it is partial and discoidal, resulting in the formation of a cap-shaped blastoderm, called the **blastodisc.** It contains about 2,000 cells and is perched high upon the noncellularized, spherical yolk mass (because the blastodisc provides some nuclei to it, the yolk is called a syncytium). As cell divisions continue and the disc is flattened out, the blastoderm expands and spreads over the yolk ball. According to the position of the edge of the blastoderm, one speaks of ¼, ½, ¾, or ¼ **epiboly.**

The early blastoderm is already composed of several cell layers. The superficial cell layer is called the enveloping or **covering layer** (some authors use the term "periblast," which, however, is used by others to designate the yolk syncytium). Below the covering layer are **deep cells.** Their number increases in gastrulation by cells immigrating at the margin of the disc.

3.9.3 Gastrulation

The process of gastrulation only remotely resembles that seen in amphibians. Cells of the blastoderm stream toward the margin and involute along the

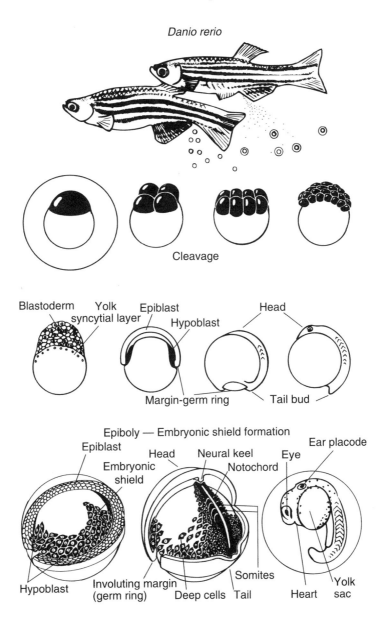

Figure 3–38 *Danio rerio*, the zebra fish. Adult fish and gametes (top); embryo development.

margin. They turn under and move toward the animal pole. Under the transparent blastoderm, a hem appears; it broadens to form the **hypoblast,** while the overlying blastoderm is called the **epiblast.** The epiblast and the hem of the hypoblast together form the **germ ring** around the circumference of the blastoderm. The hypoblast, however, is only a loose and rather transient association of cells. The cells are able to disperse themselves as amoeboid cells by migration over the surface of the yolk. These cells can also reassemble elsewhere, as do *Dictyostelium* cells in aggregation.

Deep cells converge and aggregate along a meridian that marks the future head-back-tail line. The aggregation is called the **embryonic shield** or **primitive streak,** and it is here that the primary axial organs will be formed. As epiboly continues and the blastoderm spreads over the entire yolk mass to enclose it, the primitive streak lengthens and extends over the animal and vegetal pole. Its developmental potential includes the neuroectoderm (in the epiblast) and the mesoderm (the aggregated deep cells), but neural folds do not appear. Instead, a solid neural keel detaches from the epiblast. Beneath the keel, aggregated deep cells organize themselves into the notochord and the somites. Later-arriving deep cells supplement the mesoderm by forming the lateral plates. Endodermal epithelial layers form underneath the mesoderm.

For some time, the detached neural keel persists as a neural rod. Later, a central canal transforms the rod into a neural tube. In its anterior part the tube widens and gives rise to the brain. The transparency of the fish embryo makes it possible to observe the development of the eyes in living fish and the inner ear from the ear placodes (see Fig. 4–2), and even the outgrowth of nerves, such as the spinal nerves.

3.10   Chick, Quail, and Chimeras of Both

The large size of the avian egg and its availability made it possible to observe the development of the chick even in ancient times (Egyptians, Aristotle; see Box 1), although in the first 2 days of incubation not much can be seen without a magnifying glass. Compared with amphibians, reptiles and birds have evolved some features that permit them to live completely on land and to omit a larval stage. The egg is huge—the egg cell proper is the yellow ball enveloped by the fortifying acellular, elastic, and translucent vitelline membrane. The egg cell is surrounded by the albumen (the egg white) and the whole is contained in further envelopes, including the outer calcareous shell. Moreover, within the shell the developing embryo becomes wrapped in a fluid-filled cyst, the **amnion** (Fig. 3–38). Reptiles, birds, and mammals are collectively referred to as **amniotes.**

Nowadays, fertilized eggs must be purchased from specialized farms or breeding stations, because eggs that grocery stores sell are not fertilized generally. The rooster has to do his job and contribute sperm in a timely manner, shortly after an egg is liberated from the ovary and before it is enveloped in the oviduct.

Cleavage begins immediately after fertilization in the oviduct of the hen. This is a disadvantage of working with avian embryos. Very early stages are not accessible to experimental manipulation. The morphological events of cleavage have been studied in eggs removed from the oviduct.

As in the zebra fish, cleavage is meroblastic and discoidal. Cytoplasmic division is at first by furrows produced by downward extensions of the cell membrane, leaving the blastomeres open to the yolk for some time. The progressive cellularization of the area around the animal pole results in the development of a blastodisc. As in the fish embryo, at the edge of the disc is a circular syncytial zone where the cells are still open. By continuing cell divisions, the blastodisc (Fig. 3–39) expands peripherally, while below the center a cavity—the subgerminal cavity or blastocoel—is formed. (The central, translucent area above the cavity is called the *area pellucida*, and the peripheral opaque zone, the *area opaca*.)

The cellular roof of the cavity is called the **epiblast.** At its posterior margin cells detach and settle on the floor of the subgerminal cavity, forming the **hypoblast** and covering the uncleaved mass of the yolk-rich residual egg cell. The hypoblast cells are thought to play a directing role when the endodermal and mesodermal cells immigrate, but they do not participate in the construction of the embryo.

When the egg is laid, the blastodisc has a diameter of 2 mm. **Gastrulation** begins when the egg is incubated and warmed up. During gastrulation, endodermal and mesodermal cells colonize the space between the epiblast and the hypoblast. The location where immigration happens is not an open blastopore but a **primitive groove** that remains closed, although it is homologous to the amphibian blastopore. How can cells pass through an apparently closed door? Cells of the epiblast stream toward the groove; dip down in close contact to their neighbors (therefore, the groove seems to be closed); but once inside, they migrate individually like amoebae or the deep cells in the fish embryo, to colonize the available space (Fig. 3–39). The largest free area for colonization is located in front of the primitive groove. At the anterior end of the groove is a small region, known as **Hensen's node,** where ingression is particularly active. This node corresponds to the amphibian upper blastopore lip and is endowed with a similar inductive capacity.

The germ layers are separated and the embryo is formed in front of the primitive groove. The stream of immigrating cells bifurcates into two branches; one branch advances anteriorly on the floor of the cavity, pushing aside the cells of the hypoblast. These deep-moving cells form the definitive

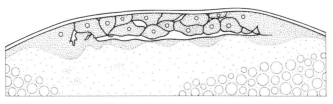

Cleavage: 64-128 Cells

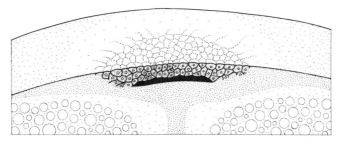

Blastodisc-Blastoderm with Blastocoel

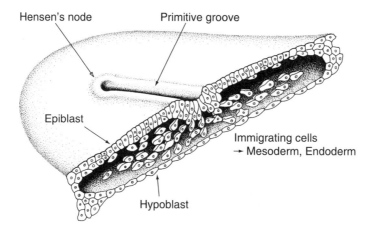

Figure 3–39 Avian egg (chick or quail). Discoidal cleavage (top), blastodisc (middle), and formation of the primitive streak (bottom).

endoderm, which will give rise to the lining of the gut. The other stream flows into the space between the epiblast and the hypoblast, broadens and forms the sheet of the **mesoderm.** As the mesoderm continues to spread anteriorly away from the primitive groove, it incorporates more and more ingressing cells and becomes densely packed. In front of the primitive groove and along the midline of the mesodermal layer, cells coalesce to form the mesodermal axial organs: the stage of the **primitive streak** is reached.

The elaboration of the large-scale organ rudiments is accomplished in a similar manner as in amphibians. A view from outside shows neural folds emerging in front of the primitive groove and above the inducing mesoderm. The neural folds merge along the midline of the embryo, forming the neural tube (Fig. 3–40). In the meantime the mesoderm subdivides itself into the

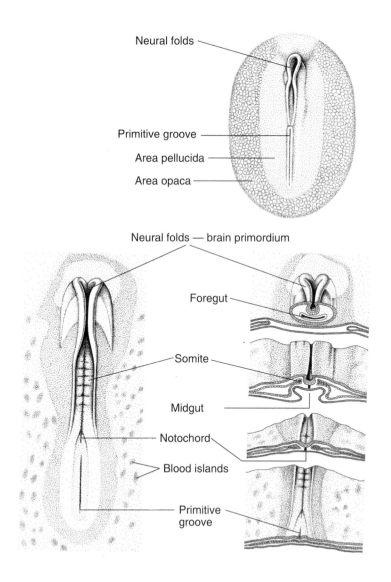

Figure 3–40 Avian embryo. Neurulation. Looking down upon the blastodisc.

notochord, the somites, and the lateral plates. The endoderm covering the yolk sphere begins to form a longitudinal fold that approaches the notochord. Later, the endodermal fold is closed into a tube: the primitive gut is formed.

The embryo has now elaborated its basic shape, but it is still open on its ventral side. Early closure on this side is prevented by the rest of the original, uncleaved egg. The endoderm, the mesoderm, and the ectoderm (epiblast) gradually grow over this residual ball. Enclosed, the yellow ball is known as the **yolk sac.** It is connected with the primitive gut, sharing with it a common endodermal lining. The body of the embryo raises itself away from the surface of the yolk sac. Thus, a partial separation is introduced between the yellow sac and the body of the embryo proper. The broad connection becomes restricted later; the embryo remains connected with the yolk sac by a stalk enclosing an endodermal canal, and later, blood vessels as well. The yolk sac gradually diminishes in size as liquified yolk is taken up by the gut and is used by the embryo.

A second canal, the urinary duct, connects the posterior intestine with an evaginated cyst, known as the **allantois** or the embryonic urinary bladder (Fig. 3–41). An egg enclosed in a shell has no means of disposing of the waste products of protein breakdown and water formed in oxidative metabolism. A solution to the problem was found by producing, instead of urea, insoluble crystals of uric acid and storing them in the allantois with a little water until the time of hatching. However, the allantois not only stores waste products but also serves as an embryonic organ of respiration. The growing allantois is accompanied by blood vessels and establishes contact with the air chamber at the blunt end of the egg.

3.10.1 Formation of the Amnion

While in the emerging embryo the neural tube is formed, dermal folds arise from the surrounding extraembryonic epiblast. They are pulled over the embryo, the free edges of the folds are welded together, and eventually the embryo is completely enclosed in a double-walled cavity—the **amniotic cavity.** In order to protect the tender embryo, the cavity is filled with fluid. The water is provided by the inner wall of the cavity; this inner wall represents the **amnion** proper. The amniotic cavity has come to be a private pond, so that embryonic development of land animals still proceeds in water. The fluids of the amnion cavity and allantois are familiar as the fluid pouring out when the chicken hatches and the thin walls of these extraembryonic containers disrupt.

Chick (and quail) embryos hatch 19 to 21 days after the egg has been laid, when incubated at 38°C.

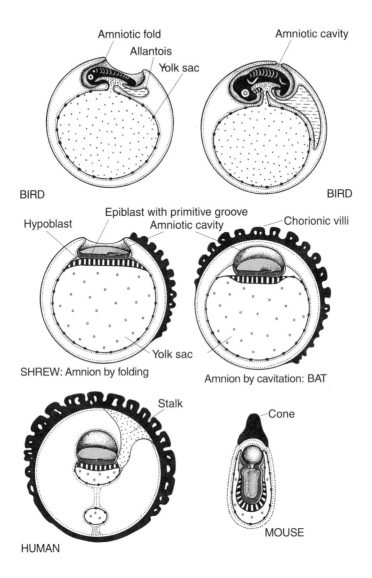

Figure 3–41   Amniotes. Formation of the amnion, yolk sac, allantois, and chorionic villi.

3.10.2 Experiments with Chick Embryos

We focus on the following three types of experiments:

1. Like all vertebrates, the chick displays a high degree of flexibility and regulation in its embryonic development. If a preprimitive streak is cut

into several fragments, most of the fragments are able to produce complete, albeit somewhat smaller, embryos. As a primitive groove has formed, the capability of complete regulation is restricted to that portion containing Hensen's node.

2. Taking advantage of features unique to avian development, the differentiation potentials of the **neural crest cells,** a group of highly versatile cells present in all vertebrates, have been most successfully analyzed in the chick. Their migration routes and roles in constructing the sympathetic nervous system and producing chromatophores are understood (Chapter 13, Fig. 13–1). In order to trace the fate of the roaming neural crest cells as well as other mobile embryonic cells, small pieces taken from quail embryos were implanted into embryos of the chick. Quail cells are tolerated without any problems and participate in the construction of a **chimeric bird,** but can readily be identified and distinguished from the host cells because the nuclei of quail cells contain considerable dense heterochromatin.

3. Pattern formation can be studied effectively on buds of the extremities, in particular, the readily accessible **limb buds.** The morphogenetic action of retinoic acid and secreted growth factors, and the significance of homeobox genes are being investigated using limb buds as model systems in vivo as well as in vitro (Chapters 9 and 10; Box 4).

3.11   The Mouse: A Proxy for Humans

The development of mammals is greatly modified compared with that of other vertebrates, due to adaptations to **vivipary.** In viviparous animals the embryo receives an adequate supply of nutrition from the mother while it is retained in the uterus, the yolk supply of the egg is superfluous and therefore is reduced, but the embryo is confronted with the need to establish intimate contact with the nourishing mother.

Mammalians are not well suited for developmental studies. Embryo development takes place hidden in the maternal body. For investigations and experimental manipulations surgery must be performed, and such an intervention is subject to authorization in many countries. The embryos are inconvenient to handle and are difficult to maintain in the laboratory for long periods of time. Only early embryos can be rinsed out from the oviduct; after implantation, embryos can no longer be removed uninjured.

In spite of all these inconveniences, the mouse is now the developmental model to which the greatest attention is being paid. Above all, it was medical interest that made the development of the mouse the paradigm for experimental mammalian embryology. Rapid, season-independent development, availability of a great number of mapped mutants, and the relative ease with which transgenic animals can be generated have made the mouse the leading model of mammalian development. However, it must be pointed out that mouse development differs in several ways from that in most mammals. With regard to oogenesis and preimplantation stages, schematic sketches of the development of the mouse and the human are fundamentally similar. Of course, the timescales are different.

For a better understanding of mammalian development, you may first read the sections on the development of amphibians and birds, and even the human, as the human embryo is not so different from a generalized mammalian archetype as is the mouse embryo. The late embryo of mammals is called the **fetus;** in the human this term is used from the 8th week onward.

3.11.1 Oogenesis and Ovulation

Five days after the birth of a female mouse, her oocytes have already duplicated their chromosomes in preparation for meiosis. The oocytes enter prophase of meiosis I. However, prophase is interrupted in the diplotene stage; lampbrush chromosomes and multiple nucleoli become visible. Of the oocytes initially present, at least 50% perish but some 10,000 remain. At 6 weeks the mouse reaches maturity. The ovulation cycle is very short: every 4 days, 8 to 12 oocytes finish the first meiotic division, extruding the polar body. The stimulus required by the egg to proceed to the second meiotic division is provided by the fertilizing sperm.

3.11.2 Embryo Development

Mice mate only when the female is in estrus. Some 50 million spermatozoa are released in coitus and seek an egg. Successful fertilization is indicated by the explosive elevation of the vitelline membrane, which is blown up by the swelling of the discharged contents of the cortical granules (Chapter 6). Once the egg is fertilized and activated, the second polar body is extruded, indicating the completion of the second meiotic division. Cleavage in mammals is extremely slow and takes days. It starts about 18 hours after fertilization (compared with 1 hour in the sea urchin) and is accompanied by early transcription of zygotic genes. There is no midblastula transition known in mammals.

The blastomeres are still totipotent through the 8-cell stage. If separated, they can give rise to eight genetically identical mice. At the transition to the 16-cell stage, **compaction** occurs, in which the blastomeres are tightly cemented together by cell adhesion molecules (uvomorulin = cadherin E). The embryo becomes a **morula.** A cavity appears inside the morula and we arrive at the **blastocyst.** In the blastocyst the first irreversible event of differentiation occurs: the cells of the outer epithelial layer, called the **trophectoderm,** undergo "endoreduplication": they amplify their entire genome and become polyploid. The trophectodermal layer surrounds a cavity, the blastocoel. In its interior, the blastocyst contains an eccentrically located **inner cell mass;** these inner cells remain diploid. The blastocyst liberates itself from its envelope (zona pellucida). Hatched, it is ready for implantation. After implantation, the trophectoderm forms **giant cells;** it is now called the **trophoblast** and will eventually give rise to the "chorioallantoic" placenta.

Compared with other mammals, the development of the mouse (Fig. 3–42, 3–43, and 3–44) displays some peculiarities that may be misleading if they are taken to be typical. For instance, the embryo takes the form of an **"egg cylinder."** This cylinder comprises an "ectoplacental cone," destined to become the trophoblastic part of the placenta, and several extraembryonic cavities. The head-back-tail line is extremely curved, giving the embryo an abnormally concave back for some time. The belly side remains open for a prolonged period after the trophectoderm in the region opposite to the ectoplacental cone has dissolved, as has its thick basement membrane, called "Reichert's membrane."

After 19 to 20 days of pregnancy, the mother gives birth.

3.11.3 Parthenogenesis, Teratocarcinomas, and Cloning

3.11.3.1 Do Mice Need a Father?   Occasionally, development can start without fertilization. This can happen in the oviduct or even in the ovary.

If precocious development starts in the ovary before the onset of the first meiotic division, the embryo will be diploid. Nonetheless, it develops abnormally, yielding a tumorlike creature called a **teratocarcinoma.** Even if a parthenogenetic embryo develops in the oviduct, it is not necessarily haploid. Perhaps the second meiotic division was omitted, or a polar body instead of a sperm cell fused with the egg. Such diploid but **maternal** egg cells lacking a paternal genome may undergo cleavage, implant into the uterus, and even develop up to the stage where the buds of the extremities appear. At this stage or earlier, the embryo dies. No living mouse has been born that definitely underwent parthogenetic development. Similarly, artificially produced **paternal** embryos do not survive. Thus, despite the parthenogenetic potential they exhibit, evidence has accumulated indicating that mice need to have a father and a mother.

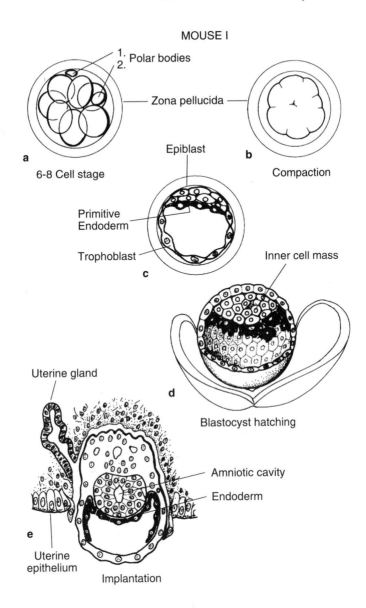

MOUSE I

Figure 3–42   Mouse I. Early embryogenesis from cleavage (a) up to the hatching of the blastocyst (d); (e) invasive blastocyst about to implant into the uterus wall.

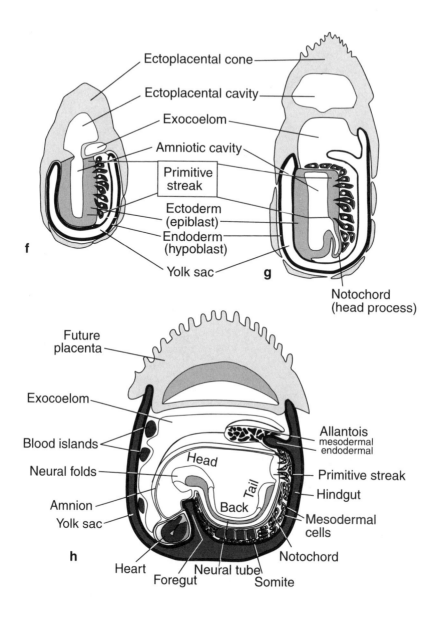

Figure 3–43   Mouse II. Gastrulation. Immigration of mesodermal cells (red) through the primitive groove into the space between the ectoderm and the endoderm. The notochord is formed by the notochordal process ("head process") that encloses a transient notochordal canal; (h) late neurulation; the embryo is approaching the phylotypic stage. Note: The embryo is curved, its ventral side largely open; the allantois and blood islands lie outside the embryo; the trophoblast (chorion) is dissolved on the ventral side; on the dorsal side the chorion with its villi is about to form the placenta.

MOUSE III

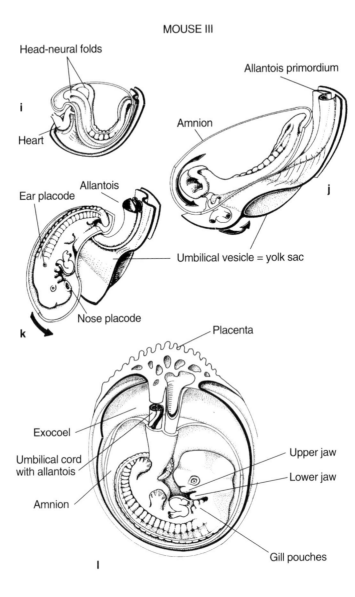

Figure 3–44   Mouse III. The concave bending of the back line is converted into a convex curvature, and the placenta is formed.

3.11.3.2 Imprinting   Related to the question whether mice (and other mammals) need to have a father is the question of whether the genomes contributed by both parents are equivalent. The answer is that they are not, not in all aspects to be more precise. Chapter 9, Section 9.5 (and Chapter 10, Section 10.10) describes the phenomenon of genomic imprinting that confers nonequivalence on the maternal and paternal genomes.

3.11.3.3 Cloned Mice?   In 1977 to 1981 sensationalized reports on cloned mice reached even the nonscientific world. Cloning was claimed to have been achieved by methods different from simple separation of blastomeres in early cleaving stages. The scientific world was informed that uniparental mice were successfully generated in preparation of a true cloning procedure. The experimental result could never be reproduced. Furthermore, the same laboratory reported that identical mice have successfully been cloned like clawed toads by transplanting somatic nuclei into enucleated eggs (Box 2). In many reexaminations by many laboratories, only transferred zygotic nuclei—but not nuclei from two-cell stages or from blastocysts—gave rise to viable mice. Under strictly controlled conditions, mice have not been cloned yet. Apparently, the early transcriptional activity of the mouse embryo coincides with early determinative events that restrict the developmental potencies of the nuclei.

However, the mouse is not a paradigm for mammalian development in all aspects and experimental results with mice must not be extrapolated precociously to other mammals or humans. In other mammals cloning was possible by transplanting nuclei from the inner cell mass of preimplantation embryos (Box 2). In no case has any cloning been achieved using nuclei from later stages or from differentiated cells.

Box 2

FAMOUS EXPERIMENTS WITH EGGS AND EMBRYOS: CLONING, CHIMERAS, TERATOMAS, AND TRANSGENIC MICE

ORGANISMAL CLONING

A **cellular** or **organismal clone** is a collection of genetically identical cells or multicellular organisms; **cloning** is the producing of such clones. Natural cloning, traditionally called **vegetative reproduction**, is known from many plants and invertebrates, and is based on mitotic proliferation. Cells arising from mitotic division are genetically iden-

Box 2   Famous Experiments with Eggs and Embryos   107

tical. In vertebrates (except in polyembryonic armadillos) cloned animals are the result of an accident or an experiment.

Miniclones consisting of identical monozygotic twins (actually two, four, or eight individuals) arise when the first two, four, or eight blastomeres fall apart or are separated experimentally. In mammals each separated blastomere may give rise to a small but complete blastocyst that implants and develops into a fetus with its own placenta. In mammals, including humans, even the inner cell mass of the blastocyst may fall apart into two or more groups: those twins are found within a common chorion (embryonic envelope, Fig. c) and have a common placenta.

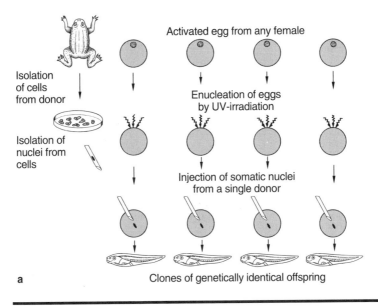

a

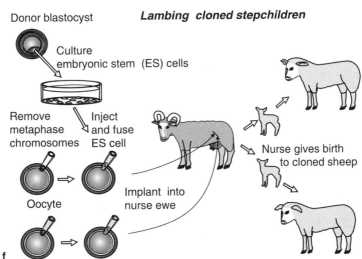

f

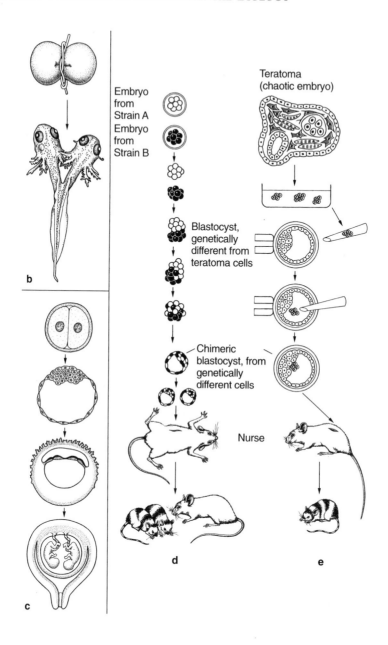

Embryo
from
Strain A

Embryo
from
Strain B

Blastocyst,
genetically
different from
teratoma cells

Teratoma
(chaotic embryo)

Chimeric
blastocyst, from
genetically
different cells

Nurse

b

c

d

e

Clones comprising many animals can be produced in the clawed toad, *Xenopus*. From activated egg cells the zygotic nucleus is removed with a micropipette or destroyed by UV irradiation. A substitute nucleus is removed from a suitable somatic cell type taken from a donor animal and transferred into the enucleated egg through a micropipette. Suitable donor tissues were first found in the intestine of tadpoles and later in several tissues of adult animals. To produce numerous clones, the procedure is repeated many times using the same donor animal: all offspring will be genetically identical with the donor and with each other.

Box 2   Famous Experiments with Eggs and Embryos   109

Cloning was only successful in mammals using this technique if the nuclei were taken from blastomeres of two-cell through eight-cell stages, before compaction. After nuclear transplantation has been accomplished, the host eggs are transferred to nurse females prepared to accept the egg by injection of gonadotropic hormones. Such nurses gave birth to lambs, cattle, pigs, and rabbits. In no case has cloning been successful if nuclei from differentiated cells were injected. In mice even nuclei from four-cell stage embryos were unable to support full development.

In experiments made known in 1996, Scottish sheep were cloned with a new variant of the procedure. Enucleated ovulated oocytes were fused with entire embryonic stem cells taken from a donor blastocyst. Previous to the transfer experiment the totipotent embryonic stem (ES) cells were multiplied through several passages in culture so that many genetically identical donor cells were available and many egg cells could be provided with a foreign nucleus. When an egg with a donor nucleus develops into a blastocyst itself, it can be used as a source of further embryonic stem cells. The morula or blastocyst stages were transferred to the uteri of ewes, and lambs were born with an identical appearance. Although the overall success rate (the ratio of live births to the number of manipulated embryos) is low at present, in sheep, and presumably in several other mammalian species as well, large-scale cloning appears to be possible.

Cloning of human beings from adult donors is performed only by science-fiction novelists.

CHIMERAS

A **chimera** is a mosaic organism composed of cells of different parental origin and, therefore, of different genetic constitution. Chimeras were produced as early as about 1920 (e.g., in H. Spemann's laboratory) by fusion of early embryos from different amphibian species. Two combined early-stage embryos (two-cell stage up to the blastula) are able to reorganize themselves, forming a giant, but uniform and viable, blastula that develops into a chimeric individual.

Mouse experiments by Beatrice Mintz used blastocysts of two different inbred strains (*albino* and *black*) that differed by their white and black coats. The blastocysts were removed from their envelopes (zona pellucida) and were fused. The fusion product was inserted into the uterus of a nurse mouse previously prepared for pregnancy by injection of hormones. The resulting tetraparental litter displayed coats with black and white stripes.

Even viable chimeras between goat and sheep have been produced. However, from such chimeras, no novel species, strains, or even genetic bastards can be bred. If chimeras composed of A and B cells are fertile at all, their egg cells or sperm cells derive from either A or B cells. The two genomes remain completely separated. Therefore, the gametes do not contain a mixture of A and B genes.

Chimeras of a different kind, with a true mosaic of genetic information, can be generated if genetic material of a different donor is introduced into the nucleus of egg cells. Transgenic organisms (see as follows) are true genetic chimeras.

TERATOMAS

Teratomas are failed, chaotically organized embryos. They derive from unfertilized cells of the germ line in the testes or the ovaries. Occasionally germ line cells start embryo development precociously. Other sources of teratomas are fertilized eggs that fail to be taken up by the tube of the oviduct and instead implant anywhere in the abdominal cavity. Experimentally, teratomas in mice are produced by removing blastocysts from the uterus and implanting them into the abdominal cavity. Teratomas may develop into **malignant teratocarcinomas,** tumors that can even metastasize. Cells derived from teratocarcinomas (such as the widely used murine 3T3 or F9) are frequently immortal and can easily be propagated as cell cultures.

As a rule, teratocarcinoma cells are still more or less intact genetically. If implanted into normal blastocysts, they may integrate into their new surroundings unobtrusively and participate in the construction of the new animal. Because in such experiments the implanted teratoma cells and the cells of the host blastocyst are usually different genetically, the resulting progeny will be a chimera.

TRANSGENIC ANIMALS

Organisms into which genes of foreign donors—even genes of a foreign species—have been introduced and which are now permanently carrying these genes are said to be **transgenic.** In mice the ability of **embryonic stem (ES) cells** to integrate themselves into a host blastocyst has been exploited. The foreign gene (**transgene**) of interest is introduced into ES cells by a process of transformation (see Box 7). The transgene is associated with a DNA sequence called a **promoter,** which allows its controlled expression in the recipient. The transgenic ES cells are pipetted into host blastocysts, and the blastocysts are implanted into the uterus of pregnant nurse mice. When the transgenic ES cells participate in the formation of the embryo, some of their descendants may give rise to primordial germ cells and eventually to germ cells. If in mating a transgenic gamete participates in fertilization, the progeny may carry the transgene. The F_1 will be heterozygous, but by repeated mating of such heterozygotes a strain of mice can be generated that is homozygous for the introduced allele. The foreign gene has now become part of the host strain's genome and will be passed on from generation to generation. It is even possible to introduce human genes into mice, for instance, to examine the involvement of a certain mutant gene in the occurrence of a disease.

3.11.4 Chimeras and Transgenic Mice

Chimeric mice with **heterogenic** tissues (tissues of different genetic origin) can be produced by a method that Beatrice Mintz developed. Using a micropipette, foreign cells are injected into the cavity of a host blastocyst. The foreign cells are derived from the inner cell mass of a donor blastocyst or from a teratocarcinoma, and are called **ES cells** (ES = embryonic stem). A teratocarcinoma results from a miscarried embryo (but not all miscarried embryos develop into teratocarcinomas). ES cells injected into a host blastocyst may be integrated into the inner cell mass of the host, proliferate, and contribute to a viable **chimerical mouse.** Such a mouse will be composed of heterogenic tissues; even primordial germ cells may originate not only from host cells but also from foreign cells. If previously a cloned gene of, for example, human origin had been introduced into the donor ES cells (by DNA injection, electroporation, or retroviral vectors, Box 7), the host embryo may include areas containing and replicating this artificially introduced human gene. In case those transgenic regions give rise to primordial germ cells, whole **transgenic mice,** carrying the human gene in all their cells, can appear among the offspring. If the human gene was randomly integrated into the genome of the donor ES cell, it will probably be present in only one of two homologous chromosomes. By inbreeding, homozygous animals can be obtained. With transgenic mice the role of a single gene in the development of a phenotype (e.g., sex) or the contribution of a gene to a hereditary disease can be studied. (See Box 2.)

3.12 The Human

Although laws as well as ethical and religious reservations forbid experimental interference, the development of the human is the subject of morphological, histochemical, molecular, and genetic investigations. In addition to aborted embryos, living cells isolated from teratocarcinomas are available for study. The following description provides information on peculiarities of mammalian development in general and human development in particular. For a better understanding, the description of amphibian and avian development should be read first.

3.12.1 Mammals Combine Novelties with Old-Fashioned Features

In mammals the embryo receives nourishment in the uterus from the mother. On the other hand, the mammalian embryo has acquired a novel feature enabling its embryo to make use of maternal sources: the **placenta.** Thus, the yolk has become superfluous, and its presence in the egg is insignificant. The human egg has a diameter of 0.1 mm and thus has returned to the size of the sea urchin egg. However, the embryo can grow. Except for the embryos of some other viviparous animals, mammalian embryos are the only ones that grow during embryonic development. From the 8th week on the growing embryo is called the **fetus.**

With the disappearance of the yolk, the mammalian egg has reverted to complete, holoblastic cleavage, but its subsequent development bears ample evidence of the former presence of the yolk. In many respects the developmental steps in mammals resemble those found in the meroblastic eggs of reptiles and birds. Thus, we will detect a primitive groove, a primitive streak, a yolk sac, an amnionic cavity, and an allantois.

3.12.2 Oogenesis: While Still Embryos, Girls Store Thousands of Eggs Which Pause in Meiotic Prophase I for Many Years

Primordial germ cells (oogonia) destined to give rise to the next generation are generated and laid aside early in embryo development; they can be identified in the prenatal female embryo in the 3rd week of development. By this time, the primordial germ cells immigrate from the yolk sac by amoeboid movements into the gonadal ridges (Fig. 5–1). In the 5th month of pregnancy the female embryo harbors some 7 million oogonia in her ovaries. By the 7th month of development the majority have perished. The surviving 0.7 to 2 million oogonia stop mitotic multiplication and become **primary oocytes.** These enter the prophase of meiosis I and pause at diplotene (Fig. 3–45). There follows an interval lasting 12 to 40 years, in which lampbrush chromosomes and multiple nucleoli (Chapter 5) are present in oocyte nuclei. Much transcription takes place, many ribosomes are produced, and the oocyte is fed with yolk material by the surrounding follicle cells. Even the relatively small human oocyte undergoes a 500-fold increase in volume.

During childhood, many more oocytes perish. At puberty some 40,000 are left; they are enveloped by nourishing **primordial follicles.** Stimulated by the pituitary gonadotropin **follicle-stimulating hormone (FSH),** 5 to 12 oocytes begin to mature in each ovarian cycle. Later, the follicles are exposed to another gonadotropin, the **lutenizing hormone (LH).** These hormones signal the egg to resume meiosis and to initiate the events culminating in the release of the mature egg from the ovary. As a rule, in one ovarian cycle, which lasts some 28 to 30 days, only one of the follicles attains the state of a mature **graafian follicle.** In the middle of the menstrual cycle, about 14 days after the last menses, one of the oocytes finishes the first meiotic division by extruding the first polar body, and the egg is released from the ovary (**ovulation**). In mammals, including humans, the second polar body is not constricted off until the egg is fertilized.

Before fertilization takes place, the egg is caught by the mouth of the oviduct. If not, a fertilized egg that falls into the abdominal cavity dies or may give rise to a monstrosity exhibiting tumorlike features, called a **teratocarcinoma.**

3.12.3 Only One of Millions of Sperm Gets a Chance

Once taken up by the mouth of the oviduct, the egg awaits the sperm with reservation: it is surrounded by a tough envelope, called the zona pellucida, and by a corona of mucous material, called the corona radiata. Both of these envelopes have been produced by the ovarian follicle and must be penetrated by the sperm.

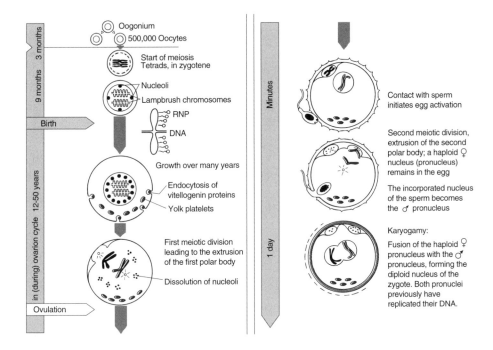

Figure 3–45 Oogenesis in the human. Oogenesis starts in the female fetus, long before the girl is born. However, meiosis I is interrupted for 12 to 40 years, during which time the oocyte undergoes growth and maturation. The appearance of lampbrush chromosomes and of multiple nucleoli indicates high transcriptional activity in the nucleus while the prophase of the first meiotic division is interrupted. Later, the oocyte increases in volume by incorporating yolk materials supplied by the liver, transported by the blood, and handed over to the oocyte by the follicle cells. Meiosis is resumed shortly before ovulation—that is, shortly before the egg is extruded from the graafian follicle and transferred to the oviduct. The formation of the second polar body is accomplished only after a sperm has contacted the egg cell.

From the 200 to 300 million sperm released into the vagina in each ejaculation, about 1% advance to reach the egg. On their long journey through the vagina, the uterus, and the oviduct, the sperm cells undergo **capacitation,** that is, they acquire the competence or capacity to fertilize through the influence of female secretory products. Only one sperm succeeds in entering the egg cell. The other sperm may help to dissolve the egg's envelopes by releasing acrosomal enzymes (Chapter 6).

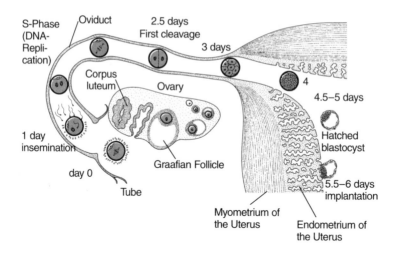

Figure 3–46 Human. From ovulation to implantation. A mature egg contained in a graafian follicle of the ovary is liberated and picked up by the tube of the oviduct. Fertilization, formation of the second polar body, and cleavage take place in the oviduct, which leads into the uterus. Once it has arrived in the uterus, the embryo has reached the blastocyst stage. The blastocyst hatches out from its envelope (area pellucida) and attaches to the wall of the uterus. Now the blastocyst begins to invade the uterine wall.

3.12.4 The Egg First Develops "Extraembryonic" Organs to Tap the Mother

3.12.4.1 Preimplantation Development (Fig. 3–46)   Although the mammalian egg lacks significant quantities of yolk and has reverted to holoblastic cleavage, cleavage in mammals is extremely slow. Not until the 3rd day is the 12- to 16-cell stage of the **morula** reached. By the 4th day the **blastocyst** stage is attained. The blastocyst looks just like a blastula, but it is not a blastula that proceeds to form an archenteron through invagination as do the true blastulae of sea urchins and amphibians. Rather, the cellular wall of the blastocyst undergoes differentiation to become the **trophectoderm.** This is an "extraembryonic" organ destined to invade the wall of the maternal uterus, absorb maternal nutrients, and dispose of waste. After implantation, the trophectoderm undergoes several rounds of endoreplication and develops polyploid **giant cells.** Part of the trophoblast will give rise to the **placenta,** a fetal organ formed to optimize all of these functions. The trophoblast encloses a cavity and the **inner cell mass** that embodies the **embryoblast** (those cells that remain diploid and eventually will generate the embryo proper). The inner cell mass is a group of cells eccentrically concentrated at a location beneath the "polar trophoblast." The blastocyst hatches, attaches to the uterine wall, and sinks into it, a process known as **implantation** or **innidation.**

3.12.4.2 Postimplantation Development (Fig. 3–47, 3–48, 3–49) The hatched blastocyst attaches with its polar trophoblast to the wall of the uterus. Entry is facilitated by lytic enzymes released from the trophoblast. The outer layer of the trophoblast proliferates into the uterine tissue. While doing so, it transforms into a **syncytial trophoblast.** The inner layer of the trophoblast, called the **cytotrophoblast,** remains cellular and encloses the blastocoel. Later, the wall of the blastocyst's cavity is supplemented by an inner layer, called the **parietal mesoderm.** The cavity of the blastocyst will not be enclosed in the embryo and is known in later stages as the exocoel or chorionic cavity.

The invasive multinucleated syncytial trophoblast prepares the space into which the embryo can grow. The solid epithelial cytotrophoblast becomes reinforced and supplemented by mesodermal cells that immigrate from the embryo: the whole organ is known as the **chorion.** From its outer surface, **primary chorionic villi** project and extend through the loose syncytiotrophoblast to establish contact with the surrounding maternal uterine tissue. The chorionic villi are the functional precursors of the much larger villi that are later formed by the placenta.

What happens inside the blastocyst? Where is the embryo formed? Before the embryo develops, the human blastocyst, like most mammalian blastocysts, makes some preparations for the benefit of the coming embryo. By the 9th day, the **amniotic cavity** is prepared and a **yolk sac** is formed. The amniotic cavity appears as a cleft in the midst of the inner cell mass. From this inner mass an epithelial layer called the **hypoblast** separates. It expands, lining the central cavity and forming the **parietal endoderm** or the **yolk sac** (a misnomer because it does not contain yolk!).

The **embryo** proper is formed from two minute epithelial areas: the floor of the amniotic cavity, now called the **epiblast,** and the roof of the yolk sac, known as the **hypoblast.** Together these two adjoining areas, one on top of the other, represent the **blastodisc.**

From this point on, generation of the actual embryo follows the path paved by our nonmammalian ancestors. Gastrulation, germ-layer formation, and neurulation are similar in many ways to embryo development in the reptiles and birds. In gastrulation epiblast cells move to a primitive groove and pass through it, colonizing the cleft between the epiblast and the hypoblast. In front of the groove, a **primitive streak** appears. Underneath, the notochord pushes anteriorly and contacts a prechordal plate that will give rise to head

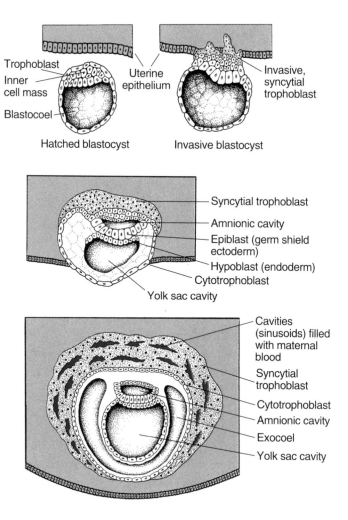

Figure 3–47 Human. Implantation (innidation), amnion, yolk sac. The invasion of the uterine wall is accomplished by an egg-derived "extraembryonic" tissue or organ, the trophoblast. It shows features of a terminally differentiated tissue, including polyploidy. The peripheral invasive trophoblast forms a syncytium by dissolving cell membranes. The amniotic cavity is not formed by amnionic folds (as in the avian and insectivore embryo) but by the separation of cells of the inner cell mass. Hypodermal (endodermal) cells surround the yolk sac cavity, forming the (empty) yolk sac. An embryo proper is not formed yet. It will be formed by the epiblast and the hypoblast by a process highly reminiscent of gastrulation (primitive streak formation) in birds (see Fig. 3–39).

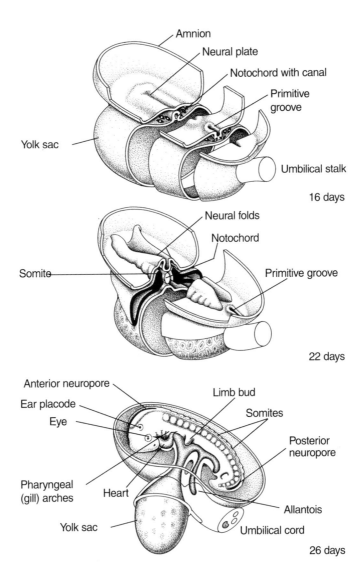

Figure 3–48 Human. Neurulation, organogenesis. The surrounding envelopes (trophoblast and chorion) are removed and the amniotic cavity is opened to show the blastodisc. The neural folds form in front of the primitive groove, just as in the avian embryo (compare Fig. 3–40). The notochord is formed by the roof of the archenteron; in an intermediate stage of its formation the tubular notochord encloses the notochordal canal. The rudimentary allantois and the yolk sac become integrated into the umbilical cord. During the entire development shown the embryo increases in length from about 1 to 5 mm.

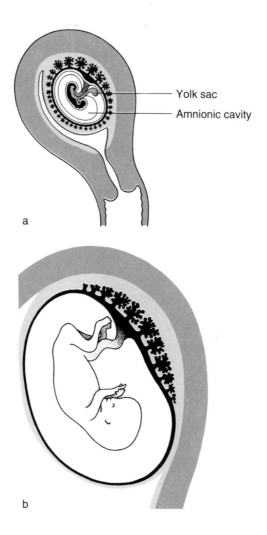

Figure 3–49 Human. Placentation. The chorionic villi (black treelike structures) are reduced on one side but enlarged on the side that faces the maternal uterine wall. The villi are provided with blood vessels (umbilical vessels) and together constitute the major part of the placenta. The maternal endometrium contributes to the placenta in a negative way as it partially dissolves. The resulting spaces (lacunae) are filled with maternal blood.

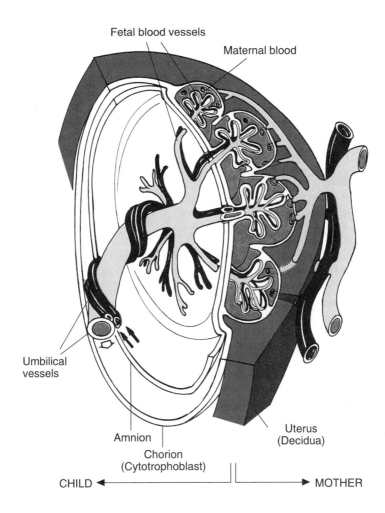

Fetal blood vessels

Maternal blood

Umbilical vessels

Amnion

Chorion
(Cytotrophoblast)

Uterus
(Decidua)

CHILD ◄ ─────────── ────────────► MOTHER

Figure 3–50 Human. Placenta. The maternal uterine wall is partially dissolved; the resulting lacunae are filled with maternal blood. The chorionic villi of the fetal placenta dip into the maternal blood.

muscles and to the pharyngeal membrane. Initially and temporarily, the noto-chord is hollow. The **chordal canal** begins with an open pore in the primitive groove close to Hensen's node and ends with another open pore in the yolk sac. A similar chordal canal (*canalis neurentericus*) is found in reptiles and some birds.

When the neural folds merge above the notochord to form the neural tube, this tube also remains open at its anterior and posterior ends. Both the **anterior** and **posterior neuropore** connect the neural canal with the amnionic cavity. The pores are closed when the heart beats and the majority of the somites are visible, as are the optic cups, the ear placodes, and the pha-ryngeal pouches ("**gill arches,**" Fig. 3–48, 4–1, 4–2).

By the 28th day of development the blood vessels have proliferated and spread inside and outside the embryo. The embryonic circulation is con-nected to the placenta through the umbilical cord (Fig. 3–49, 3–50).

3.12.5 The Placenta Is the Organ by Which the Embryo Is Anchored and Through Which the Embryo Exchanges Substances With Its Mother

The developing child lies within a double-layered cyst. Both layers are of embry-onic origin: the inner layer is the wall of the amnionic cavity and the outer layer is the chorion, largely a derivative of the trophoblast (Fig. 3–49, 3–50).

In a region of the chorion, now called the **placenta,** the tiny primary chorionic villi are replaced by larger secondary villi, and eventually by tertiary villi. The branching, tree-shaped villi grow into cavities inside the uterine wall. The villi become vascularized. Two arteries (umbilical arteries) leave the body of the fetal child at the point of the future navel, traverse the amnionic cavity inside the umbilical cord, enter the villi, and ramify and rejoin to form the umbilical vein, which returns to the embryo to supply it with nutrients and oxygen. In exchange for maternal nutrients and oxygen, waste products, such as CO_2 and urea, are transferred to the mother across the villi.

The mother is accommodating towards the embryo by removing barriers. The decidua—the outer layer of the uterine wall encasing the embryo—dissolves. In this way cavities or lacunae are created into which the branches of the chorionic villi can grow. The term "lacuna" indicates that the maternal blood capillaries open into these cavities and maternal blood flows immedi-ately around the embryonic capillaries (Fig. 3–50).

3.12.6 The Unborn Child Has a Circulatory System Similar to That of a Fish

The umbilical vein does not arrive from the lungs. Blood enriched with oxy-gen flows through the umbilical vein into the undivided heart, is distributed in the body, and is collected and pumped back through the umbilical arteries to the villi of the placenta. The fetus has a circulatory system in the form of a single circle, like a fish, with the placental villi serving as gills. Because the lungs of the fetus do not yet function, but have to take over their vital task immediately after birth, the circulatory system must be prepared for a rapid conversion into a double-circle (body and lung) system. The solution to this serious technical problem is described in Chapter 15.

3.12.7 The Placenta Is a Source of Hormones

The embryo or fetus must ensure that pregnancy continues and is not terminated prematurely by menstruation. In addition, the embryo has to prevent the immunological defenses of the mother, which are eager to reject any alien body. Evidence suggests that the embryo releases a number of signal molecules, of which only a few are known. Early in development, the trophoblast begins to release hormones signaling the mother that implantation has occurred. At least three hormones are transmitted to the mother: **chorion gonadotropin** and **progesterone** induce the maintenance of the state of pregnancy, and **chorion somatomammotropin** (also called placental lactogen) stimulates the maternal breast to produce milk.

3.12.8 Pregnancy and Birth: A Matter of Life and Death

Pregnancy and birth are a matter of life and death for both the child and its mother. The embryo is a mature fetus and is ready to be born after 38 weeks (266 days) of pregnancy. However, prior to this stage 40 to 50% of the implanted embryos perish as consequence of faulty development. Birth represents another serious risk.

It is a risk to the mother: when the placental villi are pulled out from the uterus, blood flows out from the lacunae and is lost. The mother is in danger of bleeding to death. Will the uterus succeed in sealing the large openings in time?

It is also a risk to the baby: in an instant, the lungs must be blown up and blood pumped through the lungs. The baby gets a blue color. Will it succeed in "jump-starting" the body?

In spite of all these risks, the benefits of being harbored within a protecting and nourishing mother was worth these risks. The evolutionary success of the placental mammals, and of humans, in particular, is a testament to the value of this developmental innovation.

4

Comparative Review:
The Phylotypic Stage of
Vertebrates, Common versus
Distinct Features, and
Aspects of Evolution

4.1 Observation of a Sharp Eye: von Baer's Law

Carl Ernst von Baer (1791–1876) is considered the pioneer and father of comparative embryology. von Baer discovered the notochord, and the mammalian and human egg. Comparing many embryos, he observed that early vertebrate embryos exhibit common features: "More general features that are common to all the members of a group of animals are, in the embryo, developed earlier than the more special features which distinguish the various members of the group." This statement is known as von Baer's law. Thus, the features that characterize all vertebrates, such as the dorsal brain and the spinal cord, the notochord, the somites, and the aortic arches, are seen earlier in development than features distinguishing between the various classes (limbs in tetrapods, feathers in birds, and hair in mammals). Thus, at an early stage following gastrulation and neurulation, embryos of fish, amphibians, reptiles, birds, and mammals all look alike. As development progresses, their developmental paths diverge and embryos become recognizable as members of their class, their order, their family, and finally their species (Fig. 4–1).

The common early stage may be called the **phylotypic stage** (following a proposal by Sander 1983). Its occurrence is particularly remarkable because the beginning of embryo development, such as the stages of cleavage and gastrulation, are highly distinct among the various vertebrate classes. Even among these groups—for example, among mammals—the early stages are quite different. von Baer's law is valid only if the phylotypic stage is consid-

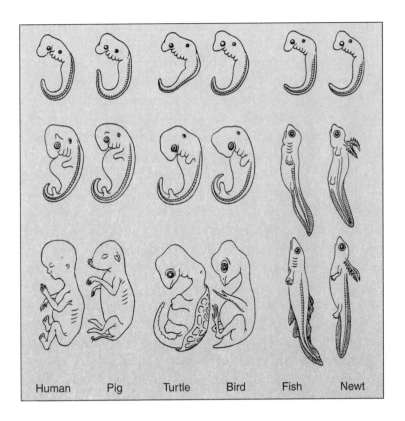

| Human | Pig | Turtle | Bird | Fish | Newt |

Figure 4–1   Embryos of various vertebrates illustrating an observation of Carl Ernst von Baer that a certain early embryo stage (now termed the "phylotypic stage") is very similar among the various vertebrates (upper row), while divergence occurs during later development (columns). The phenomenon prompted Haeckel to formulate his "biogenetic law."

ered the starting point. Apparently, this phylotypic stage is a bottleneck stage that all vertebrates must pass through.

4.2 Haeckel's "Biogenetic Law"

In 1880 Ernst Haeckel, an enthusiastic German zoologist, artist, and visionary philosopher, drafted his much-disputed **"biogenetic law."** Actually, this "law" is a hypothesis, in contrast to von Baer's law, which is based firmly on observations. In its succinct and catchy form, the law states that "ontogeny recapitulates phylogeny" in a condensed and abbreviated way. Haeckel proposed that ontogeny—the individual development of an organism from the egg to the adult—is a shortened version of the species' evolution. Multicellular organisms start their development from a unicellular egg, and pass through a blastula and gastrula stage because their common ancestry first existed as a unicellular protist, followed by a *Volvox*-like colonial hollow sphere ("blastula") and by a (putative) beaker-shaped "*gastrea*."

In his argument Haeckel made several serious mistakes. First, he equated *embryonic* stages of present-day animals with *adult* stages of ancient organisms. Second, in his view evolutionary development progressed up the steps of an ascending scale. The occurrence of a new phylum was viewed as a step towards a higher goal, eventually towards the creation of a human being. More advanced stages were sequentially added onto existing adults until a human being evolved. Third, Haeckel was not well versed in embryology and erroneously maintained that the human embryo would undergo gastrulation through invagination, thereby creating an archenteron.

In fact, the blastocyst of the placental mammals does not undergo a classic gastrulation through invagination such as occurs in the sea urchins, *Amphioxus*, and amphibians. Instead, the entire outer wall of the blastocyst is used to form a first and voluminous "extraembryonic" organ, the **trophoblast,** by which the embryo corrodes its way into the wall of the maternal uterus. Next, an epithelial **amnion** enclosing an amnionic cavity is segregated from the residual "inner cell mass," and beneath it a **yolk sac** (enclosing a nonexisting yolk) is formed.

The amnion and the yolk sac are also found in reptiles and birds, but the amnion of the sauropsids is not formed until the embryo has established its basic body architecture (Fig. 3–41). In mammals, by contrast, an embryo proper is not yet present at the time the amnionic cavity appears.

Remarkably and confusingly, the procedure by which the amnion is formed is very different among the various mammals (Fig. 3–41). In the shrew, considered by taxonomists to be a relatively archaic mammal, the amnionic cavity is formed by amnionic folds in much the same way as in reptiles and birds. However, even in the shrew an embryo proper does not yet exist at the time the amnion is formed. The embryo will arise from a small area in the bottom of the amnionic cavity (epiblast) supplemented by a corresponding area in the underlying roof of the yolk sac (hypoblast). This double-layered area develops into a blastodisc. The following processes take place in much the same way as they do in and below the blastodisc of reptiles and birds: as an expression of the beginning of gastrulation, a furrowlike depression in the epiblast indicates the downward streaming of those cells that will give rise to the endoderm and the mesoderm. The furrow is known as the **primitive groove** and is interpreted as the homolog to the blastopore in the amphibian gastrula. A Hensen's node develops at the anterior end of the groove and in front of the node a **primitive streak** emerges.

To summarize: In mammals it was very early development that underwent a considerable alteration through evolution, adapting to the protective and nutritive environment of the mother. To this extent the biogenetic law is not valid.

In principle, Haeckel and his devotees were aware of the fact that ontogeny does not only reflect ancient evolutionary steps. Terms pointing to this fact were coined by Haeckel and are still used today. "Old," recapitulated traits are called **palingenic** (from the Greek *palin* = again). Novel traits reflecting embryonic adaptations and evolutionary change are called **cenogenic** (from the Greek *kainos* = new). The distinction, however, is always relative: In mammals, including humans, there are structures, such as the amnion, the yolk sac, the allantois, the blastodisc, or the primitive streak, which are novel compared with amphibians, but ancient if one looks at reptiles. The formation of an empty yolk sac and a nonfunctional allantois in the mammalian embryo can only be understood if the development of reptiles is inserted between amphibian and mammalian embryo development.

If these observations are taken into account, Haeckel's biogenetic law merits acknowledgment as it points to the evolutionary context of developmental biology, but it must be corrected: each organism's ontogeny does not repeat phylogeny of a species but rather previous ontogenies. In each generation all species recapitulate their own ontogeny, which, compared with the ontogeny of related species, is more or less modified.

On the other hand, all vertebrates pass through a highly conserved common stage that displays a uniform basic body architecture characteristic of all vertebrates. Therefore, the biogenetic law is valid if it is modified by stating that all vertebrates recapitulate certain embryonic traits of their ancestors—in particular, a common phylotypic stage.

4.3 All Vertebrates Pass through a Highly Conserved Phylotypic Stage

All developmental pathways, although initially different in amphibians, reptiles, birds, and mammals, converge in a basic form exhibiting a neural tube, a notochord, somites, lateral plates, a foregut with gill pouches and arches, and

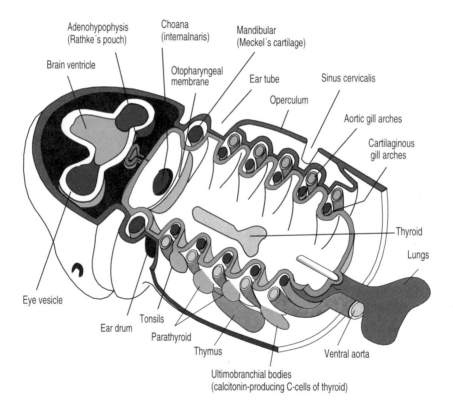

Figure 4–2 Phylotypic stage of vertebrates showing the derivation of the branchiogenic organs in the anterior body (head and pharyngeal region). The lymphatic organs (tonsils and thymus) and the hormone glands (parathyroids and ultimobranchial bodies or C cells) derive from the endodermal walls of the pharyngeal pouches. The thyroid arises from the floor of the pharynx and is considered to be homologous to the hypobranchial groove of the tunicates, *Branchiostoma* and cyclostomes. Also, the trachea and the lungs are evaginations of the foregut and are thought to have evolved from the last pair of pharyngeal pouches. The operculum, which in fishes and amphibians covers the gills, is only a transient structure in mammals.

Ectoderm and Neural Plate Derivatives

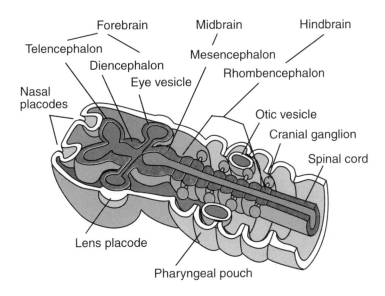

Figure 4–3 Phylotypic stage of vertebrates. Ectodermal and neural plate derivatives in the anterior body showing the segmental structure of the hindbrain and the pharyngeal region. In the hindbrain (rhombencephalon, medulla oblongata) seven segmental constrictions mark off seven to eight repeating units called **rhombomeres** or **neuromeres.** Branches originating from the first even-numbered rhombomeres (r2, r4, r6) and the associated ganglia give rise to the cranial (trigeminal, facial, vestibuloacoustic) nerves. The subsequent posterior branches and ganglia are known as the glossopharyngeal nerve and vagus; the subsequent nerves are called spinal nerves (dorsal and ventral roots, dorsal root ganglia) in that part of the central nervous system known as the spinal cord. The segmental organization of the medulla oblongata corresponds to the segmental organization of the pharyngeal pouches and organs.

a ventral heart (Fig. 3–31, 4–2, 4–3, 4–4). In vertebrates the vertebral column will be built up subsequently around the notochord.

In the phylotypic stage, the **neural system** is in the form of a closed tube, broadened anteriorly where the brain will develop. The parts of the brain are beginning to be differentiated by thickenings and constrictions of the neural tube, and the eye vesicles are bulging laterally from the forebrain-midbrain boundary. The epidermal epithelium has formed pairs of plate-shaped thickenings, called **placodes.** These become invaginated to form the **eye lens** and

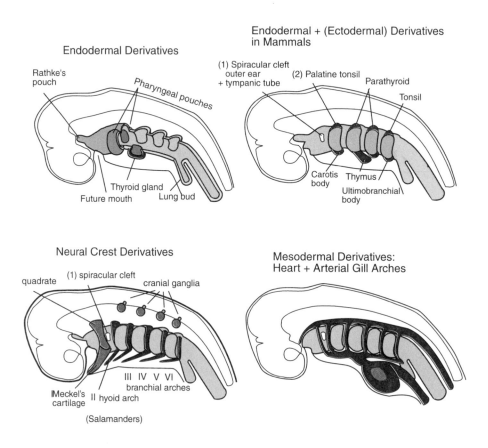

Figure 4–4   Phylotypic stage of vertebrates. Derivation of organs in the head and the pharyngeal region. The endodermal walls of the pharyngeal pouches give rise to lymphatic organs and hormone glands. The cranial and spinal ganglia as well as the cartilaginous elements of the primary jaw and the gill arches are derivatives of the neural crest. The heart and the aortic arches are of mesodermal origin.

sensory organs: the **nasal placodes** develop into the olfactory sacs; the **auditory** or **otic placodes** develop into the otic vesicles from which the inner ear with the labyrinth and the cochlea will arise. The notochord stretches underneath the neural tube from the midbrain level to the posterior end of the body. To each side of the notochord the mesoderm is subdivided into the modules of the somites. Beneath the foregut, the cardiac tubes fuse to form the ventrally located heart. Most conspicuous, however, is the structure of the pharyngeal region. In acknowledgment of its evolutionary past even the human embryo displays in its foregut outpushings known as pharyngeal (branchial) **"gill" pouches** (Fig. 3–48, 4–1, 4–2, 4–3, 4–4). They are associated with **"gill" arteries** (Fig. 4–2, 4–4, 4–6), **cartilaginous "gill" arches** (visceral arches, Fig. 4–2, 4–3), and a **primary jaw apparatus** from the first cartilaginous branchial arch.

4.4   The Phylotypic Stage May Reflect Constraints in the Mode by Which a Complex Feature Can Be Organized

Why a uniform phylotypic stage in spite of all the divergence in the earlier and later stages of development? A plausible answer recently has been found to this question: transitional structures, such as the notochord, are needed for the organization of further development. For instance, the notochord or its precursor, the midline cells in the roof of the archenteron, release signals that induce the floor plate in the spinal cord, and initiate the segregation and detachment of the sclerotome from the somite, and attract the migrating sclerotome cells toward their goal (Fig. 9–5). The signals emanating from the notochord instruct the sclerotome cells to construct the vertebrae around the notochord.

However, even when an organizing function can be assigned to transitional structures, there remains the question: Why this way and not another way? Apparently, a difficult undertaking such as the organization of development cannot easily and arbitrarily be changed and abbreviated. Solutions found in millions of years of evolution cannot easily be altered or replaced. Therefore, we do not see a succession of ancestral adult stages, as Haeckel believed, but instead we see seemingly laborious recapitulations of phylogenetically old developmental pathways.

4.5   Evolutionary Transformations Following the Phylotypic Stage

As every embryo will develop into an individual belonging to a certain species, and not into a generalized vertebrate, the common stage must subsequently be modified and elaborated further. Here only a few highlights of the comparative embryology can be outlined.

4.5.1 A First Highlight for Fans of Evolutionary Dramas: The Ear Ossicles

Even the human embryo is said to have a **primary jaw joint** as do not only the embryos but also the adults of fish, amphibians, and the nonmammal-like reptiles. Neural crest cells assemble to form the cartilaginous **mandibular arch** consisting of the **quadrate** (upper jaw) and **Meckel's cartilage** (lower jaw). The mass of neural crest cells between the first and second branchial pouches becomes the **hyoid arch** (the next becomes the first branchial arch, and so forth).

Mammals are not satisfied with the primary jaw. They develop a second one, and use the first to optimize a very different task: to optimize the transmission of sound from the eardrum to the inner ear. When the dentary and squamosal bones connect to one another in the mammalian embryo and form the secondary jaw joint, the elements of the primary jaw are freed to acquire a new function. The elements of the primary jaw (mandibular arch) are transformed into the **auditory ossicles** (Fig. 4–5): the **hammer** (malleus), **anvil** (incus), and **stirrup** (stapes). They enter the region of the middle ear to transmit sound from the eardrum to the *fenestra ovale* (oval window, inner eardrum) in the inner ear.

4.5.2 The Branchiogenic Organs and the
Derivation of the Hormonal Glands

The pharyngeal region offers some more fascinating recapitulations of old ontogenetic pathways. This applies to the endodermal derivatives (Fig. 4–4) and to the blood circulation system in the embryo (Fig. 4–6). Of course, in other regions of the body interesting transformations can also be seen. As an example, we outline the development of some hormonal glands.

The phylogenetic derivation of the hormonal glands is very diverse and sometimes curious. This also becomes apparent in ontogeny. In land vertebrates several hormonal glands arise from gill pouches or other parts of the pharyngeal foregut, tissues that are no longer needed to construct functional gills and can adopt new functions or optimize old secondary functions because respiration requirements no longer must be taken into account and compromises are no longer necessary.

In fish and in all higher vertebrates the **adenohypophysis** (**anterior pituitary gland**) develops from an area in the roof of the pharynx (known as Rathke's pouch), while the **thyroid gland** develops from a pocket in the floor of the pharynx (Fig. 4–2, 4–4). The developmental progeny of the pocket, and with it of the thyroid gland, is the hypobranchial groove (endostyle) in the branchial pharynx of the acranial chordates (ascidians and *Amphioxus*) and in the larvae of the cyclostomes.

The **C cells** within the thyroid gland and their homologous counterparts in the nonmammalian vertebrates, the **ultimobranchial corpuscles,** as well

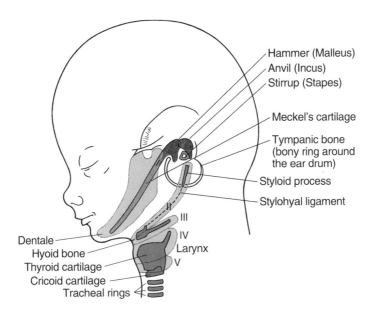

Hammer (Malleus)
Anvil (Incus)
Stirrup (Stapes)

Meckel's cartilage

Tympanic bone
(bony ring around
the ear drum)

Styloid process

Stylohyal ligament

III
IV
Larynx
V

Dentale
Hyoid bone
Thyroid cartilage
Cricoid cartilage
Tracheal rings

Figure 4–5 Human. Transformation of the visceral skeletal elements in the pharyngeal region. When the secondary jaw joints are formed and dentale contacts the squamosal of the skull, the first arch—called the **mandibular arch** (containing the Meckel's cartilage)—is modified and ossified, and gives rise to the **auditory ossicles:** the **hammer** (**malleus,** corresponding to the articular process) and **anvil** (**incus,** corresponding to the quadrate process). In lower vertebrates (fishes, amphibians, and reptiles) these two elements are bent around the mouth, forming the primary lower and upper jaws and jaw joint. The third auditory ossicle, the **stirrup** (**stapes**) is a remnant of the second or **hyoid arch.** The dorsal element, known as hyomandibular cartilage or bone in fishes, once anchored the jaws to the braincase, and braced against the otic capsule. In amphibians (or even in ancestral fishes) the hyomandibular bone was used to transmit sound waves to the ear region (and is known as the columella in amphibians, reptiles, and birds). In mammals, the hyomandibular element or columella, respectively, is used as stapes. The malleus will attach to the eardrum, transmit the sound to the incus, which in turn transmits the sound to the stapes and the stapes to the oval window of the inner ear. The ventral elements of the hyoid arch and the subsequent posterior arches are used to form the **hyoid bones** (2 + 3), and the **thyroid** (4 + 5) and **cricoid cartilage** (6) of the **larynx.**

as the **parathyroid,** develop from cells that detach from the walls of the pharyngeal pouches. In addition to hormonal glands, the pharyngeal pouches give rise to organs of the **lymphatic system,** such as the **tonsils** and the **thymus.** The thyroid, parathyroid, and thymus are shifted downward and backward until they come to lie in the neck or the thorax region.

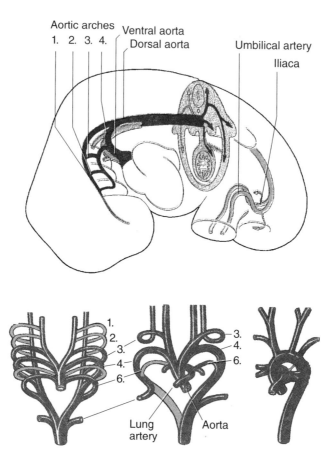

Figure 4–6   Vertebrates. Development of the blood vessels, especially of the vascular **aortic arches** (**gill arches**) in the anterior body. (Upper figure: modified after Tuchmann-Dupless.)

Very different developmental pathways create the **adrenal gland:** the central medulla is formed by aggregating sympathetic cells that once emigrated from the neural crest together with the precursors of the sympathetic nervous system. By contrast, the adrenal cortex is formed by descendants of the mesodermal lateral plates.

4.5.3 After the Phylotypic Stage, the Temporal Sequence of Developmental Events Largely Reflects the Sequence of Evolutionary Changes

Errors in Haeckel's treatises and aversions to his personality and philosophy caused the widespread refusal to acknowledge the credits of his biogenetic law. Actually, ontogeny reflects phylogeny. After all, the finished body is generally considered the result of evolution.

Some data concerning human development may point out the apparent parallelism between ontogeny and phylogeny:

- By Day 21 the somites of the neck region emerge, ear placodes are formed, the first gill slits appear, and two heart tubes form beneath the pharyngeal intestine. The embryo approaches the phylotypic stage.
- By Day 23 an embryonic circulation is established with a construction as though a fish would develop. The brain is composed of several lobes separated by constrictions, the largest and most posterior of which is the **rhombencephalon,** the next is the **mesencephalon,** and the most anterior is the **prosencephalon.** The heart lies within the cup-shaped posterior end of the branchial basket and consists of **one atrium** and **one ventricle.** A hypophyseal pouch approaches the midbrain; the **thyroid gland** sinks down. The intestinal tract has **gill pouches** and continues to the **cloaca** into which the ducts of the **pronephric kidney** also lead. The notochord is present but no other skeletal elements: jaws are absent. The putative fish embryo apparently will become a lamprey or slime eel.
- By Day 27 the "fish" apparently prepares to immigrate into shallow freshwater areas, often with low oxygen levels and seasonal droughts. A lung develops. The kidney transforms into a mesonephros. In addition, jaws and a tongue are forming. Apparently, a lungfish or an amphibian species is about to develop.
- By Day 34 we see a **pectoral** and a **pelvic limb** developing a **palm.** The intestine is associated with a pancreas that is needed to digest "dry" terrestrial food. By Day 38 the heart is divided into **two auricles** but still has **one ventricle.** Its position is shifted posteriorly. Apparently, an amphibian is about to conquer land. Its travel on land is facilitated by some other features: the nasal sacs are shifted from a lateral to a frontal position and the eyes are shifted from a dorsal to a lateral position (ultimately, in the human fetus the eyes will be shifted anteriorly).
- By Day 42 the **lung** is fully developed; the trachea is separated from the esophagus. The gill slits disappear with the exception of the first slit, which will become the middle ear cavity and eustachian tube. The human fetus has a long tail and the tail has now reached its longest length. Hands and feet are directed inwards, while the elbow and the knees are directed outwards, as is the case in newts, salamanders, and reptiles.
- By Day 45 the nasal sacs are connected with the pharyngeal cavity through **choanae** (**internal nares**), an achievement first made by the amphibians.

The secondary bony palate, which separates the nasal from the mouth cavity, is finished by Day 57. Such a secondary palate is characteristic of crocodiles, mammal-like reptiles, and mammals. A **four-chambered heart** is characteristic of crocodiles and mammals as well, and is now developed gradually.

* The stage of a putative reptilian is further indicated by the appearance of finger nails ("claws") by Day 60. Features of mammal-like reptiles are, for example, longer, rotated extremities with elbows pointing posteriorly and knees pointing anteriorly; and the development of different types of teeth, the ear ossicles hammer and anvil, and an external ear canal.

* By Day 45 the genitalia begin to develop, but a penis or a vagina with a clitoris is recognizable only at the end of the 4th month. Amphibians and reptiles do not have such devices. Sexually differing outer genitalia have been developed only by mammals. Concomitantly, the gonoducts are now separated from the hindgut (only the egg-laying mammals preserve a cloaca).

* Other mammalian features also develop late. From about Day 100 onward the fetus sucks its thumb. By Day 110 it develops a—transient—dense hairy coat, called the **lanugo.** The disappearance of this coats marks the beginning of the final human phase of the fetal development. It includes the legs growing longer than the arms.

However, it must be pointed out that the time course of a few ontogenetic events deviates from that of the evolutionary course. Such deviations are known as **heterochronies.** An example is the reduction of the tail in the human fetus. It occurs from Day 42 to 70—that is, in the "reptilian phase" of the fetal development, whereas in evolution the tail was reduced only when the higher primates emerged from the level of monkeys.

4.6   Evolution Can Also Be Seen in the Ontogeny of Invertebrates

A look beyond the borders of the various animal phyla discloses common features. A **spiral cleavage** is observed in several protostomial animal phyla (plathelminthes, nemertina, annelids, and mollusks); a **radial cleavage** is observed in several deuterostomial groups (echinoderms, acrania, and vertebrates). Developmental pathways starting with spiral cleavage often pass through a **trochophore** (Fig. 3–11) or a trochophore-like larva. Ontogeny and phylogeny are connected intimately with each other, for phylogeny is nothing but an uninterrupted series of slowly modified ontogenies.

4.7   Are Vertebrates Upside-Down Insects?

In 1822 the French zoologist Geoffry St.-Hilaire proposed a fantastic view of present-day and ancient animals. In his view vertebrate and arthropod embryos have a common body plan, but they have an inverted dorsal-ventral axis. The insect larvae resemble upside-down tadpoles. One or the other of these two groups crawled around on its back rather than on the belly of a common, wormlike ancestor. Therefore, the main nerve cord extends along the ventral midline in insects and along the dorsal midline in vertebrates, while the heart beats in the dorsal body in insects and in the ventral body region in vertebrates.

This thesis fell into disrepute, but recent progress in the molecular analysis of development has revived the debate (see Bibliography for Chapter 4). Data on the expression pattern of genes found in both groups might vindicate the old fantastic idea.

Speculations are permitted in science if the following three criteria are met: (1) they must not be presented as established facts, (2) they must not be contradictory in themselves, and (3) they must not contradict established facts. At present, a definitive answer is not possible.

4.8   An Intricate Matter: Homologous Organs, and Orthologous and Paralogous Genes

Evolutionary changes have their origin in changes in the genome. Evolutionary relationships and changes should be reflected in the genome and in the spectrum of proteins coded by the genome. The discovery of genes containing the DNA motif known as the **homeobox** in all animal groups down to the coelenterates, indicates genetic continuity and common developmental principles at the molecular level in spite of all the anatomical and morphological diversity exhibited by individual developmental pathways.

Morphologists speak of **homologies,** referring to characters with a common evolutionary origin. For example, homologous organs in various organisms derive from a common ancestral organ, though evolutionary changes may hide the derivation. The concept of homology distinguishes **general homologies,** correspondences between structures found in different species, from **serial homologies,** corresponding structures (e.g., forelimb, hindlimb) or repeating structures (e.g., somites) in the same animal.

In a similar but less precise manner, molecular biologists refer to **homologous genes,** meaning nonrandom sequence identity in DNA sequences from different organisms. Clearly, a high degree of sequence identity usually results from common ancestry: for example, genes of a gene family certainly originated from a common ancestor gene. Parallel to the terminology morphologists use, molecular geneticists speak of **orthologous genes** when genes with high sequence identity are found in various species, and of **paralogous genes** when several genes with high sequence identity are found in the genome of a single organism, because the original gene has given rise to several genes by gene duplication.

Does homology at the level of the genome allow us to read off homology at the level of the phenotype, and vice versa? Unfortunately, it does not!

Cells, tissues, and organs are composed of large sets of molecules. Now consider a bone. In ontogeny and phylogeny a skeletal element is often formed first with cartilage; later the element becomes modified and optimized by ossification. In the course of such a conversion to bone, all molecular components of the cartilage are removed and replaced by new and different components. Even the collagens, which are among the few macromolecular components found in both structures, are exchanged by other members of the family: the cartilage contains Type II and XI collagen, whereas the bone contains Type I (and V), as do the skin, the tendon, and the cornea.

On the other hand, macromolecules whose genes exhibit sequence similarity and are thus "homologous" to the molecular geneticist are often found in tissues or organs that are clearly nonhomologous in the morphologist's view. For instance, the **Hox-D13 protein** is expressed in the mouse embryo in the tip of the outgrowing tail as well as in the tips of the outgrowing arms and legs (Fig. 10–2). Like Hox-D13, **wnt-3a** is found in the tip of the tail but also in somite number 7.

Irrespective of the inconsistent use of the term "homologous," there are genes that have a dominant function in directing position-dependent elaboration of cell types, tissues, and organs. Such genes often code for transcription factors that in turn direct whole sets of subordinate genes. Such controlling genes are called **selector** or **master genes.** Orthologous and paralogous master genes found in various organisms indicate common principles in the control of development. These common principles are emphasized in the following chapters.

5

The Egg Cell and the
Sperm Get a Dowry

⟨⟩⟨⟩⟨⟩

5.1 Gametogenesis: Primordial Germ Cells Are Often Set Aside for Storage Early in Development.

Frequently in embryo development cells that are not used to construct the **soma**—the body of the new individual—are put aside but remain in stock as **primordial germ cells.** Such cells are known in nematodes, insects, and vertebrates.

When observing the chromosomes in the cleaving embryo of the horse roundworm, *Ascaris*, (now, *Parascaris*), in 1904, Theodor Boveri observed unusual behavior in those chromosomes. This nematode has only two pairs of large chromosomes. After the first cleavage, the chromosomes in one of the two blastomeres fragment into many pieces and many of the fragments are lost. This phenomenon of **chromatin diminution** occurs in the cell destined to participate in the construction of the body. The task of giving rise to germ cells is assigned to the second blastomere and in this cell the chromosomes remain intact. During the next divisions the story is repeated: only the cells on the direct route to the primordial germ cells (compare *Caenorhabditis elegans*, Fig. 3–9) preserve the entire chromosomes, whereas all future somatic cells lose chromatin. By experimental interference, such as centrifugation, Boveri (1910) was able to demonstrate that localized "cytoplasmic determinants" protect chromosomes of the germ line cells from being fragmented. Moreover, these cytoplasmic factors were thought to commit their host cells to become germ cells.

In *Drosophila* the primordial germ cells are the first cells to be formed in the embryo. They are known as **pole cells** because they arise at the posterior pole of the egg (Fig. 3–14, 3–15). As the pole cells form, they enclose **pole granules.** These are fibrous conglomerates containing considerable protein and RNA, and they embody **germ cell determinants.** An unusual component of these granules is RNA of mitochondrial origin, which has been released by mitochondria. However, this mitochondrial RNA probably is not the germ cell determinant itself. The true determinants either are derived from the maternally active gene *oskar* or they are allocated to the posterior

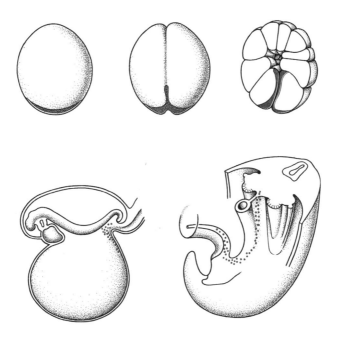

Figure 5–1   Origin and migration of primordial germ cells. The upper row shows the distribution of the RNA-rich "germ plasm" in *Xenopus*. Cells containing this germ plasm constitute the "germ line," leading to the primordial germ cells. The lower row shows the migration of germ cell precursors in the mammalian embryo. (After Gilbert.)

region of the egg under the organizing influence of *oskar*. When injected beneath the anterior pole of the egg, the *oskar* product causes ectopic development of pole cells at the anterior end of the embryo (**ectopic** from the Greek *ek*, *ekto* = outside, *topos* = place; outside the normal place). These, however, will not find their way into the gonads. Only pole cells properly located at the posterior pole are incorporated eventually into the gonads.

Earlier we discussed the remarkable work that has made it possible to follow the developmental heritage of every cell in nematodes, but even in vertebrates the lineage of the primordial germ cells can be traced back to the uncleaved egg. In *Xenopus* certain staining procedures demonstrate a particular "**germ plasm**"—granular cytoplasmic constituents close to the vegetal pole of the egg (Fig. 5–1). Like the pole cytoplasm in *Drosophila*, this region harbors RNA-loaded granules. Later these granula are found in the primordial germ cells that arise in the region of the embryonic hindgut and migrate along the mesenteries (a device to suspend the intestine) into the gonadal ridges (the rudiments of the ovaries or the testes).

In mammals the first primordial germ cells are dispersed over the posterior yolk sac. From here they emigrate, taking a similar route as in amphibians.

5.2 Only in the Gonad Is the Commitment to Eggs or Sperm Made in Vertebrates

In the early vertebrate gonad the immigrating primordial germ cells colonize the peripheral layer of the "female" **cortex** as well as the internal "male" **medulla** (Fig. 20–1). At this stage the embryo's sex has not yet been determined definitively. In case the gonad develops to a testis, directed by the master gene *sry* on the Y chromosome (Chapter 20), the primordial germ cells in the cortical layer perish, and those of the medulla differentiate into **spermatogonia**. In females there is no Y chromosome and therefore there is no *sry* gene. In the absence of *sry* the gonad develops into an ovary by default. Only the primordial germ cells of the cortex survive and become **oogonia**.

5.3 In Many Animals, Including Vertebrates, the Nuclei of the Oocytes Exhibit Lampbrush Chromosomes, rDNA Amplification, and Multiple Nucleoli

Oogonia undergo a phase of mitotic proliferation (Fig. 3–45). After these rounds of division have ceased, the germ cells are termed **oocytes.** The ovary of a female human fetus contains about 500,000 oocytes. After the birth of a girl, the proliferation ceases, in contrast to spermatogenesis in a boy, where proliferation resumes in puberty. With beginning puberty about 90% of the oocytes die.

As early as in the 3rd to 7th month of human female fetal development, oocytes enter prophase of meiosis (Fig. 3–45). The chromosomes duplicated earlier during S phase. In zygotene of prophase the homologous chromosomes, each consisting of two chromatids, pair along their long axes, held in register by a **synaptonemal complex.** But then, remarkably, prophase is interrupted for a prolonged time period, lasting 12 to 40 years in humans. The chromosomes remain paired but decondense and form lateral loops, taking on the appearance of "lampbrushes"—brushes once used to clean oil lamps.

5.3.1 Lampbrush Chromosomes

The lateral loops extruding from the paired meiotic chromosomes represent stretches of largely uncoiled DNA and indicate busy engagement in RNA synthesis. The transcripts are packed in ribonucleoprotein (RNP) particles and are shipped into the cytoplasm. Eventually, the egg cell is filled with RNP particles, enclosing many species of mRNA, in preparation for the unfolding of the program of early embryo development.

5.3.2 rDNA Amplification and Multiple Nucleoli

A large quantity of nucleoli (**multiple nucleoli**) appear, indicating production of an enormous number of ribosomes. Their appearance is the manifestation of a previous selective amplification of ribosomal genes (**rDNA amplification**).

In the genome of *Xenopus* the genes for the ribosomal 18S, 5.8S, and 28S RNAs are clustered together in a single transcriptional unit. This transcriptional unit has already been multiplied in evolution by repeated gene duplications, resulting in a tandem array of up to 450 copies of the rDNA genes, even before the rDNA becomes amplified additionally in oogenesis. The array is sometimes referred to as the **nucleolar organizer** because the factory to produce ribosomes is constructed along this region of the chromosome. Normally, one factory is built on each of the two homologous chromosomes in somatic cells. Thus, two nucleoli can be seen in the nucleus. Because *Xenopus* is tetraploid, up to four nucleoli can be expected in somatic cells. However, in oocytes many more nucleoli are seen. This is because an additional multiplication, called **amplification,** takes place. Through selective replication by way of a rolling circle of DNA, from each of the preexisting 4 ($\times$ 450) copies, about 250 copies are generated. Eventually, 1,000 additional pieces of rDNA are present. Each of these rDNA copies detaches from the coding DNA strand, is closed into a circle, and is supplemented by proteins. Thus, 1,000 factories for producing ribosomes are generated. The 1,000 nucleoli each contain 450 copies of three (18S, 5.8S, 28S) of the ribosomal genes. The missing 5S rRNA is procured as transcripts of the 24,000 copies of the linearly multiplied 5S gene present in the regular genome. These 24,000 copies are clustered on another chromosome and have been multiplied in evolution. In the additional 1,000 nucleoli about 10^{12} ribosomes are assembled. Without gene multiplication, the frog would require 500 years instead of a few months to produce such an amount! The egg is prepared to undergo rapid development because it has enormous potent machinery to produce proteins.

5.4   Somatic Cells Often Assume Additional Nursing Duties

The oocyte of vertebrates produces mRNA and ribosomes by itself. However, yolk proteins and yolk lipids, which serve as energy stores or building materials, are not made by the oocyte. **Vitellogenins,** the protein precursors of the yolk materials, are produced by the liver, transferred via the blood to the ovary, taken up by follicle cells, and handed over to the oocytes. Vitellogenins are found only in the female blood. Incorporated by endocytosis, the vitellogenins are split in the oocytes and are converted into the heavily phosphorylated **phosvitin** protein and the lipoprotein **lipovitellin.** Encased in membranes, these two materials constitute the main content of the **yolk platelets.** In addition, the maturing egg deposits **glycogen** granules. The oocytes develop into the largest known cells. The egg cell of the chick (the yellow sphere in the center of the egg) has a volume 9×10^9 times greater than that of a normal somatic cell! The egg cell in the ostrich is the largest animal cell known today.

5.5   In *Drosophila* All the Needs of Oocytes
Are Provided by Nurse Cells

In insects such as crickets, which give themselves time to complete oogenesis, we observe lampbrush chromosomes and multiple nucleoli like those in vertebrates. This is not so in *Drosophila;* this fly lives for only 14 days and produces an egg in 12 hours. The oocyte is supplied entirely with all it needs by nurse cells.

The oogonia are packaged in tubes termed **ovarioles** (Fig. 3–13). Each oogonium divides mitotically four times to generate an association of 16 cells. The cells remain connected through thin tubes called **fusomes.** Two of the centrally located cells are connected with four neighbors; one of those two cells becomes the oocyte, whereas the other 15 sister cells are destined to become nurse cells. These amplify their genome, become polyploid, and eventually contain 500 to 1,000 copies of the genome. This multiplication permits a high level of transcriptional activity.

The products of their synthetic activity, including RNP particles and protein, are directed into the swelling oocytes. Among the gene products are development-controlling molecules such as *bicoid* mRNA, deposited at the anterior pole of the egg, and *nanos* and *oskar* mRNAs, translocated to and accumulated at the posterior pole. These RNAs will then specify where the head and the abdomen are to be made (Chapter 3, Section 3.6, and Chapter 9). All of these products are of maternal origin, as are the products procured from the **follicle cells** in the wall of the ovarioles.

As in vertebrates, yolk proteins in *Drosophila* are manufactured outside the ovary in the form of vitellogenins. In insects the factory is the fat body; the vitellogenins are released into the hemolymph and forwarded to the oocyte by the follicle cells.

5.6   Oocytes Become Polar (Asymmetrical) and Are Enclosed by Extracellular Membranes and Envelopes

The constituents of the yolk are deposited asymmetrically in the oocyte. In *Drosophila* the nucleus stays in the center of the egg. However, anteroposterior and dorsoventral polarity are manifested by the distribution of yolk materials and the elliptic form of the egg, which is stabilized by the hardy envelope of the **chorion** secreted by the follicle. But the apparent bilateral symmetry is not decisive for the bilateral organization of the future body. In certain mutants (*bicoid, dorsal*) or as a result of experimental interference (injection of *bicoid* mRNA into the posterior region of the egg), the architecture of the body can be altered fundamentally in an ostensibly normal egg (Chapter 3, Section 3.6, and Chapter 9).

Most animal egg cells, including those of vertebrates and sea urchins, appear spherical; but in their internal structure they are endowed with an **animal-**

vegetal polarity: yolk platelets and glycogen granules are accumulated in the vegetal hemisphere. The huge nucleus of the oocyte, traditionally called the **germinal vesicle,** is located near the animal pole. The extent to which later development is influenced by the nonuniform distribution of the oocyte constituents must be tested experimentally (Chapters 8 and 9).

Finally, the egg cell is surrounded by an acellular, strengthening **vitellin membrane** and by additional enveloping layers of various consistency. In mammals these layers are designated as the **zona pellucida** and the **corona radiata.** In reptiles and birds the albumen and the egg shell are wrapped around the egg cell by the wall of the oviduct only after fertilization.

5.7   In Vertebrates Hormonal Signals Initiate the Formation of the Polar Bodies and the Final Maturation of the Egg

The mammalian oocyte undergoes a 500-fold increase in volume, a process that can take several years (12 to 45 years in human females). When the oocyte has reached its final size, meiosis is resumed. The chromosomes condense into a form suitable for transport. The **germinal vesicle breakdown** indicates the decomposition of the nuclear membrane. Subsequently, the first meiotic miniature sister cell of the egg cell, the first **polar body,** is extruded. The mammalian egg segregates the second polar body only after fertilization.

Resumption and completion of the interrupted meiosis is controlled hormonally. Stimulated by an internal clock in the brain, or in a few mammals such as rabbits, by the act of intercourse, the pituitary gland first releases the **gonadotropic** (gonad controlling) hormone **FSH (follicle-stimulating hormone).** In response to this hormone, maturing follicles undergo the last phase of growth and granulosa cells express receptors for the second gonadotropic hormone **LH (luteinizing hormone).**

FSH and LH stimulate the production and release of a second set of hormones: the steroids **estrogen** and **progesterone.** These steroids prepare the uterus for implantation. A certain species-specific ratio of FSH, LH, and steroid hormones induces the extrusion of the first polar body and the subsequent **ovulation:** the release of the egg from the ovary. In amphibians progesterone plays the dominant role in breaking the dormancy of the egg. In human females LH induces 10 to 50 oocytes to resume meiosis in the first half of the **menstrual cycle,** after years of arrest. However, as a rule, only one mature egg is liberated. Ovulation occurs in the middle of two menstrual bleedings. In most other mammals ovulation takes place in the period of

estrus ("heat"), which must not be equated with bleeding, but indicates the time of ovulation and therefore readiness for conception.

5.8   The Sperm: A Genome with a Motor

Primordial germ cells become spermatogonia in the seminiferous tubules of the testis (Fig. 20–1). Spermatogonia are stem cells that retain the ability to divide: one daughter cell remains a stem cell, while the other becomes a **spermatocyte.** The stem cells lie along the inner surface of the tubules; the segregated spermatocytes are displaced in the direction of the inner lumen of the channels. The spermatocytes derived from one stem cell remain interconnected and undergo their further development synchronously. Together, the oocytes cease mitotic division but grow only slightly, unlike oocytes. At puberty, the spermatocytes undergo meiosis. Each spermatocyte gives rise to four equal spermatids. In a process of terminal differentiation four spermatids give rise to four sperm cells. During their entire development the descendants of a primordial germ cell remain attached to each other by cytoplasmic bridges (just like the oocyte remains connected with the nurse cells in the ovarioles of *Drosophila*). Only the finished sperms separate from each other. While sperm cells accumulate in the center of the tubules, the stem cells along the tubule walls provide a fresh supply, continuously (in the human) or during the mating period.

   A typical animal sperm cell, such as that of the sea urchin (Fig. 6–1, 6–2), possesses an acrosomal vesicle beneath its tip, or an **acrosome;** proceeding posteriorly are the highly condensed nucleus, the neck with a pair of centrioles, the midpiece with the mitochondrial power station, and the propulsive flagellum. The sperm is ready for delivery and fertilization.

5.9   An Invisible Dowry: Imprinted Methylation Patterns from the Father and the Mother

The "maternal" genome, supplied by the egg, and the "paternal" genome, supplied by the sperm, do not always contribute equally to the programming of the characteristics of the new organism. In the mouse embryo, for instance, the paternal genome predominates in the extraembryonic structures (the trophoblast, the placenta), while in the embryo proper the maternal genome appears to have greater influence. On the other hand, in the human the expression of the *chorea huntington* gene is stronger if the (dominant autosomal) congenital disease is inherited from the father. This phenomenon, known as **genomic imprinting,** is due to differing methylation patterns in the DNA of the egg and the sperm (Chapter 10, Section 10.10). Other additional mechanisms of genomic imprinting have been proposed.

As a rule, the DNA of the sperm and the DNA of the egg cell are methylated differently; in mammals the overall amount of methylation is less in the egg cell. The methylation pattern influences the expression of genes from each gamete. In the gametogenesis of the next generation it appears that the methylation pattern can be extinguished partially and a new pattern can be imprinted, whereby the DNA of the sperm becomes more extensively modified by methylation than the DNA of the egg cell.

5.10   Mutations and Genetic Manipulations Are Transmitted to the Next Generation Only If They Affect the Germ Line

The germ cells convey genetic information accumulated over millions of years of evolution to the next generation. In the ontogenetic development of multicellular organisms a line can be drawn from the egg cell to the germ cells. This lineage is called the **germ line.** It is a matter of discussion whether a germ line exists in all organisms that definitively separates somatic cells from future primordial germ cells, as happens in *Caenorhabditis elegans* (Fig. 3–9). However, even if somatic cells can be converted to primordial germ cells, a linear series of DNA replications undoubtedly connects the egg cell with the germ cells. Only mutations and genetic manipulations that happen in the germ line, and only genes that are introduced into the germ line, get into the next generation. This point is relevant to any informed discussion about the capacity to alter the genetic constitution of organisms and introduce alien genes permanently.

6

The Start: Fertilization and Activation of the Egg

$\Longrightarrow\!\!\!\bullet\!\!\!\Longleftarrow$

6.1   When Does Life Begin?

Beginnings are often difficult to pinpoint, particularly in life cycles, for life is continuous. Nevertheless, each individual life has two discrete boundaries: fertilization and death. Sperm cells and egg cells are not capable of starting an independent life on their own. Once released from the testes or the ovary, the lifetime of sperm cells and egg cells is restricted to a few minutes or hours. Only the fusion of sperm and egg results in a cell with the potential to survive and to give rise to an independent individual as well as to a new generation. In biological terms, a human is a human from fertilization until death. No stage of development can be omitted, no nonhuman stage can be defined.

6.2   Are Insemination and Fertilization the Same?

Maternal and paternal genomes are brought together in the process of fertilization. Fertilization is preceded by insemination, but these two events are

146

often not so clearly distinguished as some terminological purists want. According to their definition, **insemination** leads to the encounter of the sperm cell and the egg cell, and culminates in the fusion of the sperm cell membrane with the egg cell membrane; **fertilization** is the fusion of the two haploid genomes to form the diploid genome of the zygote. A distinction between insemination and fertilization is meaningful and necessary in those organisms where the sperm enters an egg cell before the egg nucleus has completed meiosis. This is the case in mammals as well as in the nematode *Ascaris*. In such cases the haploid sperm nucleus has to wait until the egg nucleus is also haploid before fertilization proper can take place. Usually, however, all processes, starting with the first contact of the sperm cell and the egg cell and ending with the fusion of the two nuclei, are grouped under the heading "fertilization."

6.3 The Egg Cell Attracts the Sperm; the Sperm Can Undergo a "Capacitation"

The best-investigated process of fertilization is that of the sea urchin. The freshly spawned egg emits an attracting **gamone** in order to guide the sperm to its goal. In mammals this may happen as well. In mammals the sperm becomes capable of fertilization only if it is primed on its way through the uterus and the oviduct by substances secreted by the female. This effect is called **capacitation** of the sperm.

6.4 The Acrosome: A Chemical Drill

The contact of the head of the sperm with components of the egg envelope induces the **acrosomal reaction.** The acrosomal vesicle in the head of the sperm cell opens and releases a collection of hydrolases, such as proteases and glycosidases. The sperm possesses a chemical drill head and enzymatically lyses a path through the jelly layer and vitelline membrane, driven by its flagellum. In the sea urchin sperm the drill head elongates to a long boring rod or finger: from the bottom of the opened acrosomal vesicle a finger like structure, the **acrosomal filament,** projects through the enzymatically bored channel, penetrating the jelly envelope. The stretching out is accomplished through a rapid polymerization of globular G actin molecules to F actin filaments in the evaginating acrosomal finger (Fig. 6–1).

FERTILIZATION: SEA URCHIN

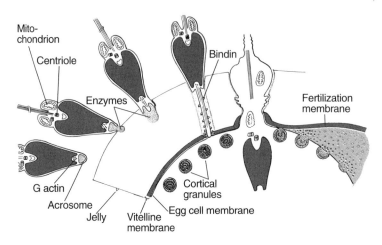

Figure 6–1   Fertilization in the sea urchin. Contact of the sperm with the jelly coat of the egg induces the **acrosomal reaction:** an acrosomal filament perforates the jelly coat and the vitelline envelope. A canal is formed and the nucleus of the sperm, together with the centrioles, is drawn or injected into the egg cell. As the cortical vesicles are discharged (cortical exocytosis), the swelling contents of the opened vesicles displace and inflate the vitelline envelope. It is now known as the **fertilization envelope.**

6.5   Species-Specific Receptors of the Egg Envelope Evaluate the Captured Sperm

6.5.1 Sea Urchins

In the free seawater of natural marine environments sperm of several species of sea urchin may contact an egg. It can happen that sperm of foreign species may attach to the jelly coat and drill their way through the outer envelope of the egg. The jelly coat contains components that trigger the acrosome reaction in sperm from various species. As a final control, the sperm and the egg check each other out by displaying their identity card: on the outer surface of the acrosomal filament **bindin** molecules serve as a distinguishing sign. Bindins from different species differ in structural details. If the correct bindin is present on the surface of the sperm, it will fit onto corresponding **bindin receptors** in the egg membrane (the extracellular domain of this receptor probably is integrated into the vitelline membrane that encloses the egg cell). This sperm or bindin receptor exists as a disulfide-bonded homotetramer of a transmembrane glycoprotein. The bindin and the bindin receptor allow a mutual, species-specific recognition: only sperm of the same species can establish intimate contact with the receptor and thus with the cell membrane of the egg.

6.5.2 Mammals

Why is the species-specific identity card of the sperm required at all in mammals? One might expect that males and females of different species would never copulate. Be that as it may, there are species-specific sperm-binding proteins in the zona pellucida, termed ZP1, ZP2, and ZP3. The ZP3 protein is considered to be the primary sperm-catching molecule. ZP3 is a glycoprotein that is in a form allowing the head of the sperm to couple with it only in the zona pellucida of unfertilized eggs. After fertilization, the sperm-binding capacity of the zona pellucida decays.

In mammals the sperm-binding proteins in the zona pellucida—in particular, ZP3—not only check the species of the sperm but also trigger the acrosomal reaction (in the sea urchin two different molecules of the egg envelope were responsible for the induction of the acrosomal reaction and the subsequent species-specific control and binding). However, the sperm-catching proteins of the zona, including ZP3, are not integral membrane proteins and therefore are not the ultimate sperm receptors. Present hypotheses assign the role of the ultimate sperm receptor to a transmembrane protein with internal tyrosine kinase activity. The following sequence of events has been proposed: the sperm first communicates with ZP3; ZP3 attaches onto the sperm head and stimulates exocytosis of the acrosomal vesicle. By this process an inner acrosomal protein, functionally comparable to the bindin in sea urchins, becomes exposed on the tip of the sperm. This newly exposed protein is the ligand that couples the sperm cell to the ultimate sperm receptor. This attachment enables fusion of the membranes of the sperm cell and the egg cell.

Following membrane fusion, the internal contents of the spermatozoon—its nucleus, mitochondria, and centrioles—are drawn into the egg. The egg's surface bulges outward into a **fertilization cone,** and long microvillar processes erupt from the surface toward the sperm. The processes stretch out and grasp the head of the spermatozoon. It becomes engulfed.

6.6 As a Rule, One Sperm per Egg

When the cell membranes of the sperm and the egg fuse, entry of further sperm should be prevented. The biology of fertilization distinguishes between two

FERTILIZATION: MAMMAL

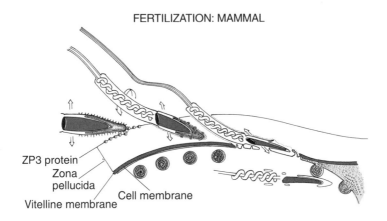

ZP3 protein
Zona pellucida
Vitelline membrane
Cell membrane

Figure 6–2   Fertilization in a generalized mammal. Upon contact with the ZP3 protein of the egg envelope (zona pellucida) the acrosomal membrane is fenestrated. Zona-lysing enzymes released by the fenestrated acrosomal vesicle digest a slit into which the rotating spermatozoon moves, propelled by the flagellum. As the advancing sperm establishes contact to the egg cell, the membranes of the spermatozoon and the egg cell fuse and give way to the nucleus and the mitochondria of the sperm. Mitochondria supplied by the sperm are believed to be destroyed and degraded in the egg cell. A cortical reaction similar to that seen in the sea urchin egg (Fig. 6–1) provides a fertilization envelope that hinders other sperm from entering the egg cell.

mechanisms that both serve to block **polyspermy** (the entrance of multiple sperm). The primary, fast polyspermy block is mediated by the inactivation of the sperm-binding receptors on the egg, which now prevent additional sperm from attaching. The secondary and permanent block is realized by the rapid inflation of the so-called **fertilization membrane** (Fig. 6–1, 6–2, 6–3). Both blocking mechanisms are switched on during activation of the egg (Section 6.7).

Often, inactivation of the binding molecules is vital. In most animals polyspermy is disastrous and causes early death of the embryo. Amphibian and avian eggs, however, appear to tolerate polyspermy; supernumerary sperm are destroyed within the egg.

6.7   Only Maternal Mitochondria Survive

Sperm contain mitochondria that are introduced into the egg cell along with the sperm nucleus. However, genetic analyses indicate that only the maternal

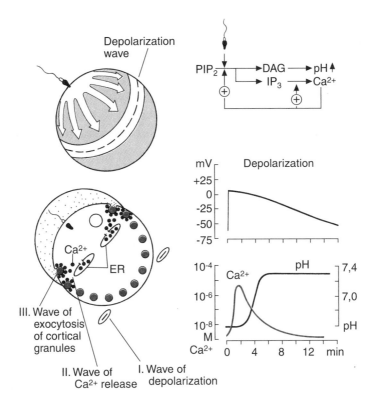

Figure 6–3   Activation of the egg (see text).

mitochondria survive. Therefore, mitochondrial genes are transmitted only in the female lineage. The exclusivity of maternal mitochondrial inheritance has prompted population geneticists to construct matriarchal phylogenies.

6.8   Activation of the Egg: Sleeping Beauty Awakened with a Kiss

The unfertilized egg sleeps: transcription, protein synthesis, and cell respiration are at or near zero levels. Contact of the acrosomal filament with the vitelline membrane triggers a cascade of dramatic events (Fig. 6–3):

1. The membranes of the sperm cell and the egg cell fuse at the site of contact. A passageway is established through which an activating factor is injected into the egg by the sperm. Subsequently, the **nucleus,** the **cen-**

triole, and the mitochondria of the sperm are directed into the egg cell through this passageway.

2. The egg membrane becomes depolarized electrically. A **wave of depolarization** starts at the point of sperm entry and spreads over the egg surface in the form of an action potential. In sea urchins, voltage-dependent alterations of the binding receptors are thought to provide the **early, fast polyspermy block:** sperm that arrive late cannot establish contact with the voltage-deformed sperm receptors.

3. At the sperm attachment site the **phosphatidylinositol phosphate (PI) signal transduction pathway** (Box 3) is initiated. Within seconds the second messengers, **inositol triphosphate (IP$_3$)** and **diacylglycerol (DAG)**, are produced. Additional second messengers such as cGMP and cyclic ADP-ribose, appear transiently. Because fertilization is a vital prerequisite for development, presumably there are redundant mechanisms to ensure signal transduction and egg activation.

4. The second messengers, IP$_3$ and cyclic ADP-ribose, or an activating factor supplied by the sperm, cause the release of **calcium ions** from the endoplasmic reticulum (ER) into the cytosol. A positive feedback loop results in an explosion of free calcium: the released Ca^{2+} ions evoke the release of further Ca^{2+} in the neighboring parts of the ER. The calcium reserves in the ER behave as an assemblage of distinct compartments. Calcium released from the first compartment stimulates calcium release in the second compartment, and so forth. The process of calcium release spreads explosively across the egg. Because the liberated calcium ions are pumped very quickly back into the ER, a circular propagating Ca^{2+} wave begins at the sperm entry point and travels across the egg towards the opposite pole. In the hamster and the mouse eggs the first calcium wave is followed by a burst of secondary calcium oscillations; repetitive waves occur at 1- to 10-minute intervals.

5. The calcium waves induce an explosive exocytosis of the **cortical vesicles** (also called cortical granules). Beneath the egg membrane are thousands of cortical vesicles. Upon activation by calcium ions the vesicles release their contents into the space between the egg cell membrane and the shielding vitelline membrane.

   The released material has an extremely high swelling capacity. Water rushes from the environment through the vitelline membrane, driven by the high osmotic value of the substance released from the vesicles. On the other hand, the vitelline membrane is highly elastic and has an extreme capacity to expand, like a balloon or an air bag. The pumped-up

Box 3   The Pi Signal Transduction System   153

━━━━━━━━━━━━━━━━━━━ Box 3 ━━━━━━━━━━━━━━━━━━━

THE PI SIGNAL TRANSDUCTION SYSTEM

Among the basic properties of life is the ability to be stimulated—to take up external information and respond to it. From the perspective of the individual cell, controlling signals emanating from other cells also belong to the category of external stimuli. Signals conveying information within the cellular community of an organism control cell division, determination and terminal differentiation, or direct cell movement.

The ability of a cell to respond to a particular signal molecule depends on its having specific antennae, called **receptors,** made of polypeptides. Many receptors are membrane anchored and are exposed on the exterior cell surface. The message picked up by the receptor through binding of the corresponding signal molecule (ligand), for example, a growth factor or inducer, must be amplified and transduced across the cell membrane into the interior of the cell. Inside, the message is converted into a **second messenger,** the first messenger being the extracellular signal itself. Frequently, more than one second messenger is generated or released from storage (among them, as a rule, calcium ions). The second messengers are directed into the diverse compartments of the cell (e.g., the endoplasmic reticulum or the nucleus). By cascades of secondary events (often phosphorylation reactions), the signals are distributed among various molecular targets such as ion channels in the membrane, the cytoskeleton, or diverse enzymes and transcription-regulating factors. For all these tasks, animal cells are equipped with signal transducing systems.

In development, but also in the physiology of the adult organism, **PI-PKC (phosphatidylinositol–protein kinase C)** systems play particularly diverse roles. Such systems mediate the following:

- **Activation of the egg** upon contact with the sperm (Chapter 6)

- **Reception of the neuralizing induction signal** in the future neural plate (Chapter 3, Section 3.8; Chapter 9, Section 9.5)

- **Generation of positional information and memory** in *Hydra* (Chapter 9.9)

- **Stimulation of cell division by growth factors** (Chapter 17)

- **Release of prolactin and thyroid-stimulating hormone** in amphibian metamorphosis (Chapter 19) and vertebrate vegetative physiology

- **Stimulation of B lymphocytes by antigens**

- **Signal transmission at certain synapses** (e.g., those operating with muscarinic acetylcholine receptors)

The following description of the PI-PKC system is simplified from a range of specific cases.

The external signal molecule binds to a **receptor** inserted into the cell membrane and exposed on the surface. Frequently, each polypeptide receptor contains five to seven transmembrane domains. The receptor is coupled through a molecular switch, a **trimeric G protein,** to an enzyme located on the interior surface of the cell. In PI-based systems this enzyme is usually **PLC (phospholipase C, phosphoinositase).** When the

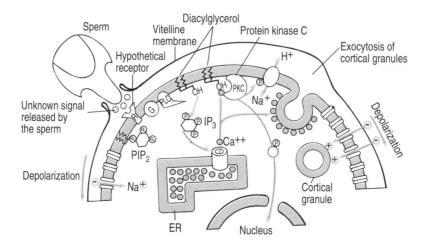

receptor is occupied by a ligand, PLC is activated and splits a special, membrane-associated phospholipid, **PIP$_2$ (phosphatidylinositol-biphosphate),** into two second messengers. One second messenger is the highly hydrophilic and electrically charged **IP$_3$ (inositol triphosphate);** the other is the lipid **DAG (diacylglycerol).** These second messengers serve to initiate the following two cascades:

1. The polar, water soluble IP$_3$ diffuses into the cytosol, binds to IP$_3$ receptors located along the endoplasmic reticulum (ER), and stimulates the release of calcium ions from the ER into the cytosol. In turn, calcium ions mediate a large variety of subsequent reactions.

2. The apolar DAG remains integrated into the membrane and now serves as a link onto which a specific key enzyme attaches. The enzyme, known as **PKC (protein kinase C),** is translocated from the cytosol, where it was present in an inactive form, onto the membrane, where it becomes activated by DAG (and Ca^{2+}). The activated PKC transfers phosphate from ATP onto specific serine and threonine residues of a large array of target proteins.

Gated ion channels in the cell membrane are among the potential target proteins that become loaded with phosphate: K+ channels, Na$^+$/K$^+$ ATPase, and the Na+/H+ antiport. Other substrates are elements of the cytoskeleton and a series of other protein kinases. These protein kinases mediate cascades of phosphorylation reactions. One such pathway leads into the nucleus. **Tertiary messengers** appear in the nucleus of various animal cells, for example, transcription factors coded by the *c-fos* proto-oncogene. Some cells respond to tertiary messengers with DNA replication and cell division, while other cells respond with cell differentiation.

Most cells are equipped with an array of diverse receptors that enable them to respond to various signals and various combinations of signals. Correspondingly, these cells have several transducing systems, and elements of the various systems interact with each other so that a variety of signals can be integrated. Frequently, the PI system is coupled with the **phosphatidylcholine cycle (PC)** and with the liberation of another second messenger, the **arachidonic acid (AA).** In turn, AA is the mother compound of a large array of signal molecules, collectively called **eicosanoids.** These include prostaglandins, leukotrienes, thromboxanes, and hydroxy fatty acids. All of these eicosanoids are potentially intracellular second messengers but they can also

Box 3   The Pi Signal Transduction System   155

cross cell membranes. Thus, one or a few external signal molecules may start cascades of secondary signaling systems.

---

balloon is known as **fertilization membrane** or **envelope** and represents the **second, permanent block to polyspermy.**

In mammals the explosion of the cortical granules not only sets free swelling gelatinous materials but also sets free enzymes that destroy the sperm-binding proteins of the zona pellucida. In any event, the inflated vitelline membrane has lost its sperm-binding capacity.

6. The Ca^{2+} signal stimulates the **metabolic activation** of the egg. This metabolic activation is also mediated by DAG, although it remains in the egg membrane. A protein kinase C (PKC) is translocated from the cytosol to the membrane, associated with DAG and activated. Subsequently, PKC stimulates a Na^+/H^+ antiport in the egg membrane by transferring phosphate onto serine and threonine residues of the antiporter. By the activity of this ion exchanger, H^+ is extruded from the cell and Na^+ is taken up instead. The pH in the cytoplasm of the egg rises. Present hypotheses assume that ion exchange and increased pH result in conditions under which the mRNA stored in ribonucleoprotein (RNP) particles can be liberated and used for translation. Among the newly produced proteins are histones, which are needed for reduplication of chromosomes in the course of cleavage.

7. Another cascade of events starting from activated PKC eventually results in the start of **DNA replication.** DNA replication is initiated in the haploid nucleus of the sperm and the egg, and is completed before the two nuclei meet each other to fuse. Directed into the egg, the sperm nucleus decondenses and is called the **male pronucleus.** It migrates to the **female pronucleus,** guided by microtubules that emanate from the centrioles. In mammals the female pronucleus first must complete the second meiotic division before it can mate the male wooer. Completion of meiosis is manifested by the extrusion of the second polar body. With the encounter and fusion of the male and the female pronuclei, fertilization is concluded.

Without delay the first cleavage is initiated. To organize microtubules for construction of the spindle apparatus, the centrosome, which usually contains a pair of centrioles, is needed. In some eggs the centrosome provided by the sperm is duplicated, in others the maternal centrosome, and in still others both the maternal and paternal centrosomes are used to direct and accomplish the distribution of the duplicated chromosomes. The first cell division can start.

7

Precisely Patterned Cleavage Divisions Are Driven by an Oscillator

❖

7.1 Cell Cycles in the Early Embryo Only Consist of S and M Phases

The fertilized egg is a single cell that must give rise to a multicellular organism; and this should happen as fast as possible because the embryo cannot evade predators (if it is not carried away inside a viviparous mother). The egg uses stored molecules to jump-start development. As a rule, until gastrulation, the early embryo can abstain from transcribing genetic information, because all needed mRNA is already present in the maternal ribonucleoprotein (RNP) particles. From these particles, mRNA is liberated and used for translation. Chromosomes are not needed for directing protein synthesis and the embryo can focus on their multiplication. In rapid succession S phase (replication of the DNA and duplication of the chromosomes) and M phase (mitosis) alternate. Mammalian eggs are exceptions to this rule, and the cell cycle in cleaving mammalian eggs is unusually long. The eggs of sea urchins, *Xenopus*, and the clam *Spisula* are popular for research because of their rapid and synchronous cleavage.

Because mRNA for histones and other chromosomal proteins is present in the cytoplasm in many copies and because these proteins can therefore be supplied rapidly, chromosome replication is finished in 20 to 30 minutes. Immediately thereafter, the chromosomes are condensed to a form suitable for movement during mitosis. The early embryonic cell cycle actually consists only of S and M phases. In *Xenopus* the G1 and G2 phases are inserted only in the blastula stage, when new transcripts of zygotic genes are needed. This stage is known as the **midblastula transition.** Now the cell cycles become longer and the divisions of the various cells become asynchronous. Increasingly, the start of a new cell cycle becomes dependent on external cues.

7.2   The Mitosis-Promoting Factor Oscillator

The internal oscillator that drives the cell cycle in the early embryo uses the same molecular components that underlie the controlling circuitry of any cell cycle.

The protein kinase **p34** or **cdc2** (complete designation p34^{cdc2}) is present continuously, whereas the protein **cyclin B** is produced periodically in preparation of the mitosis (Fig. 7–1). Cyclin B accumulates and is destroyed repeatedly in synchrony with the cell cycle. However, these oscillations do not represent the complete clock. Both components are engaged primarily in the initiation of the mitotic division. For mitosis to occur, cyclin B and cdc2 must associate with each other to form **mitosis-promoting factor, MPF** (synonyms: **meiosis-promoting factor,** or **maturation-promoting factor**).

MPF is subject to regulatory modifications—to phosphorylation and dephosphorylation. With the activation of the egg, the first round of cyclin B synthesis is initiated. The newly synthesized cyclin B couples to cdc2 to form MPF. However, the complex is still inactive; it becomes activated only after DNA replication has finished. Activation of MPF is mediated by an alteration in the pattern of its phosphorylation. For MPF to be activated, threonine-161 (i.e., the amino acid threonine at position 161 of the cdc2 moiety) must be loaded with phosphate (positive activation), and threonine-14 and tyrosine-15 must have their phosphates removed (release of inhibition). In this state MPF has come to be the true mitosis-promoting factor. After mitosis has finished, cyclin B is destroyed by proteases.

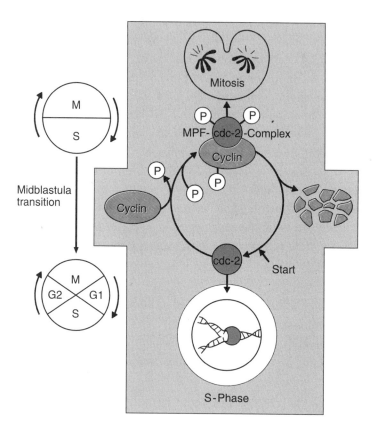

Figure 7–1 The embryonic cell cycle (left) and the mitosis-promoting factor (MPF) oscillator. Binding of cdc2 to cyclin is followed by several steps of phosphorylation and dephosphorylation, indicated by the circled P. The P circles merely symbolize these events. There are actually multiple sites where threonine, serine, and tyrosine are loaded with phosphate. Activation of the MPF complex (cdc2 + cyclin) demands both phosphorylation of certain amino acids (threonine-161) and dephosphorylation of others (threonine-14; tyrosine-15) in the cdc2 moiety. In the S phase cdc2 is coupled with cyclin A (not shown).

Mitosis is only a short phase of the entire cell cycle. In the fully developed cell cycle, DNA replication is prepared in G1 phase and actually is accomplished in the following S-phase DNA. In G2 the following mitosis is prepared. A cell must proceed through the whole cycle in an orderly sequence. For example, mitosis must not start until DNA replication is finished. Therefore, several key events in the cell cycle are regulated to ensure an orderly progression. At decisive checkpoints, cyclins and cyclin-dependent kinases (CDKs) play key roles again. Both exist in multiple variants (isoforms). The association of cyclin E with CDK2 is thought to control the commitment of the cell to DNA replication, an event called **START** in yeast and **restriction point R** in animal cells. During the S and G2 phase, cdc2 and CDK2 associate with cyclin A. Only then is active MPF formed as outlined previously.

Much remains to be learned about the details of the oscillating machinery. Why are so many diverse cyclins and CDKs present in animal cells? The entire oscillating apparatus also contains many other components: several sets of kinases, phosphatases, and cofactors. Their orderly interaction leads to the repeated construction and destruction of MPF as well as sequential ordering of events, so that reduplication of the chromosomes is finished before cell division takes place.

7.3   Tightly Patterned Sequences of Cell Divisions Allow Cell Genealogies to Be Traced in Some Species

In the small embryos of small organisms careful observation has revealed fixed patterns in the temporal progression of the timing of cell divisions and the spatial arrangement of daughter cells. The diverse lineages always lead to the same types of cells, tissues, or organs. In nematodes, such as *Caenorhabditis elegans*, for example, the P lineage always terminates in the primordial germ cell P4 (Fig. 3–9). In the spiralian embryo the D line leads to the primordial mesoblast d4, which in turn gives rise to all mesodermal tissues (Fig. 3–10). In the ascidian embryo a certain cell line leads to the musculature of the tail and another line leads to the notochord (Fig. 3–23). However, even in *C. elegans*, where a particularly precise and rigid order governs the sequence of cell divisions and subsequent events of cell differentiation, not all muscle cells are descendants of one and the same founder cell and not all nerve cells are descendants of another founder cell. Instead, several different founder cells contribute to the final inventory of muscle cells and nerve cells. Most tissues are thus said to have a **polyclonal provenance.**

A cell lineage faithfully repeated from generation to generation does not permit a priori reliable statements about the time course of determination. When is the fate of the cells programmed, and when is their destiny irreversibly fixed? However, experimental evidence suggests that exact genealogy is correlated with early determination.

8

Determination:
Cells Are Programmed and
Committed to Their Fates

———◦—◦—◦———

8.1 With the Allocation of a Certain Task,
Alternative Developmental Potencies Become Curtailed

In the course of embryogenesis all newly generated cells, with very few exceptions, receive complete genomic information. In mitotic cell cycles identical and complete strands of DNA are apportioned to both daughter cells. Initially, the cells of an embryo have **genomic equivalence:** each cell is thought to have the same genome as every other cell and therefore is capable in principle of becoming any cell type. All cells are believed to be **initially equivalent** and **totipotent,** and experiments with transplanted nuclei (Box 2) confirm this basic notion.

Determination is the process by which specific tasks are allocated to cells and cells become programmed differentially. The alternative term, "**commitment,**" is frequently used when attention is focused on a particular cell type rather than on an organ or body region.

The term "**specification**" also is used to indicate allocation of tasks. "Specification" does not imply strict and definitive programming in the sense of "determination." Often the term "specification" can be read in the context of pattern formation in areas of cells.

A cell fated to take a distinct path cannot be distinguished from its previous uncommitted state on the basis of its appearance. A practical instruction for using the terms "specification" and "determination" is given later in Section 8.3.

8.2 Two Basic Mechanisms of Determination Are Asymmetrical Cell Division and Cell Interaction

8.2.1 Asymmetrical Cell Division and Autonomous Development

In the cytoplasm of a founder cell (an egg cell, a blastomere, or a stem cell in later stages of development) developmental determinants, such as ribonucleoprotein (RNP) particles loaded with mRNA for transcription factors, are distributed nonhomogeneously. The determinants become segregated in cell division and distributed unequally among the daughter cells. As a consequence, a cell division gives rise to two differently fated cells.

If the developmental pathway of a cell and its descendants is solely the result of its internal determinants, the cell will be able to acquire its terminal form and function independently of its surroundings. Thus, the founder cell of the tail muscles in the ascidian larva is informed of its task by maternally generated determinants laid down in the egg. The developmental pathway that this founder cell and its descendants will take is autonomous. When such cells are transplanted into a foreign environment or maintained in cell culture, they will adhere to their program regardless of the altered environment.

8.2.2 Cell Interactions and Dependent Development

Daughter cells are initially equivalent not only genetically but also (more or less) with regard to their determinative components. The cells become different because they influence each other mutually. Their development responds to environmental influences.

Cell interactions are manifold (Chapters 9 and 17). Founder cells whose fate is also determined by their neighborhood are flexible in their response. For example, cells from the animal hemisphere of the amphibian blastula give rise to the epidermis, the nervous tissue, or the musculature, depending on the influences acting upon them (Fig. 3–36).

8.3 Mode and State of Determination Must Be Ascertained Experimentally

The observation that some blastomeres receive certain cytoplasmic components while others do not, does not tell much about the mode of determination. For example, the frog egg contains black pigment granules in its animal cortical layer; these are allocated only to animal blastomeres. Yet these granules are not significant with respect to determination, as albino eggs develop normally.

Whether determinative components are present must be determined experimentally. Cytoplasmic components can be displaced by pressure or centrifugation; or cytoplasm can be sucked off and reinjected at another location.

The state of determination is tested with the following procedures:

1. **Isolation assay.** Embryo cells or associations of cells can be grown in petri dishes on suitable media. The culture medium must be neutral with respect to the developmental pathway of the cells. If the cells in the dish always develop autonomously (or after the addition of certain differentiation-promoting substances) into the same cell type, they are considered to be committed. Following Slack (1991), cells whose commitment is stable in a neutral environment are said to be **specified.**

2. **Transplantation assay.** Cells, associations of cells, or larger pieces of an embryo are transplanted microsurgically into another embryo at another place. The **donor** cells may be implanted into a **host** embryo of the same species (**autologous** or **homoplastic transplantation**) or into host embryos of another species (**heterologous** or **xenoplastic transplantation**). Heteroplastic transplantations can be performed without complication between various newts and salamanders, frogs and newts, or chick and quail.

Following Slack (1991), transplants that retain their state of commitment even in an alien environment are said to be **determined.** If transplants are already determined irreversibly, they will behave "**according to their provenance**" (Spemann 1938)—according to the internal program brought along with them as dowry—and they will behave autonomously, without being led astray by their new surroundings. By contrast, cells that may only be "specified" but are not yet determined irreversibly, will develop "**in accordance with their location,**" if determining signals emanating from their new neighborhood act upon them.

In a famous experiment from the school of Spemann, an ectodermal piece from a frog blastula, which would have formed skin of the belly at its original location, develops teeth when transplanted into the mouth region of a newt. The piece responds to local cues. However, the teeth are horny because frog cells have genes to make horny teeth but lack genes to make hard teeth (Fig. 3–33).

The ability of transplanted tissue to adopt the identity of its new surroundings is considered to be evidence of its developmentally uncommitted state and also demonstrates the phenomenon of **positional information** (Box 5). The results of the isolation and transplantation assay are not always consistent, because specification and determination may have taken place before transplantation, but may be unstable and retuned upon transplantation.

3. **Nuclear transplantation assay.** With commitment to a distinct task, the potency to develop into other cell types is lost, as a rule. The developmental potencies become restricted. Whether such a restriction is associated with an irreversible genetic alteration can be tested by transplantation of the nuclei.

In famous experiments first carried out by Briggs and King (1952), and Gurdon (1962), and subsequently continued by others, *Xenopus* nuclei were removed from somatic tissue, such as tadpole gut cells or adult lung cells, and were injected into enucleated egg cells. Remarkably, the eggs gave rise to normal clawed toads. With increasing experimental experience, the fraction of eggs displaying complete developments increased to large numbers. The nuclei of the donor cells were still **totipotent** (synonym: **omnipotent**); determination in these cells did not result in an irreversible loss of genetic potency (Box 2).

Provided the transplantated nuclei are all taken from a single individual, the offspring are genetically identical to the donor and to each other. The toads have been **cloned** (a clone is a collective of genetically identical individuals, usually generated by asexual reproduction).

8.4 "Mosaic" and "Regulative" Development Differ in the Time Course of Determination

In the embryos of ascidians, nematodes, and spiralians determination takes place very early. As a consequence, the regulatory capacity of injured embryos is very low and removed founder cells can often not be replaced or compensated by reprogramming other cells. Such reprogramming is possible in many other developmental systems and is known by the term "transdetermination" (Chapter 3, Section 3.6.5, and Chapter 21). In the groups of animals listed previously, transdetermination was unknown for a long time. Frequently, the development of particular cells or cell lineages is "**autonomous**" and is continued upon isolation of the cells and maintenance in the cell culture. Such observations have prompted the view that a mosaic of determinants is laid down in the egg of those animals, and the term "**mosaic development**" was coined. However, closer analysis revealed that even in these embryos many cell interactions take place; however, they occur very early.

In the embryos of hydrozoans, sea urchins, and vertebrates determination is mainly based on interactions among cells (Chapter 9). In addition, the period of programming is long. Correspondingly, the ability to regulate development persists for a long time in these organisms.

In many organisms certain stem cells, such as those of the blood-forming system in vertebrates, retain pluripotency for life. **Pluripotency** refers to the ability to give rise to several different cell types. On the other hand, in the eggs of sea urchins and vertebrates as well as in eggs of organisms exhibiting mosaic development, more and more maternally encoded gene products are being identified, and distinct functions are assigned frequently to the cells that receive such gene products. For instance, in sea urchins the micromeres and the founder cells of the larval skeleton, as well as the primordial germ cells in the amphibian embryo, have their functions assigned by maternal gene products. Thus, the differences between organisms exhibiting "mosaic" or "regulative" development are merely gradual.

8.5   Through Determination Selector Genes May Be Turned On Constitutively

Prompted by the example of the muscle-specific selector gene *MyoD1* (Chapter 10, Section 10.3), hypotheses have been proposed to explain how a cell-type-specific program of determination might be installed and passed from cell generation to cell generation in proliferating progenitor cells. Selector genes, also called master genes, hold in check batteries of subordinate genes. In order to realize a distinct differentiation program, a distinct set of genes must be activated. Selector genes select and control sets of genes required for cell-type- and region-specific development. Therefore, the activity of selector genes is pivotal for the initial selection, actual realization, and further maintenance of a particular differentiated state. Chapter 10 discusses this topic further.

9

Epigenetic Pattern Formation: New Patterns Are Created During Development

9.1 How Is Positional Information Acquired and Used in Differentiation?

Cells of the embryo must behave in accordance with their location. Here they construct the nervous system, there they form a muscle, at this place they must jointly produce an element of the skeleton giving it a distinct shape, and at that place they must commit suicide to create a cavity. How do cells know where they are? Do their genes somehow tell them what their position is at each moment? This is not possible, for the DNA of the nucleus does not contain a street map, and the cells of the embryo cannot look into their nuclei to find out where they are. Somehow position-dependent cell differentiation must be initiated from the outside. By "from outside" we mean by signals having their origin outside the nucleus of the cell in question.

9.1.1 Substantiated Working Hypotheses Help to Trace the Origin of Positional Cues

Several hypotheses have been proposed to explain how a cell gets information on its position:

1. **External cues.** Information comes from the external environment. For example, gravity or light could provide cues to guide orientation.
2. **Patterns of maternal determinants laid down in the egg cell.** Maternal cytoplasmic determinants (represented by RNA or protein) are assigned differentially to the cells. They function like airline tickets, specifying the route and the destination. However, various tickets are needed and must be allocated to the cells in an orderly spatial pattern.

    Two mechanisms for the creation of spatial order can be imagined and both have been identified. (1) The cytoplasmic determinants, whether synthesized by the oocyte itself or supplied by neighboring nurse cells, are deposited in the egg at distinct places during oogenesis. (2) A process of sorting out and internal patterning, called **ooplasmic segregation,** takes place within the egg after fertilization.
3. **Cell interactions and concerted behavior.** Ballet dancers can create beautiful patterns guided by mutual arrangement and exchange of gestures. Cells may also "look" at their neighbors and make agreements with them to create coordinated patterns.

9.1.2 Pattern Formation as an Epigenetic Process

The pattern of cytoplasmic determinants can only be a first guide to orientation (Fig. 9–1). Think of a *Hydra:* by budding, the parental animal gives rise to hundreds of offspring; hundreds of offspring give rise to thousands of offspring, and so forth, until after thousands of years vast numbers of polyps have been produced without intervention of an egg cell and sexual reproduction. How could the detailed development and shape of all these hydras be specified by cytoplasmic determinants laid down in the one original egg cell many years ago? Or imagine apple trees of the variety Golden Delicious. Trees all over the world have been grown by grafting and are the cloned offspring of one egg cell that existed decades ago. How could the pattern of cytoplasmic determinants in the founder egg cell determine in advance the exact pattern of all these trees with all their branches, leaves, and blossoms?

Pattern formation is a reproducible epigenetic accomplishment. **Epigenetic** refers to processes that are determined by events "above" (Greek: *epi* = on, above) the level of genetic information. The processes that are decisive for pattern formation are interactions of cells with other cells (Fig. 9–1; Box 6) or

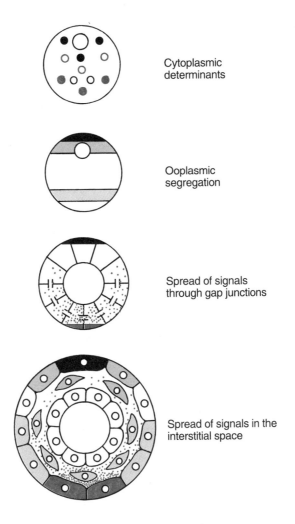

Figure 9–1   Steps of determination. A hypothetical scheme based on classic experiments. First, specification or commitment is accomplished by cytoplasmic determinants (top). These are ordered spatially in the course of an egg internal patterning
process, known as ooplasmic segregation, and differentially allotted to the blastomeres (daughter cells of the egg). Later in development, pattern formation and
commitment are based on the exchange of signals between different cells.

with extracellular substrates (extracellular matrices). Of course, genetic information is required to produce those molecules that mediate cell interactions.
For example, genes specify the signal molecules that can be exposed on the
cell's surface or secreted into the surroundings, and receptors that can be used
to receive and recognize such signals. However, all these interactions also are

determined significantly by nongenetic principles, such as the physical laws of diffusion or the physical forces of adhesion.

9.2   External Cues Guide Determination of Spatial Coordinates

First, a decision is made where the head, the tail, the back, and the belly will be located. Most animals are bilaterally symmetrical. Perpendicular to an anteroposterior axis is a dorsoventral axis. In developmental biology such axes of asymmetry or anisotropy are called **polarity axes.**

Egg cells are always organized in a polar manner. Frequently, oocytes in the ovaries are not surrounded completely by nourishing nurse cells. The oocyte is fed predominantly from one side (Fig. 3–13, 3–18), which suggests a source of cues for asymmetric organization. Even gravity has an influence in certain cases. Thus, in the oocyte various substances, whether produced by the egg itself or provided by nurse cells, are not deposited uniformly. Heavy yolk granules often accumulate near the vegetal pole, while the nucleus of the oocyte comes to lie near the animal pole, where the polar bodies are constricted off later in the course of the meiotic divisions (actually, the location of the polar bodies defines the animal pole).

In most eggs it is not difficult to recognize the animal-vegetal axis under the microscope. However, in the vast majority of egg cells only one unipolar axis is visible, not a complete bilaterally symmetrical architecture. The egg of *Drosophila* is exceptional, displaying a bilateral shape and internal organization. Most animal eggs, however, display neither a bilaterally symmetrical shape nor bilateral organization in their internal constituents.

The animal-vegetal axis may coincide with either the future anteroposterior or dorsoventral body axis, or with neither of them. In mollusks, ascidians, and birds the animal-vegetal axis of the egg coincides with the dorsoventral axis of the embryo.

9.2.1 In Vertebrates Environmental Cues Are Significant

In **amphibians** (and in the zebra fish) the animal-vegetal axis enables an observer to predict approximately where the head will be formed. The head will be located within a short radius of the animal "North Pole." The spine will extend over the animal hemisphere, cross the equator, and end at about the Tropic of Capricorn. Initially, however, the actual longitudinal line along which the spine will extend is not determined. The position of this "zero meridian" is determined at the moment of fertilization. Two external parameters are involved: gravity and the point of sperm entry.

The sperm attaches randomly at a point anywhere in the animal hemisphere. Under the additional influence of gravity the **gray crescent,** or its functional equivalent, appears diagonally opposite to the sperm entry point. The gray crescent harbors several important determinants of development and marks the site where the blastopore will appear in gastrulation (see Chapter 3, Section 3.8). By experimentally rotating an egg, the location of the gray crescent and hence of the future blastopore can be displaced. The gray crescent (unfortunately not visible in *Xenopus*) marks the future location of the spine. The zero meridian is now fixed. The head-back-tail line will start at the animal pole (urodeles such as newts and salamanders) or near the Tropic of Cancer (anurans such as frogs and toads) and extend along the zero meridian down to the gray crescent at the Tropic of Capricorn (Fig. 3–27). An internal architecture, the animal-vegetal asymmetry, determines bilateral symmetry in cooperation with external cues (point of sperm entrance and gravity).

In the **avian egg** the dorsoventral polarity is predetermined by the structure of the yolk-rich egg. The animal pole marks the center of the upper surface of the blastodisc and defines the dorsal side. The location of the head-tail line is thought to be specified by the combined effect of gravity and rotational movements: as the egg descends along the oviduct and uterus, it rotates. However, only the shell and some internal constituents rotate, not the yellow egg cell. Thus, the blastodisc remains at the top of the egg cell but its plane is tilted; the angle to the horizontal is 45° and therefore the bastodisc is subjected to shearing forces. The direction in which the egg is rotated is believed to specify the direction of the head-tail line.

Mammalian and **human embryos** appear to leave the site where the inner cell mass will segregate to chance. The position of the inner cell mass defines the future dorsal side. How the head-tail polarity is specified is unknown.

9.2.2 In *Drosophila* the Mother Decides the Future Polarity Axes of Her Child in Advance

In the invertebrate model *Drosophila* symmetry determination is markedly different than described previously for vertebrates. A significant difference is that the process of specifying bilateral symmetry is under the control of the maternal genome. As *Drosophila* has a dominant position in developmental biology, there is the risk that peculiarities found in *Drosophila* are taken as paradigmatic for animal development in general. However, even in *Drosophila*, the decisive factors originate outside the egg cells.

The **anteroposterior polarity axis** is specified by the spatial distribution of particles enclosing maternally generated mRNA [ribonucleoprotein (RNP) particles]. *bicoid* message is deposited at the front pole, the mRNAs of *nanos* and *oskar* are deposited at the tail pole. Viewed from the egg cell, the origin of these gene products is external, but the determinants are now internalized. When *bicoid* mRNA is translated into protein, this protein acquires the function of a transcription factor controlling gene activities in the early embryo (Chapter 3, Section 3.6, and Chapter 10). In addition, the embryo produces the TORSO receptors using maternally supplied *torso* mRNA; these receptors are exposed on the surface of the egg cell and recognize external cues. The cues are factors produced in the ovary by follicle cells and stored in the space between the egg cell and the egg envelope, in front and behind the egg cell.

Specification of the **dorsoventral polarity** is mediated by signal molecules that have been secreted by follicle cells at the future ventral side, stored in the "perivitelline" space between the egg cell and the egg envelope, and picked up by receptors (coded by the maternal gene *toll*) of the egg cell membrane. The signals trigger a mechanism that redistributes the determinative cytoplasmic factor: the maternal factor DORSAL, initially homogeneously distributed in the precleavage egg cell, migrates into the ventral nuclei of the blastoderm (Fig. 3–17, 3–20).

Because the messages for each of these factors (derived from *bicoid, nanos, oskar,* and *dorsal*) and each of these receptors (derived from *torso* and *toll*) are products of genes, the *Drosophila* genome is involved directly in the establishment of the coordinates of the body. It is, however, the **maternal genome** that is involved. From the perspective of the oocyte, even in *Drosophila* the decisive cues for orientation come from outside, from the nurse cells and follicle cells of the maternal ovary.

9.3   Ooplasmic Segregation: In the Egg Cytoplasm Epigenetic Patterning Results from the Redistribution of Determinants

In many animal egg cells a redistribution of cytoplasmic components is observed before the first cleavage plane divides the egg into separate compartments (Fig. 9–1). This **ooplasmic segregation** is seen, for example, in the eggs of spiralians and ascidians (Fig. 3–26).

In *Drosophila* regulatory gene products are displaced before the cellular blastoderm stage. For example, the DORSAL protein is uniformly distributed in the freshly laid egg but eventually accumulates in the nuclei of the ventral blastoderm (Fig. 3–17). This redistribution of cytoplasmic constituents is an expression of internal patterning within the egg.

9.4 Pattern Formation by the Exchange of Signals between Adjacent Cells

With the completion of the first cleavage (in *Drosophila* with the completion of the cellular blastoderm), pattern formation is based on the exchange of signals between cells. Cells make arrangements with their neighbors: "If you make this, I will make that." Such agreements between adjacent cells are found even in classic "mosaic" embryos, such as those of *Caenorhabditis elegans* and ascidians. The following two examples are from *Drosophila*.

9.4.1 Eye Contact: Photoreceptor Cell Development

The compound eye of the *Drosophila* comprises about 800 small eyes called the ommatidia. Each **ommatidium** is composed of 20 cells, including eight **photoreceptors.** These cells are arranged in a precise, consistent pattern. The first photoreceptor to develop is the central cell, called R8; it is the BOSS cell. BOSS refers to the gene *bride of sevenless.* The protein encoded by this gene is exposed on the surface of R8 and is shown to the adjacent cell R7 as an identity card. R7 in turn is equipped with a membrane-associated receptor to check the identity card of its neighbor. The receptor is a transmembrane tyrosine kinase derived from the gene *sevenless.* If photoreceptor R8 is unable to offer the correct signal due to a mutation, or R7 is unable to recognize the message, the cell R7 will not develop properly. Instead of developing into a photoreceptor, it makes a lens. The lens cell is the default option if the photoreceptor program is not turned on (Fig. 9–2).

9.4.2 Nerve Cell or Epidermis?

In the *Drosophila* blastoderm the cells segregate into two groups: **neuroblasts** and **epidermal** cells. How is the decision made? Initially, all of the cells are nearly equivalent and have the potential to become either the epidermis or the nervous system, although in the ventral side there is an initial bias for becoming neuroblasts. However, the cells are not yet committed irreversibly to one path or the other. If transplanted into a more dorsal position, ventral cells adopt the features of a dorsal cell and participate in forming the epidermis. Likewise, if displaced ventrally, prospective epidermal cells comply with the social rules and instructions valid at the new location, and become nerve cells.

Both groups segregate definitively from each other by the exchange of signals while they still are clustered more or less in intermingled groups. In this process of separation signal molecules exposed on the surface of the cells

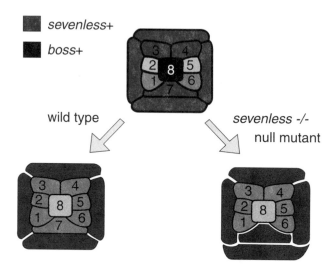

Figure 9–2 Pattern formation and commitment in the *Drosophila* ommatidium by close-range interactions. The photoreceptor R7 is formed only when the cell R8 expresses the gene *boss*⁺, and the cell R7 expresses the gene *sevenless*⁺. The proteins coded by these genes are exposed on the cell surface and presented to the neighbors. If either of the two gene products is absent or defective, photoreceptor R7 is replaced by an additional lens cell. Lens cell formation is the default option if the photoreceptor program is not turned on.

are of particular significance. One of the surface molecules is encoded by the gene **Notch**. When the NOTCH protein is defective due to a mutation in the *Notch* gene, all cells develop into neuroblasts instead of a mixture of neuroblasts and epidermoblasts. NOTCH interacts with another membrane protein exposed by neighboring cells and encoded by the gene **delta**. Both NOTCH and DELTA are transmembrane proteins having multiple functions: they serve cell adhesion, and simultaneously play the roles of both ligand and receptor. NOTCH is the ligand for the DELTA receptor and DELTA is the ligand for the NOTCH receptor (Fig. 9–3).

NOTCH is needed by the cells to become the epidermis. In order to reach this end, the NOTCH receptor—once stimulated by the DELTA ligand—initiates a cascade of events resulting in the switching on of a gene complex called *enhancer of split* (ESPL-C). When NOTCH is defective, the intrinsic preference to take the neurogenic path is not suppressed properly.

DELTA is needed by neurogenic cells. It enables them to suppress the neurogenic tendency in their neighbors by showing them their DELTA-

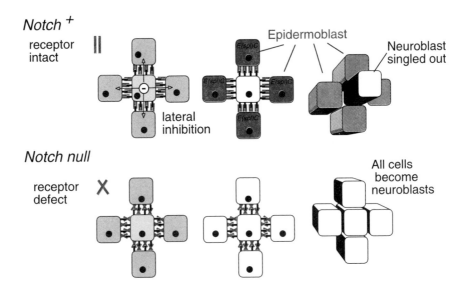

Figure 9–3   Segregation of a neuroblast from epidermoblasts in the ventral blastoderm of a *Drosophila* embryo. Initially, in the ventrolateral blastoderm all cells are (almost) equivalent, with a bias toward becoming neuroblasts. This bias is indicated by the presence of transcription factors that are encoded by the *achaete scute* gene complex. Local inhibitory cell interactions mediated by cell surface proteins restrict the expression of *achaete scute* factors to the future neuroblasts. The surface proteins in question are encoded by the *Notch* and *delta* genes. The particular cell that is fated to become a definitive neuroblast is thought to be specified by random fluctuations in some signaling activity. Once a definitive neuroblast is specified, it suppresses the neurogenic bias in adjacent cells by DELTA/NOTCH-mediated signaling. Finally, the neuroblast is singled out and delaminated into the interior of the embryo. In *Notch* and in *delta* null mutants all cells of the ventral blastoderm continue to express the *achaete scute* gene complex and adopt the neuroblast cell fate, which leads to lethal hypertrophy of the nervous system.

decorated cell surface. To realize their own neurogenic potency, the DELTA-decorated cells must continue to express another complex of genes, called *achaete scute-complex (AS-C)*.

The two cell-type programming gene complexes (the epidermis-specific *ESPL-C* and the nerve-cell-specific *AS-C*) cannot be activated simultaneously: their expression is mutually exclusive. Eventually, the differently programmed cells separate mechanically. The nerve cell precursors are shifted into the interior of the embryo (Fig. 9–3), whereas the epidermal cells remain outside and secrete the shielding cuticle.

9.5 The Principle of Embryonic Induction: Cells Emit Signals into Neighboring Areas

9.5.1 Induction, a Matter of Great Significance

In induction signal-emitting (inducing) cells instruct or permit signal-receiving (responding) cells to take a specific developmental path. Such instructive or permissive interactions are among the fundamental events leading to the cooperative formation of tissues and organs. The term "induction" is used in biology in diverse contexts (e.g., *induction* of an enzyme refers to turning on transcription of a gene whose product is an enzyme). To distinguish between the various meanings, developmental biologists often speak of **embryonic induction** (although inductive influences are not restricted to embryogenesis).

Transfer of inductive information can be accomplished by signal molecules exposed on the cell surface. Section 9.4 described two examples: in the *Drosophila* ommatidium the cell R7 is instructed to develop into a photoreceptor by the adjacent photoreceptor R8, and cross talk between neuroblasts and epidermoblasts in *Drosophila* facilitates their segregation from each other. Both are examples of induction. However, the term usually is associated with signal molecules capable of spreading into a larger area by diffusion, though spread by diffusion is not a strict criterion defining inductive molecules.

By transplanting cells or pieces of tissue, and by eliminating putative emitters of inductive signals, inductive interactions have been revealed in all animal organisms that have been investigated with adequate methods. Attempts to cause the development of structures in the wrong places have been successful in sponges, hydrozoa, sea urchins, and vertebrates. Even in embryos previously considered to represent a genuine mosaic type of development, such as *Caenorhabditis elegans* and ascidians, more and more inductive interactions between adjacent cells are being detected.

A structure that is induced artificially in the wrong location is called **ectopic.** The first discovery of ectopic induction was in 1909 in *Hydra:* a small piece taken from the mouth cone (hypostome) or the subtentacular region and inserted into the gastric region organizes the formation of a second head and body axis (Fig. 9–10). For many years inductive phenomena have been studied most extensively in amphibian embryos (Chapter 3, Section 3.9, and Section 9.5.4).

9.5.2 The Receiver Must Be Competent

In radio broadcasting a message is heard only if the receiver is tuned to the transmitting station sending the message. Likewise, in embryonic induction tne response to a signal requires competence of the recipient cells, where **competence** is defined as the ability to respond to an inductive signal. The acquisition of competence implies the production and display of **receptors** ("antennae") for the inducing signal. Competence often is restricted to a narrow window of a few minutes or hours.

9.5.3 Inducing Signal Molecules Are Difficult to Identify

Evidence for inductive events is usually deduced from transplantation studies: when host tissue touches transplanted tissue, it responds by developing a structure that would not otherwise form. Although such experiments have been performed many times, identification of the inducing signals turned out to be very difficult. Embryos are tiny, surgical intervention is difficult, and signal molecules are present in only minute quantities, so that tons of embryos would have to be extracted to obtain chemically analyzable traces of the inducing substance.

Despite these unfavorable circumstances, impressive progress has been made in *Drosophila* and *Xenopus*. In *Drosophila* mutagenesis helped identify genes for signal molecules and their receptors. Using the methods of molecular biology (Box 7), eventually the genes of several inducing factors and receptors were sequenced, and more signal molecules are still being identified. Inducing factors and receptors can be expressed in large amounts in appropriate experimental systems, such as transfected bacteria or animal cells.

In vertebrates progress also depended on methods from molecular biology, although classic biochemical methods have also been employed with some success. For example, activins (see as follows) were first isolated from chick and amphibian embryos using chromatography guided by bioassays.

9.5.4 Induction by Diffusible Signal Molecules: The Great Topic in Amphibian Development

9.5.4.1 History   In 1924 Hans Spemann and Hilde Mangold drew the attention of the scientific world to the astonishing inductive power of the upper blastopore lip of the amphibian gastrula: a blastopore lip inserted into a blastula at an ectopic site organizes the surrounding cells into a supernumerary gastrulation, resulting in a secondary body axis (Chapter 3, Section 3.8). This discovery stimulated an intense race to identify the putative inducer. In fact, the amphibian inducers proved to be remarkably elusive and the first were isolated only in 1987.

In the hunt for inducers appropriate bioassays are crucial. The "**animal cap assay**" was both the source of many pitfalls and of the final success. Caps are excised from the animal hemisphere of blastulae (Fig. 3–36) and maintained in cell culture. For decades the caps were taken from newt blastulae. Pieces of gastrulae, that were excised from other regions of donor embryos and were supposed to be the source of inductive substances, were attached onto the caps. Alternatively, the animal caps were bathed in buffered solutions that contained the putative inducers. After some time of contact or incubation, the caps were examined microscopically for the presence of differentiated cells and complex structures. Even eyes or brain structures could be identified in the isolated and treated caps. Thus, the beginning was promising.

However, the hunt for inducers, in particular, the hunt for neural inducers, was soon confounded by the fact that just about any culture condition triggered neural development, an effect that became known as "autoneuralization." Fortunately, *Xenopus* ectoderm or "animal cap" tissue does not "autoneuralize" readily or develop mesodermal cells, such as muscle cells, blood cells, or cells typical for the notochord. In the end, two lines of research converged.

The first path of successful investigation began with a happy coincidence: animal caps excised from *Xenopus* blastulae (Fig. 3–36) were maintained in culture. When their medium was supplemented by certain growth factors used in the culture of mammalian cells, the caps differentiated mesodermal cells. Both **FGF** (fibroblast growth factor) and **TGFβ** (transforming growth factor beta) showed inductive properties.

The second line of research was based on hard work. From tons of incubated chick eggs and other biological sources, a mesoderm-inducing substance was extracted, tested using the animal cap assay, and purified. Animal caps, which develop into ciliated epithelia in the absence of factors, formed mesodermal cells, such as muscle cells when treated with traces of the substance. A similar substance, called **XTC-MIF,** was found in the supernatant of a cultured *Xenopus* cell line. Both substances turned out to be related to each other and to the known growth factor TGFβ. The mesoderm-inducing members of this protein family are now known as **activins.** These findings suggested that other known "growth factors" should be assayed for inductive properties, and that similar molecules should be sought in the amphibian embryo using molecular probes (Box 7).

9.5.4.2 Summary of Embryonic Induction   The embryonic development of *Xenopus* is governed and controlled by **cascades of inductive processes.** The cascade starts as early as fertilization, or, at the latest, in the early blastula.

Primary Induction.   The processes of primary induction take place first (Fig. 3–35). These include the following:

1. **Mesodermalizing induction** (also called **vegetalizing induction**). A broad ring of equatorial cells in the blastula is specified to develop mesodermal tissues by signals emanating from the vegetal pole region.
2. **Dorsalizing induction** (also called **caudalizing induction**). The dorsal side of the embryo and the head-tail polarity are specified by signals originating under the cortical layer known as the gray crescent (or more ventrally in the "Nieuwkoop center"). The sources of these signals are allocated to this region of the egg during the sperm-induced rearrangement of the cytoplasmic constituents (ooplasmic segregation), while other determinants are confined to the prospective anterior region. Once the embryo begins to gastrulate, the "**upper blastopore lip**" forms in the area of the former gray crescent. This lip is a rich source of inductive signals and also is known as the "**Spemann organizer.**" The signals are often referred to collectively as dorsalizing inducers but also as neuralizing inducers because the bundle of signals emitted by the organizer include signals that inform the adjacent animal ectoderm to begin the expression of neural features. The dorsalizing induction coincides with the beginning of neuralizing induction.
3. **Neural induction.** Signals emitted by the Spemann organizer and subsequently by the roof of the archenteron (dorsal mesoderm) cause the overlying animal ectoderm to develop the central nervous system.

Many inductive actions can be attributed to secreted soluble factors. Such factors are encoded by **maternal RNA** and are produced in the early embryo by cells in the vegetal hemisphere of the egg. The factors are thought to be released into the interstitial spaces via exocytosis.

The bioassay in which animal caps are exposed to solutions of those factors (Fig. 3–36) has demonstrated that several protein families exhibit inducing capacity:

- **Mesodermal induction** is attributed to factors belonging to (1) the FGF family; (2) the TGFβ family, including VG1; several variants of activins; and BMP (bone morphogenetic protein).
- **Dorsalizing induction** is assigned to (1) activin B, (2) NOGGIN, and (3) CHORDIN.

Several factors even have the capacity to mimic the inducing capacity of a living blastopore lip. These include certain activins (first found in mammalian tissues but also present in *Xenopus*), representatives of the WNT protein family, and both NOGGIN and CHORDIN, which were obtained by introducing *Xenopus* mRNA into "expression vectors" (Box 7). When injected into a blastula at appropriate sites, they induce twinned axes and eventually development of a more or less complete second embryo (like in Fig. 3–34). Such results suggest that natural induction is accomplished not by one factor but by a collection of synergistically acting factors. In fact, there may be redundant signals that ensure that the important events of induction take place. However, it is not known with certainty whether the powerful organizing activity emanating from the upper blastopore lip is initiated by one particular factor being the sole trigger of a cascade of secondary processes, or whether the inductive power is brought about by simultaneously and synchronously acting factors as suggested previously. In any event, the Spemann organizer (Fig. 3–34, 3–35) is the source of a surprisingly large number of potential signal molecules. Besides activins, WNT, NOGGIN, and CHORDIN, the mRNA messages for FOLLISTATIN and SONIC HEDGEHOG, and the nonpolypeptide signal molecule **retinoic acid** have been identified. While the function of these substances in the region of the Spemann organizer remains to be elucidated, we will discuss the function of SONIC HEDGEHOG and retinoic acid in later development in Section 9.5.7.

In birds and mammals the part of the amphibian upper blastopore lip is played by **Hensen's node** at the anterior end of the primitive groove. Hensen's node is homologous to the upper blastopore lip, is endowed with similar inductive power when transplanted into an amphibian blastula, and generates similar signal molecules. In addition, all these "organizers," the amphibian upper blastopore lip and Hensen's node in birds and mammals, express certain homeobox-containing genes such as *goosecoid* (Chapter 10). These genes confer the ability to produce inducers. Note, however, that the final structures induced by these organizers, such as a forehead or a second head-trunk structure, can only be initiated by the primary inducing signals. Many events must follow to realize such complex structures.

The molecules used for **neural induction** are largely unknown, although candidate molecules have been nominated. Promising examples are NOGGIN, and CHORDIN although they probably need synergistically acting helpers. These two secreted proteins have been nominated previously as candidates for the dorsalizing induction as well. This double candidature reflects the fact that dorsalizing and neuralizing induction overlap, both spatially and temporarily. Both activities emanate from the Spemann organizer. Subsequently, neuralizing induction is continued by signals sent out from the roof of the archenteron (the chordamesoderm) and initiate the formation of the neural plate and hence the central nervous system in the overlying ectoderm. In addition, "planar" signals emitted from the blastopore lip and traveling through the future neural plate are believed to contribute to the **regionalization** of the central nervous system: to its subdivision into the various brain regions and the spinal cord.

9.5.5 How to Create Region-Specific Diversity

How is diversity generated during development? Why is a head formed in the anterior region of the embryo and a tail formed in the posterior region? Evidence is accumulating that signals emanating from the posterior region (from the region of the blastopore) are responsible for region-specific subdivision not only of the neural tube but of the entire body axis. A candidate molecule fulfilling the criteria of a "posteriorizing" factor or morphogen is retinoic acid, whose putative role in patterning processes will be discussed in Section 9.8.

In the present section we address more generally the basis for developmental determination. For example, what specifies whether brain structures (e.g., telencephalon or forebrain, diencephalon or midbrain, metencephalon or hindbrain) or structures of the trunk (spinal cord, notochord) are made, and whether heart muscle cells or muscles of the trunk are made? At least four possible answers have been proposed as follows:

1. The **particular quality** of the locally acting substances is responsible. Factors may be similar but not identical in the various body regions; they might even belong to the same protein family, but the different members are expressed in a region-specific pattern. For example, several paralogue genes of *hedgehog* are present in vertebrates and are expressed in different spatial patterns: *banded hedgehog* is expressed throughout the neural plate and in the dermatome part of the somites, *cephalic hedgehog* is expressed in ectodermal and endodermal structures of the head, and *sonic hedgehog* is expressed in the notochord and limb buds (see Section 9.5.7).
2. The **local concentration** of factors is responsible. The concentration may take the form of a gradient rather than being spatially uniform. Thus, activin is thought to stimulate the development of different cell types, depending on its local concentration (see Section 9.6).
3. The **locally changing proportions in the mixture** of inducing factors is responsible. As with medications, mixtures differing in their quantitative composition may have different effects.
4. The **locally different competences** of the responding tissue is responsible. As a consequence of their history, various recipients of the same signal molecule may respond differently.

9.5.6 Primary Induction Is Followed by Secondary and Tertiary Inductive Events

The classical textbook example of a secondary embryonic inductive event is the induction of the eye lens (Fig. 3–37).

9.5.6.1 Development of the Eye
In its inner "kernel" the eye is a derivative of the brain and thus of the neural tube, whose development is initiated by the "primary" induction emanating from the roof of the archenteron. Two bulges of the lateral wall of the midbrain enlarge, giving rise to the **optic vesicles.** By a process of invagination, the vesicle transforms into the double-walled **optic cup.** The inner wall gives rise to the layers of neurons and photoreceptors, collectively called the **retina,** whereas the outer wall will form the pigmented layer. Mesodermal cells enclose the optic cup and supplement the eyeball with the layers of the vascular chorioid coat and the sclera. Among the solid structures, only the lens and the cornea are still missing.

9.5.6.2 Inductive Events
The optic cup comes into contact with the overlying ectoderm. In a process known as secondary induction, the cup stimulates the formation of an ectodermal **lens placode** (*placode* = thickening; Fig. 4–2) in the right time, at the right place, and in the right size. As the lens placode detaches from the ectoderm, it becomes covered by the surrounding ectodermal epithelium. A tertiary signal emanating from the lens causes the ectodermal epithelium covering the lens to transform into a translucent cornea. The molecular nature of the inducing stimuli is still unknown, although some evidence points to the involvement of FGF-2.

Surprisingly, it was discovered that in some vertebrates, and in fact in some amphibian species, lens formation does not depend on having received an inducing stimulus from the eye cup. Apparently, the cells of the future optic vesicle begin to emit inducers before the optic vesicle is detectable. The finished optic cup has a subsidiary, amplifying influence that is still strong and required in some species, whereas it can be omitted in other species.

9.5.7 Signal Molecules Such as SONIC HEDGEHOG May Be Used in Different Organisms and Organs for Different Purposes

The molecular techniques known collectively as "reverse genetics," particularly the use of heterologous gene probes (Box 7), allow the rapid determination of whether a gene known from, say, *Drosophila*, is also present in another organism. Such investigations yielded many surprising results, for example, in the case of the potential signaling molecules coded by the *Drosophila* gene

hedgehog and the signal molecules coded by a corresponding gene in vertebrates, *sonic hedgehog*.

The mutation *hedgehog* makes a *Drosophila* larva look like a hedgehog with a
crew cut; the back is covered by chitinous spines. In the *Drosophila* embryo the
gene is expressed in various stripes and areas. First, *hedgehog* is expressed at the
anterior border of the parasegments in those cells whose nuclei contain the
transcription factor ENGRAILED. Unlike ENGRAILED, HEDGEHOG is not a
transcription factor but a transmembrane protein that is presented to the adjacent cells. These sense the HEDGEHOG cue and in turn emit the signal WING-
LESS. WINGLESS spreads by diffusion and is picked up by neighboring cells
with membrane-associated receptors (Fig. 9–4). Adjacent emitters of HEDGE-
HOG and WINGLESS stimulate each other to continue signaling and to preserve
their own program of gene expression.

In other regions of the larval fly, such as the imaginal discs, another variant
of HEDGEHOG is produced. This variant detaches from the cell surface.
HEDGEHOG is not only a signal molecule but also a proteolytic enzyme and
cleaves itself into a domain remaining in the membrane and a component that
floats away. The detached HEDGEHOG domain serves as a signal molecule,
capable of reaching distant targets by diffusion. HEDGEHOG is used for short-
distance and long-distance communication in vertebrates (zebra fish, *Xenopus*,
chick, mouse) as well.

In vertebrates *sonic hedgehog* is expressed in the upper blastopore lip and
subsequently in the derivative of the invaginated lip cells, the notochord. The
notochord presents SONIC HEDGEHOG protein as an inducing cue to the adjacent neural tube in a contact-dependent manner. In response to this cue the
ventralmost midline cells of the neural tube form the **floor plate**. In the

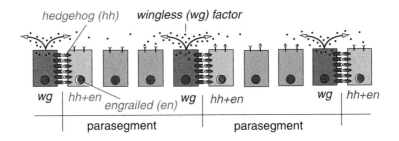

Figure 9–4 *Hedgehog* in the blastoderm of *Drosophila*. In the particular area shown
the HEDGEHOG protein is a membrane protein that is produced by cells also
expressing the transcription factor encoded by *engrailed*. The exposed HEDGEHOG
protein interacts with a matching membrane protein of adjacent cells. This interaction occurs at the boundaries of the parasegments.

neural tube the induced floor plate takes over the production of HEDGEHOG. The signal stimulates the development of **motoneurons** on either side of the floor plate (Fig. 9–5).

The soluble diffusing component of SONIC HEDGEHOG reaches the somites. In the somites the HEDGEHOG signal induces the sclerotome cells to detach. They migrate towards the emitter of the signal, enclose the notochord, and construct the **vertebrae** (Fig. 9–5).

SONIC HEDGEHOG gets another chance to demonstrate its versatility when the limbs are formed. In the limb bud of the chick embryo HEDGEHOG is expressed at the posterior border, where a signal emitter known as **ZPA**

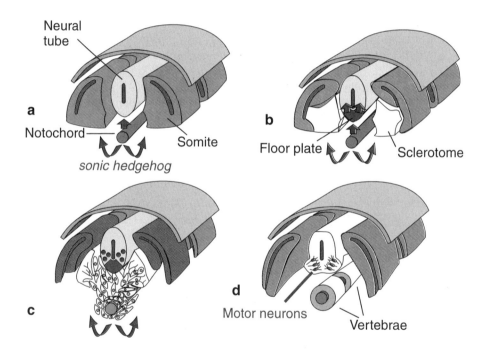

Figure 9–5 *Hedgehog* in vertebrates. The protein encoded by the gene *sonic hedgehog* can appear in a membrane-associated form or in a form that detaches from the cell surface and spreads into the surrounding spaces where it reaches distant target cells. SONIC HEDGEHOG directly presented by the notochord to the adjacent neural tube induces the formation of the floor plate in the neural tube (a). Subsequently, the floor plate also produces HEDGEHOG (b), inducing neuroblasts to differentiate into motoneurons (c, d). SONIC HEDGEHOG released by the notochord into its surroundings induces the sclerotome cells to detach from the somite (b), migrate around the notochord (c), and form the vertebral bodies (d).

(**zone of polarizing activity**) is located. When cells expressing HEDGEHOG are transplanted to the anterior border of the limb bud, supernumerary fingers are formed. This intriguing phenomenon is discussed further in Section 9.8.

9.6   Inducers, Morphogens, and Morphogenetic Fields

Inducers can be morphogens as well. By definition, a **morphogen** acts locally to organize the spatial pattern of cell differention by virtue of spatial concentration differences. A high concentration of a morphogen at location 1 specifies cell type or structure A, a lower concentration at location 2 specifies cell type or structure B (Box 4).

Several inducing factors—if not all of them—in the early embryo not only act upon the neighboring area but also have a patterning function in the area where they are produced. In at least some instances the pathway of differentiation is a function of the concentration of the inducing factor. This has been shown using the animal cap assay (Fig. 3–36). When exposed to very high concentrations of **activin B,** the uncommitted cap cells develop into endoderm-type cells. Intermediate concentrations of activin B cause cap cells to differentiate into the dorsal mesoderm, such as the notochord and somite-derived muscle cells, and very low concentrations cause them to form the epidermis. Thus, activin B is also acting as a morphogen.

A **morphogenetic field** is an area whose cells can cooperatively bring about a distinct structure or set of structures. Another way to define a morphogenetic field is as an area in which signal substances—"inducer," "morphogen," or "factor"—become effective and contribute to the subdivision of the field into subregions. An example of a morphogenetic field is the circular area in the trunk wall from which the forelimb originates. The area is subdivided into concentric rings, an outer ring destined to become the shoulder girdle, and an inner area destined to become the free limb. A secondary morphogenetic field in the limb is the hand field, which subdivides itself into the palm and the fingers.

Morphogenetic fields define developmental potencies; these are not necessarily exactly coincidental with developmental fates. Usually fields are initially larger than the area fated to construct a particular organ. Fields exhibit the faculty for regulation. When cut into two, each half will give rise to a complete structure, albeit of half the size. The processes of subdivision, spatial restriction, and developmental restriction are phenomena of **self-**

organization, pattern formation (Box 4), and **commitment.** The entire process of organ formation consists of sequences of such phenomena.

Box 4

MODELS OF BIOLOGICAL PATTERN FORMATION

Pattern formation is a central topic of developmental biology. Pattern formation refers to spatially ordered (nonrandom) cell differentiation that results in ordered arrays of structures. Complex patterns are the synergistic outcome of many interacting molecules and cells. Because the complexity of reality often surpasses the faculties of our intuition, simplified models have been constructed that allow computer simulations. As examples, two simple, historically significant models are outlined here.

POSITIONAL INFORMATION ACCORDING TO WOLPERT

An emitter releases a signal in the form of a soluble substance, the hypothetical **morphogen S.** The concentration of S diminishes with increasing distance from the source. A continuous gradient may be set up from the source to a sink. The morphogen source may be located at one end of a string of cells, with the "sink" destroying the morphogen at the other end. In this configuration, under equilibrium conditions, the concentration decreases linearly along the string. Alternatively, all cells along the string remove S, in which case the concentration decreases exponentially. The resulting gradient of S provides **positional information.** Cells are capable of measuring the local concentration and, by doing so, are capable of locating their position along the gradient. The positional information thus deduced is "interpreted" by the cells in that they behave in accordance with their position (and their previous history).

In addition, the S gradient is a guiding principle and is used to adjust a second gradient of **positional value, P.** Positional value is a relatively stable tissue property that serves as positional memory in case of need, for example, regeneration. The concentration profile of S determines the shape of the P gradient; the local S value acts as the inhibitor, determining the upper limit of the local P value. When the source of S is removed and S drops below a critical value, P automatically rises. Cells acquiring the maximal P value first begin to emit S, thus suppressing an increase of P elsewhere. S and P are involved in pattern formation. If there are threshold values above which the cells respond by becoming, say, "*red*," and below which they respond by becoming "*white*," a sharp segregation into two populations, *red* and *white*, is possible.

The hypothesis does not explain how a gradient is established for the first time, and it does not propose a mechanism as to why and how P values increase when no S brake is applied.

REACTION-DIFFUSION MODELS

The primary aim of reaction-diffusion models is to propose mechanisms by which patterns could arise from homogeneous or chaotic initial conditions. Ideas as to how patterns of orderly, yet different, cells might arise in initially uniform populations of

Box 4   Models of Biological Pattern Formation   185

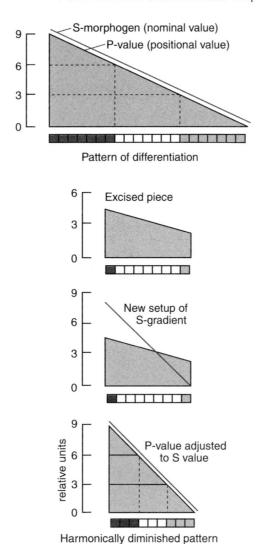

POSITIONAL INFORMATION after Wolpert

S-morphogen (nominal value)

P-value (positional value)

Pattern of differentiation

Excised piece

New setup of S-gradient

P-value adjusted to S value

relative units

Harmonically diminished pattern

genetically identical cells have been developed mainly by mathematicians (Turing, Murray) or physicists (Prigogine, Haken, Gierer, Meinhardt). The basic concept of reaction-diffusion models is the **morphogenetic field,** in which several biochemical reactions create **prepatterns** in the distribution of chemicals, generally called **morphogens.** Cell determination and differentiation would follow these chemical prepatterns.

The simplest versions of reaction-diffusion models create stable, nonuniform concentration patterns by combining the production of at least two interacting substances that feed back onto their own production as well as the production of the other substance. In a basic model developed by Gierer and Meinhardt on ideas that

REACTION-DIFFUSION-SYSTEM
after Turing, Gierer and Meinhardt

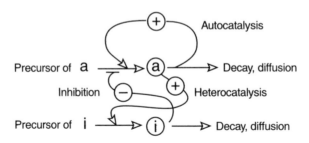

$$\frac{\delta a}{\delta t} \quad = \quad \varrho a \quad + \quad \frac{a^2}{i} \quad - \quad \mu a \quad + \quad D_a \, \frac{\delta^2 a}{\delta x^2} \; ;$$

Change of a-concentration	Source density	Auto-catalysis	Decay	Diffusion
in time		Inhibition		

$$\frac{\delta i}{\delta t} \quad = \quad \varrho_i \quad + \quad a^2 \quad - \quad vi \quad + \quad D_i \, \frac{\delta^2 i}{\delta x^2} \; ;$$

Turing originally proposed, the generation of an activator, *a,* is an autocatalytic event. Through a process of nonlinear amplification, *a*'s presence generates the production of more *a*. The explosive rise in *a* is limited by the decay of *a*, its diffusion into neighboring areas, and by the production of an inhibitor *i*, which is initiated heterocatalytically by *a*. The inhibitor *i* sets limits to the production of *a*. Furthermore, it is assumed that the diffusibility of the two substances is different: the activator *a* diffuses slowly and only over short distances, while the inhibitor *i* diffuses rapidly and has a long range. Due to these properties, *i* allows *a* to increase in concentration up to a limit, but suppresses the start of an autocatalytical *a* increase outside the first peak. Hence, *i* prevents the occurrence of a second, competing peak (**lateral inhibition**).

The behavior of the two substances in time and space is described by two partial differential equations on which computer simulations are based. By selecting appropriate parameters (basic rates of production, rates of decay, diffusion constants, size of the field), and, if necessary, by introducing more interacting substances, a multitude of patterns can be created with a computer. For example, gradients are generated that remain stable and, after experimental disturbance, regenerate. Periodic patterns, such as fields of cuticular spines, patterns of stripes, or branching patterns, have been generated with remarkable fidelity to those observed in nature. Periodic patterns arise in these models if the range of the inhibitor is shorter than the length of the field, and when a relatively high level of activator initiates a new, self-enhancing activator peak outside the inhibited area. Many patterns have been simulated successfully, such as patterns on mollusk shells.

However, chemical reactions have not yet been identified experimentally that would be as simple as recent models propose. It is likely that reaction-diffusion models

Box 4   Models of Biological Pattern Formation   187

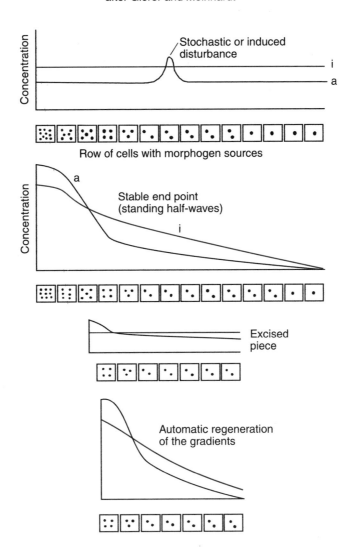

REACTION-DIFFUSION PREPATTERN
after Gierer and Meinhardt

describe some general formal principles, such as the principle of self-enhancement (autocatalysis), rather adequately, but not actual mechanisms.

EXTENDED AND ALTERNATIVE MODELS

Pattern formation can be the result of a multitude of physical forces and chemical processes as well as result from the behavior of motile cells. Accordingly, a multitude of explanatory models have been designed, including mechanical and mechanochemical models (Murray 1989, see Bibliography). Many models share formal basic proper-

Computer simulation using an expanded
model from Gierer and Meinhardt

Bristle field

Patterns from
mollusk shells

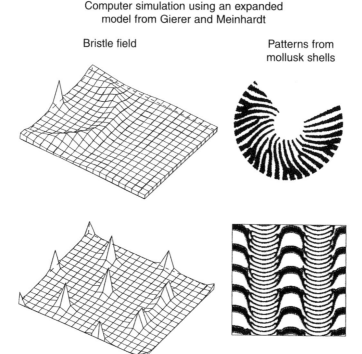

ties: they embody a process of **self-enhancement** ("autocatalysis", positive feedback)
and a process that sets limits to the self-enhancement. Restrictions on unlimited
increase could be achieved by production of an inhibitor, depletion of a substrate or
cell type, or saturation. Models of biological pattern formation are reminiscent of
models used in ecology to describe prey-predator relationships and the spread of epi-
demics, populations, or genes.

9.7   How to Create a Field, Subdivide It, and Define a Point within It

To pinpoint a location, such as a spot on the wall where we wish to drive a
nail, we often draw vertical and horizontal lines that cross at the desired point.
Nature can do the same.

In the *Drosophila* embryo the future development of a leg is prepared by
the formation of an **imaginal disc,** a circular group of cells that is specified to
form an appendage. The imaginal disc is invaginated subsequently and is
stored inside the larva until the larva undergoes metamorphosis. How is the
location of a leg disc in the blastoderm of the embryo determined?

The blastoderm is already subdivided into compartments called parasegments.

At the border of two parasegments two vertical stripes of cells producing signal molecules are juxtaposed: a stripe secreting WINGLESS(WG) and a stripe of cells expressing *engrailed* and exposing HEDGEHOG (HH) on its surface (Fig. 9–6). Due to another process of patterning, a second, horizontal coordinate subdivides the anterior compartment into a dorsal and a ventral area (the molecular tools used to draw this line are unknown). The stripe of cells secreting HEDGEHOG induces the adjacent stripe of cells in the dorsal anterior compartment to secrete a third signaling molecule, DECAPENTAPLEGIC (DPP).

The site where the boundaries of the three compartments (anterior dorsal, anterior ventral, and posterior) intersect is the only site where cells expressing *DPP* contact those expressing *WG*, and this site defines the center of the disc (Fig. 9–6). A working hypothesis proposes that the central point

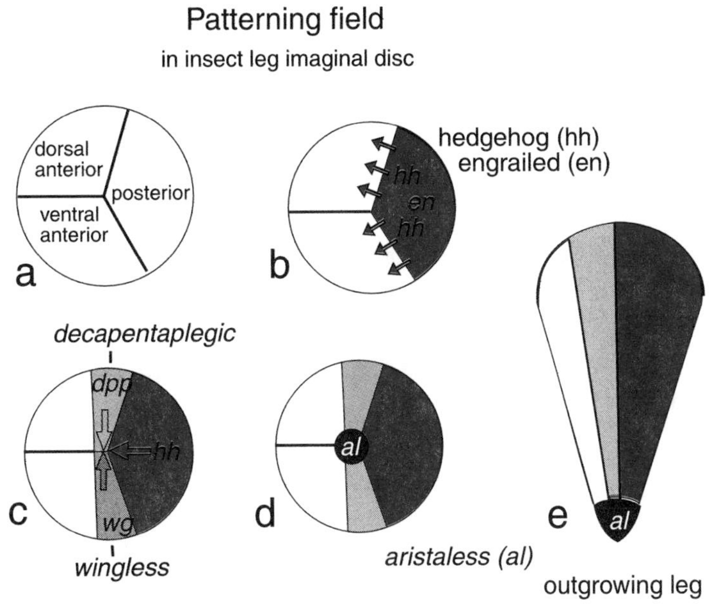

Patterning field
in insect leg imaginal disc

a dorsal anterior / ventral anterior / posterior

b hedgehog (hh) engrailed (en) hh / en / hh

c decapentaplegic dpp hh wg wingless

d aristaless (al) al

e al outgrowing leg

Figure 9–6 Specification of a developmental field in a circular area of the blastoderm fated to become an imaginal disc (leg or wing). The disc is specified where compartments expressing different genes border on each other. In the center is a small region where cells expressing *wingless* are in close association with those expressing *decapentaplegic*. At this site, the (homeobox) gene *aristaless* is activated and a proximodistal organizing center is established that controls the growth of the leg during metamorphosis.

becomes the source of a morphogen (DPP itself?); a high concentration in the center of the disc specifies the future "fingers" (tarsalia).

9.8   The Avian Wing as a Model Limb

The development of limbs is a highlight in developmental biology. The avian wing bud is particularly accessible to surgical manipulations by the experimenter; the outcome of disturbed pattern-forming processes can readily be read off from the pattern of the skeletal elements.

As a three-dimensional structure, the wing has three polarity axes:

1. The proximal-distal axis from the shoulder to the digits.
2. The anteroposterior axis from the second to the fourth finger (fingers 1 and 5 are missing in the wing).
3. The dorsoventral axis from the upper side to the lower side (this axis is not considered here).

9.8.1 Along the Proximal–Distal Axis a Temporal Program Is Translated into a Spatial Pattern

A limb originates from a limb bud. Shielded by an ectodermal epithelium, the limb bud incorporates mesodermal cells that emigrate from the lateral plate (these cells give rise to the skeletal elements) and cells that emigrate from the somites (these cells give rise to the muscles). The immigrating cells reaggregate to form a "mesenchyme." This mesenchyme, initially unstructured, must be properly organized so that the humerus, the radius, and the ulna; the elements of the palm; and the elements of the phalanges are laid down in the correct sequence.

In birds and mammals the limb bud is covered by an ectodermal cup over which an **apical ectodermal ridge** extends. The ridge produces growth factors such as **FGF-2** and secretes them into the underlying mass of mesenchymal cells. Upon removal of the ridge, the outgrowth of the bud stops.

The signaling range of the growth factors defines the limit of the **progress zone** in which cells proliferate. The resulting cells gradually become displaced out from the shelter of the apical cup and thus from the signaling range of the growth factors. As a consequence, proliferation ceases.

The first cells to leave the space of the sheltering cup and cease proliferation form the upper arm with the humerus. The next group forms the lower arm with the radius and the ulna, the third group forms the carpals, and the last group forms the phalanges. Various experiments have shown that a **tem-**

poral program is run in the growing limb. When a very early bud with its young progress zone is transplanted onto the stump of a later-stage bud, the young bud begins with the humerus according to its own autonomous program, and it does so even when the stump already contains cells specified to form a humerus. A medium-stage progress zone continues its program by lying down the radius and the ulna; a late-stage progress zone grafted onto a young stump merely adds fingers without taking into account that the removed progress zone of the young bud did not yet have the chance to provide cells for the ulna and the radius; these are now missing (Fig. 9–7).

The nature of the temporal program is unknown, though several indications point to a gene-controlled mechanism: in the growing limb a series of homeotic genes are activated sequentially (Chapter 10). In the course of the proximal-distal pattern specification, continuously changing **positional values** are assigned to the cells (Box 4). These values can be used in amphibians to supply patterning information to regrow an amputated limb (Chapter 21).

9.8.2 The Order of Fingers 2 to 4 Is Specified by an Emitter of Signal Molecules

At the posterior margin of the limb (where in our hand the little finger would appear), an emitter of signals is located, the **ZPA (zone of polarizing activity).** A small group of cells near the posterior junction of the limb bud and the body wall is the source of one or more secreted signal molecules. Two hypotheses have been advanced to explain finger development.

A traditional hypothesis sees the signal intensity from the ZPA decreasing from the posterior margin to the anterior margin. This gradient of signal strength specifies the sequence and characteristics of each finger. If an additional ZPA is implanted into the anterior margin where signal intensity would ordinarily be low, it elicits the appearance of an additional set of fingers. A mirror-image duplication of the finger pattern results in the sequence of digits 432234 or 4334 (Fig. 9–8).

Retinoic acid (RA, vitamin A acid) can substitute for a transplanted ZPA. A porous bead soaked with RA and implanted into the anterior margin induces a mirror-image duplication of the digits just like a ZPA does. Because RA, derivatives of RA, receptors for RA, and even a graded distribution of bound ^{3}H-RA have been found in the limb bud, RA has been identified by some researchers as the first true morphogen.

However, this position has been disputed by the suggestion that the bead soaked with RA causes cells in the anterior marginal area to adopt features of ZPA cells. These artificially induced ZPA cells would emit the natural signal

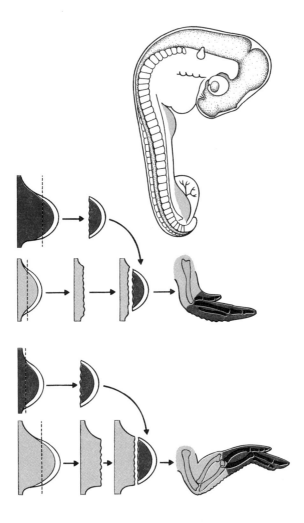

Figure 9–7   Chick wing bud I. Patterning along the proximodistal axis during growth of the wing bud. A cap out off from a bud and grafted onto the stump of another bud continues its developmental program irrespective of the pattern elements that are still lacking, or already laid down, in the stump. (Redrawn and modified after Alberts et al.)

that is not RA, and thus would trigger a cascade of secondary events. SONIC HEDGEHOG has been proposed as the natural signal. Indeed, SONIC HEDGEHOG is expressed in the ZPA, and cells secreting SONIC HEDGEHOG can replace a ZPA in its polarizing activity. A cascade of subsequent signaling processes could follow. SONIC HEDGEHOG might trigger the release of further signal substances (e.g., peptide growth factors such as FGF-4 and BMP).

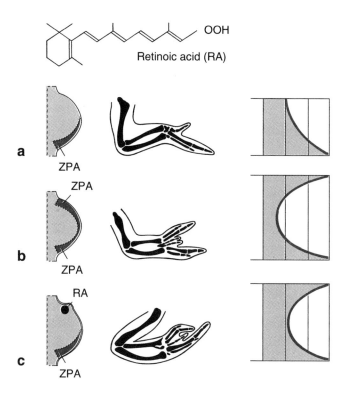

Retinoic acid (RA)

Figure 9–8   Chick wing bud II. Pattern specification along the anteroposterior axis. (a) The type of fingers and their sequence are specified under the influence of a zone of polarizing activity (ZPA) that is located at the posterior margin of the bud. (b) Implantation of posterior tissue into the anterior margin of a host bud results in the mirror-image duplication of the sets of digits. (c) A resin bead (black dot) soaked with retinoic acid mimics the effect of an implanted ZPA. The results have been interpreted in terms of a concentration gradient of a hypothetical morphogen that is released by the ZPA or the implanted bead, respectively (curves). Above certain threshold concentrations (parallel lines) the cells respond by forming the digits 2, 3, or 4. (Redrawn and modified after Wehner and Gehring: Zoologie, Stuttgart, 1990.)

On the other hand, *sonic hedgehog* is expressed rather late in the development of the limb bud. Thus, its expression could also be interpreted as an intermediate event in a cascade started, perhaps, by RA or some members of the FGF family.

Whether or not RA is the primary signal in the specification of anteroposterior polarity in limb buds, the numerous morphogenetic effects of RA are remarkable. RA influences proliferation, pattern formation, and cell differentiation in many rudimentary organs. It does so presumably by stimulating gene expression in a way known from steroid hormones (Box 5). As pointed out previously in Section 9.5.5, RA contributes to the regional subdivision of the central nervous system.

9.9   Positional Information and Positional Memory in *Hydra*

The freshwater polyp *Hydra* needs positional information not only in embryogenesis but for later life as well. It needs information to guide the perpetual renewal of its body from stem cells and to be ready to replace body parts whenever regeneration is required (Chapter 21). Accidents resulting in damage to the body can easily happen to these unshielded and soft animals.

Wherever the *Hydra* body column is bisected by a transverse cut, a head is restored at the upper end of the lower fragment, and a foot is restored at the lower end of the upper fragment. If we cut at various levels, we will see that the entire body column between the existing head and the foot has the potential to form both the head and the foot. Why are these potentials not expressed in the intermediate gastric region? Traditional views assign to the head an organizing power that includes a suppressing activity, comparable to the **apical dominance** phenomenon known from plants.

Experimentally, the system of pattern control has been shown to have the following properties:

1. An existing head suppresses the formation of another, competing head, while it promotes the formation of a foot at the opposite end of the body column. Additional heads grafted onto the body column can even evoke the development of supernumerary feet, while additional feet do not evoke additional heads.
2. Though in principle heads and feet can be made along the entire trunk column, the capacities are not distributed uniformly. They are graded: a position close to the head confers a high capacity to form head structures but a low capacity to form a foot. In contrast, tissue located near the foot end has a low potential for head formation but a high potential for foot formation.

This gradient specifies polarity as well. In an excised section of the body column, whether from the upper, middle, or lower part of the column, the head-forming potential is always greatest at the upper end, while the foot-forming potential is greatest at the lower end.

The relative positional values along the body column can be measured in the following two ways:

1. **Dissociation-reaggregation.** Pieces of tissue are excised from the trunk, dissociated into single cells, and allowed to reaggregate. These arrange themselves in new orders, as shown in Fig. 9–9. In such aggregates those cells that had occupied the position nearest to the head in the animal will form head structures such as tentacles, whereas the cells from the position nearest to the foot will form feet. Dissociated cells thus preserve a memory of their previous position. The newly formed structures are a function of the relative position the cells had in the previous body. Polarity is determined by a scalar parameter: the slope of positional value.

2. **Transplantation.** Pieces of the body wall are excised and transplanted into another body region. In a surrounding with the same positional value the pieces simply preserve their identity. In a surrounding with moderately different positional value the pieces fit into the new community by adopting the identity of the new neighbors (position-dependent behavior). However, in a neighborhood with significantly different positional value, they behave strangely and change their own state completely. In an environment of significantly lower positional value the pieces form ectopic heads, and in an environment of significantly higher value they form feet (Fig. 9–10). The frequency of ectopic head and foot formation reflects the sign and degree of the differences in positional value.

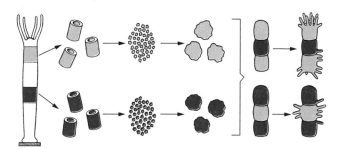

Figure 9–9   Experiments with dissociated cells in *Hydra*. Sections of the body column taken from two different levels are dissociated; from the resulting cell suspensions aggregates are prepared, and subsequently the aggregates are combined in triplet arrangements. Tentacles are formed preferentially by those cells that originally possessed the relatively highest positional value.

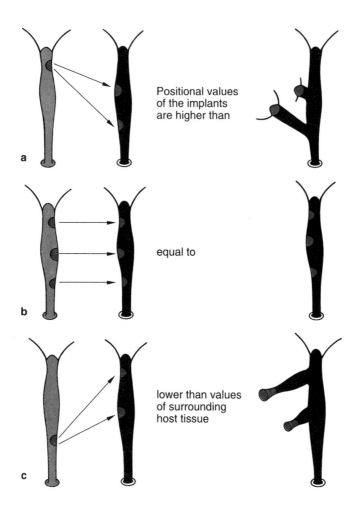

Figure 9–10   Synopsis of transplantation studies in *Hydra*. Pieces of tissue taken from donor animals are implanted into host animals at various positions along their body axis. If the positional value of an implant is higher than that of its new surroundings (a), it may form a head; if it is lower (c), it may form a foot. Implants whose positional value corresponds to that of the surrounding host tissue (b), integrate themselves into the host unobtrusively.

The molecular bases of positional information and positional memory are largely unknown, as are the mechanisms of pattern control. The following two hypotheses have been put forward to interpret the results:

1. Gradients of morphogens provide positional information (Box 4). In particular, in the head region a head activator would dominate, in the trunk a head inhibitor would dominate. Both morphogens are thought to be pro-

duced predominantly in the head and destroyed in the trunk. A mirror-image countersystem comprising a foot activator and a foot inhibitor also exists. Positional values would reflect the density of morphogen sources.

2. An alternative hypothesis considers the head to be the dominant competitor for resources of limited availability. A head would suppress the development of another head, and simultaneously promote foot formation, by attracting precursor cells (such as neuroblasts) and especially by binding and removing head-promoting hormonal factors. Factors needed for head formation would be produced in the gastric region and distributed in the interstitial spaces throughout the body. The complement of receptors to pick up these factors would be a function of positional value: cells in the upper body would be equipped with many receptors and thus capable of binding a large quantity of factor; they would form the head, while cells at the lower end of the body would possess only a few or no receptors, would lose out in the competition for these factors, and therefore would form the foot. An autoregulatory feedback loop between the amount of bound factor and the number of newly expressed receptors would restore the gradient of receptor complement in regeneration.

The significance of soluble extracellular factors is indicated by the pivotal role that the **phosphatidylinositol phosphate (PI) signal transduction system** (Box 3) has in pattern formation. Stimulation of the key enzyme PKC (protein kinase C) with diacylglycerol results in the development of **multiple supernumerary heads,** whereas suppression of the PI system by long-term exposure of the animals to lithium ions (which block several enzymes of the PI cycle) results in the formation of multiple ectopic feet.

9.10   Intercalation in Insect Appendages

Positional values as an expression of positional memory also play a role in the extremities of vertebrates and insects. In the legs of insects discontinuities in positional values are generated by grafting. For example, a leg may be bisected experimentally, a piece removed, and the distal leg fragment grafted onto the remaining stump. Or, an additional piece of a leg taken from another specimen can be inserted in either the correct or the wrong orientation. In any case, the confrontation between cells with different positional values results in the intercalation of leg structures. Dissonances stimulate cell proliferation, and the missing positional values are assigned to the additional leg structures until discontinuities are smoothed and the positional values are no longer

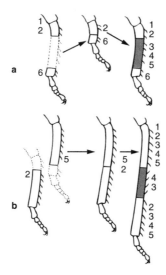

Figure 9–11   Intercalation of missing parts in the insect leg. Disparities in positional values are generated by (a) removing a segment or (b) replacing the distal part of the leg by a longer part taken from a donor animal. Disparities induce the insertion of parts (red) during molts. After one or several molts, the structure of the intercalated segment is such that the sequence of adjacent positional values no longer shows a gap. Note: In (b) the intercalated segment displays a reversal of polarity.

interrupted by a gap. The experimenter's trickery may even force the insect to intercalate a piece with reverse polarity (Fig. 9–11).

9.11   How to Make Periodic Patterns

Patterns composed of repeating units are frequent in organisms. In colonies of hydrozoa and corals, polyps (hydranths) are arranged in regular and beautiful patterns. Periodic patterns are displayed in the segments of articulated animals and by many segmental units in vertebrates such as somites and spinal ganglia. Repeating units include scales, feathers, and hairs. The following three mechanisms of periodic patterning, which may also occur in combinations, can be imagined (Fig. 9–12):

1. **Transformation of temporal rhythmicity into a spatial periodic pattern.** In growing systems internal oscillators (internal clocks) trigger the

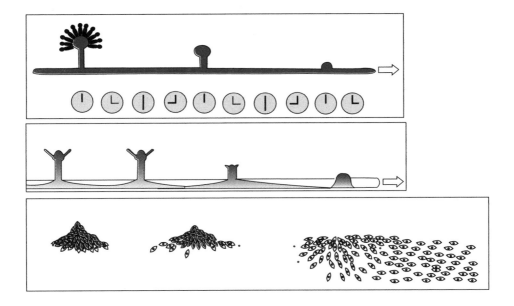

Figure 9–12   Formation of periodic patterns. Top: A new structure is formed at regular time intervals. An example is the periodic development of spore-forming conidiophores in the fungus *Neurospora*. New conidia are added in daily intervals (circadian rhythm). Middle: The spacing of hydranths (polyps) in colonial hydroids is specified by inhibitory influences that originate in existing hydranths and decrease in intensity with distance. The inhibitory influence may be mediated by a substance emitted by the existing hydranth into the stolon; alternatively, inhibition may be based on the depletion of an essential factor that is taken up by the hydranths and removed from the stolon. Bottom: Spacing of aggregates of, say, *Dictyostelium* cells is (in part) due to depletion of cells in the vicinity of the aggregates. An aggregate emits attracting signal molecules. The distance between aggregates depends on the range the chemoattractant can spread with sufficient concentration. Beyond the signaling range of an aggregation center, a new center is established by the autonomous activity of pacemaker cells.

generation of new units at regular intervals. Such oscillators might be coupled to the cell cycle or to an internal circadian clock.

2. **Lateral inhibition.** Existing structures are sources of inhibitory signals. As the signals spread into the neighboring area, their intensity decreases with distance, due to enzymatic degradation, spontaneous decay, or dilution. Outside the range of the inhibiting signals, new structures are made spontaneously or induced by stimuli acting from outside.

3. **Spacing by depletion.** A developing structure uses up essential substances or incorporates irreplaceable precursor cells. The area around the

emerging structure becomes depleted. Farther away, such substances or precursor cells are still available and another structure can be made. This type of periodic patterning is found in competing aggregates of *Dictyostelium* amoebae and in similar multiple cell aggregations (Fig. 9–12).

Box 5

SIGNAL MOLECULES ACTING
THROUGH NUCLEAR RECEPTORS

Despite wide differences in their chemistry, the signal molecules retinoic acid, steroid hormones, and thyroxine are all received by their target cells in very similar ways and they release similar intracellular mechanisms of response.

The retinoids derive from carotene B and are found in three oxidative derivatives: retinol (vitamin A), retinal (vitamin A aldehyde), and **retinoic acid (RA,** vitamin A acid). All-trans-retinoic acid and its metabolic derivatives, 9-*cis*-retinoic acid and 3,4-didehydro-retinoic acid, are known for their morphogenetic actions.

Being lipophilic, retinoic acid is said to penetrate the lipid bilayer of cell membranes by simple diffusion. To be sure, RA molecules can infiltrate cell membranes easily, but there is little reason for them to leave the membrane to enter the hostile water of the cytosol. From the membrane, RA is collected by a specific protein known as **CRABP (cytoplasmic retinoic acid binding protein)** and is transported into the cytoplasm. Within the cell, RA can function as a cofactor in the glycosylation of proteins, or it is transferred to a second class of intracellular receptors. Bound to their matching ligand, these second receptors, **RAR (retinoic acid receptor)** are eventually found in the nucleus and function as **transcription factors.** Stabilized by zinc fingers, the RARs bind to gene-controlling DNA regions (enhancers, promoters) containing sequences called **RARE (RA-responsive elements).** The RARE sequence upstream from a gene marks it as a target gene. To function as a transcription factor, two receptors must collaborate and aggregate along the RARE, forming a dimer. One of the two partners of a dimer is a similar, but not identical, receptor **RXR,** which binds another hormonal signal molecule X (e.g. thyroid hormone). Together, RAR, loaded with RA, and RXR, loaded with, say, thyroid hormone, form a functional heterodimer. More factors may attach and join the aggregate, modifying its activity.

Likewise, the thyroid hormone **thyroxine** and the **steroid hormones** (such as the insect-molting hormone **ecdysone,** the mammalian sexual hormones **testosterone, estradiol,** and **progesterone**) infiltrate the cell membrane. The mechanism by which they are detached from the membrane is largely unknown. Eventually, they are found in the nucleus, bound to receptors. Surprisingly, the receptors for RA, steroid hormones, and thyroxine belong to the same protein family, meaning that these receptors derived from a common polypeptide precursor. In fact, heterodimers between a receptor loaded with a steroid and one loaded with thyroxine have been found (like the RAR/RXR heterodimers mentioned previously).

Very different, however, are the effects of all these signal molecules at the organismal level: the biological contexts in which RA, steroids, and thyroxine are used to convey developmental signals are highly diverse, as are the specific groups of genes

Box 5   Signal Molecules Acting Through Nuclear Receptors   201

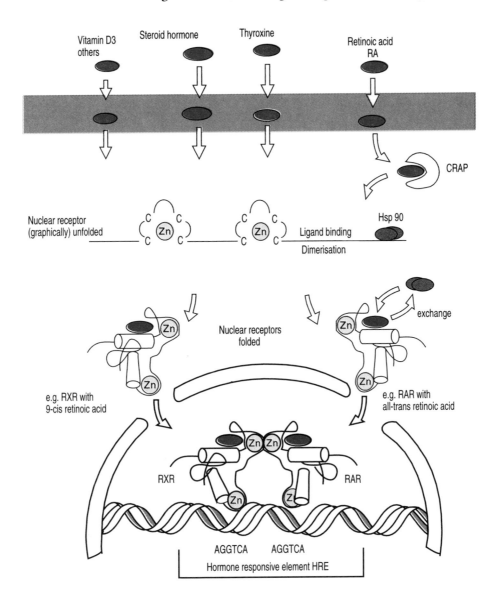

activated by them. RA strongly influences patterning in the limb bud (Chapter 9, Section 9.7; Fig. 9–8) and in the axial organs, including the neural tube. Overdosage elicits severe teratogenic effects (embryonic malformations). For example, feathers may form instead of scales on the legs of birds.

Besides stimulating the basic energy-extracting and energy-transferring metabolism, thyroxine is used to control metamorphosis in amphibians (Chapter 19) and molting in birds. Steroids have a pivotal role in the control of molting and metamorphosis in insects (Chapter 19), and in the control of sexual development in vertebrates (Chapter 20).

10

Differentiation Is Based upon Differential Gene Expression that Is Programmed during Determination

---

10.1 Initially Cells Are "Genomically Equivalent," but They Become Programmed Differently

In sexual as well as in asexual reproduction of a multicellular organism, all cells arise by mitotic divisions. Cleavage of a fertilized egg is a series of mitoses. If developing systems follow traditional textbooks and undergo nothing but regular mitotic divisions, the genomic information of all cells is identical as long as nothing is changed after DNA replication. In fact, initial genomic equivalence appears to occur for the vast majority of embryonic cells. Yet, although every cell contains the entire set of genetic instructions, in each particular cell only a small fraction of the total genetic information actually is being used and expressed in a collection of proteins specific for the respective cell type. In different cell types different sets of genes are expressed. Which parts of the genome will used and which will not is programmed in the process of **determination.**

The program specifying the genes to be activated does not need to be run immediately. Frequently, cells committed to a distinct developmental path undergo many rounds of cell division without revealing their identity and destiny in their appearance. The daughter cells inherit the program from their parent cell. This phenomenon is called **cell heredity.**

Frequently, the inherited program is actually executed after a considerable lapse. Only while undergoing **terminal differentiation** are the cells actually provided with a characteristic molecular inventory and acquire their characteristic shape and function. For example, myoblasts synthesize muscle-type molecules such as sarcomeric actin, myosin, tropomyosin, and troponin; using these molecules, the developing muscle fiber constructs its contractile machinery. Erythroblasts acquire hemoglobin, carbonic anhydrase, and spectrin while they adopt the task and shape of red blood cells. Neuroblasts produce, among a multitude of other proteins, neural cell adhesion molecules N-CAM, neurofilaments, voltage-gated ion channels, and synaptophysin. The retinal photoreceptors form the discs of photoreceptive membranes inserting rhodopsin into them and attaching transducin molecules. The number of identified cell-type-specific proteins is huge and growing daily.

Potentially, the acquisition of cell-type-specific molecular equipment can be regulated at many sites of the cell's biochemical machinery. The equipment with active enzymes can be regulated at the level of transcription, at the level of translation, and through posttranslational modification of the synthesized proteins (e.g., by phosphorylation, methylation, acetylation, and glycosylation). However, only modifications that are effective for long periods of time can bring about cell differentiation. As a rule, once a program is installed and the terminal differentiation is attained, the state of differentiation is maintained for life. **Transdetermination** (change of a program) and **transdifferentiation** (exchange of the particular type of differentiation or change of cell type) are infrequent (Chapters 11 and 21).

The developmental biologist is presented with several intriguing questions:

- How are the genes to be expressed activated in a coordinated manner, and how is the expression of other genes silenced?
- How are developmental decisions *"frozen"*; how can the state of determination be transmitted faithfully to the daughter cells during cell division?
- Is differentiation accompanied by secondary qualitative and quantitative changes in the structure or accessibility of the genome?

10.2 Chromosome Puffing in Flies:
Genes Activated but Equivalence Lost

Once upon a time, textbooks and teachers told the story of "puffing" activities along the giant chromosomes found in the salivary glands and excretory organs (Malpighian tubules) of larvae of dipterans such as *Drosophila*. The story was, and still is, presented as a classic example proving the theory of differential gene activity. In fact, the giant chromosomes display gene activity in such an obvious form that it can easily be seen under the light microscope. Giant chromosomes are called **polytene** because they contain more than a thousand aligned replicated DNA strands that remain attached to each other. Each polytene chromosome has a characteristic series of bands (chromomeres) and interbands. The *Drosophila* genome comprises about 5,000 bands, each containing about 10 genes on average. During molting and metamorphosis, several sites along the chromosomes become enlarged and puffed out. Puffing results from the conversion of the DNA into a decondensed form; it indicates regions where genes are transcribed and the newly synthesized RNA is packed into ribonucleoprotein (RNP) particles. Using isolated salivary glands, it has been shown that one single regulatory factor, the hormone **ecdysteron** (hydroxyecdyson), causes transcriptional activation of several genes and the inactivation of other genes. The pattern of puffing changes during fly development, and is different in salivary glands and Malpighian tubules. Thus, puffing indeed reflects organ-specific gene expression.

On the other hand, salivary glands and Malpighian tubules are already differentiated; they merely change their functions. Also, the chromosomes are polytene, which means that an irreversible amplification of the entire genome has taken place during the development of those organs. Polytene chromosomes originate through repetitive rounds of DNA replication. At the end, about 1,000 DNA strands stick together and form a twisted cable. Thus, polytene chromosomes also represent an example of an irreversibly altered genome and loss of **genomic equivalence**. Genomic equivalence is not always maintained during differentiation (see Chapter 11).

10.3 A Practically Perfect Paradigm:
The *MyoD/myogenin* Family Programs a Myoblast
and Its Descendants Take Over the Program

How is a particular cell type programmed? One peculiar type of cells has yielded its developmental secrets more readily than any other: the **cross-**

striated muscle fiber. Skeletal muscle fibers are large, multinucleated syncytial cells formed by the fusion of many myoblasts. Myoblasts are cells of mesodermal origin. They become programmed step by step: before and during gastrulation they become committed to take the mesodermal path, and in the somites they constitute the part termed myotome, where they are fated definitively to become the precursors of the striated muscle fibers. This gradual programming takes place under the influence of inducing factors (Chapter 9). In the course of the programming central **myogenic key genes** are switched on.

The first myogenic ("muscle generating") gene to be identified was the *Myoblast-Determining gene 1, MyoD1.* This gene is a **master gene** that brings other, subordinate genes needed for terminal differentiation under its own control. Through transfection with *MyoD1* mRNA, fibroblasts (precursors of connective tissues and cartilage) and adipoblasts (precursors of fat cells) can be reprogrammed and prompted to become stably and heritably committed myoblasts. This is an example of a **gain-of-function** experiment.

In order to control other genes, **MyoD1** protein is provided with a **basic helix-loop-helix (bHLH) domain.** This domain attaches to the control region, called E box, of the genes to be ruled. MyoD1 protein acts as a transcription factor. It not only controls other genes, but it also autocatalytically regulates its own production. This is the cleverest aspect of the whole matter: MyoD1 physically binds to the upstream controlling region of its own gene, keeps it active, and thus amplifies and maintains its own production. This is referred to as "autocatalysis" or "positive feedback."

During DNA replication transcription factors may detach but this does not cause problems. After completion of cell division, transcription factors return into the nucleus searching for their target promoters. MyoD1 will be among them, will find the E box in front of its gene, and increase its own production again. MyoD1 is not diluted out—both daughter cells remain myoblasts and the state of determination survives cell division. The muscle cell is committed stably to its task.

From such an important gene one expects that a loss-of-function mutation would result in offspring without muscles and incapable of life. Yet, to the surprise of researchers who conducted this experiment, targeted mutagenesis (Box 7) did not have this effect, even when the defective gene was homozygous due to inbreeding (**knockout** or **null mutation**).

This result, while initially disappointing, paved the way to surprising new results and understanding. Perhaps because the gene is so important, there exists not only one key gene but several similar genes that can replace one another, just as we usually have more than one key for our house and car. This phenomenon is referred to as **genetic redundancy.** In addition, various myogenic selector genes might be of significance in the programming of muscle subtypes.

At present, four myogenic selector genes are known in mammals: *MyoD1*, *myf-5*, *MRF-4*, and *myogenin*. All belong to the same gene family—that is, they display a high degree of sequence similarity, and all four genes code for DNA-binding transcription factors containing a bHLH domain. When injected into fibroblasts, they are each able to start the muscle differentiation program.

During embryo development, *myf-5* and *MRF-4* are expressed first, but only transiently. Subsequently, the *MyoD1* and *myogenin* genes are activated. The MYOGENIN protein is exceptional: it only appears when the myoblasts fuse to form myotubes and start constructing their contractile machinery. MYOGENIN is produced constitutively, and it is indispensable: *Myogenin* null mutants are lethal. In this case there is clearly no genetic redundancy. Genes belonging to the same family cannot always replace each other, and the organism is not insured completely against all disastrous mutations.

10.4   Genes Directing Development Are Often Master Genes Regulating Batteries of Subordinate Genes

Many genes directing and executing fundamental decisions in development, like *MyoD1*, are **master genes,** also called **selector genes.** They code for proteins having "transactivating" gene regulatory function. That means that they can turn on genes located on other chromosomes. Long-distance control is rendered feasible by the ability of these proteins to bind specifically to control sequences located upstream of those genes that are put under their command. Because their DNA-binding activity serves to determine whether the gene is transcribed into a message, these regulatory proteins are classified among the **transcription factors.** With the attachment of such factors onto the promoter region of subordinate genes, some of the genes become activated, others become suppressed. The control sequences upstream or within the promoter region of the regulated genes are called **enhancers,** or **responsive elements** (RE), or simply *xyz* **boxes.**

Some Examples of DNA-Binding, Gene-Regulatory Domains are as follows:

1. **Homeodomain.** In the development of the basic body architecture of *Drosophila* and probably of all eumetazoans, **homeotic genes** have a pivotal

role (see Section 10.6). In particular, they have a highly conserved motif called the homeobox, or *HOM* motif, in common. This box is the basis of a **lock-and-key system.**

The *HOM* sequence provides the protein for which it codes with a structure called a **homeodomain.** This is the key structure of the protein, enabling it to recognize particular lock sequences on the DNA. The homeodomain motif comprises 60 to 70 amino acids and displays an angled **helix-loop-helix** or **HLH** structure (also called helix-turn-helix). The angled HLH key fits into the major groove of the DNA, but it does not fit into it anywhere on the DNA. Instead, it is able to recognize specific sequences called **responsive elements** (RE elements) in the promoter region of those genes put under its control. Small differences in the homeodomain, on the one hand, and the RE elements, on the other hand, ensure that different keys fit into different locks. Because the many variants of the homeodomain proteins bind to different promoters, the metaphor of molecular address labels also has been used to characterize their function.

Today, the homeoboxes identified originally in *Drosophila* genes exemplify a large class of related genes. Some of them show extended sequence similarity to the *HOM* genes of *Drosophila* (see Section 10.6), others are very different, but all share a homeobox coding for an HLH domain.

2. **Basic helix-loop-helix domain (bHLH).** At first glance, the bHLH is similar structurally to the HLH homeodomain. However, in order to control genes two bHLH domains must attach to the promoter or enhancer region of the respective gene, forming a dimer (as a rule, a heterodimer between two similar but not identical domains). The best-known examples of transcription factors with the bHLH motif are the members of the muscle-cell-determining *MyoD1/myogenin* gene family (see Section 10.3).

3. **Pax domain.** Other genes controlling developmental events are equipped with a Pax motif, coded by the *paired box* sequence of the corresponding genes. An example is the master gene controlling eye development, *Pax-6* (see Section 10.5).

4. **Zinc-finger domains.** In *Drosophila* more than 30 proteins have a controlling function in the establishment of the basic architecture of the body. Many are equipped with a homeodomain, but not all. Some of the gene-regulatory transcription factors are equipped with zinc fingers, including the protein coded by the segmentation gene *hunchback*.

Among the proteins with zinc-finger domains are the **receptors** of the **steroid hormone family.** The members of this family are related closely to each other—that is, they display high sequence homology. The family comprises the receptors not only for steroid hormones but also for **thyroxine,** retinoic acid, and vitamin D₃. Interestingly, all of these hormones are used to control both metabolism and development. Loaded with their ligand, the receptors move into the nucleus and become transcription factors controlling gene expression (Box 5). Differences in the ligand-binding region enable these receptors to bind different hormones, and small differences in the zinc-finger motif enable them to bind to different promoters and thus to control different sets of subordinate genes.

5. **Other domains in transcription factors regulating development.** All known DNA-binding motifs are found among transcription factors controlling development. For example, the gene for the hormone prolactin is under the control of a factor exhibiting a **POU domain.** Prolactin is involved in the control of amphibian metamorphosis (Chapter 19).

Finally, the mammalian **testis-determining factor TDF** contains a DNA-binding domain of the **high mobility group HMG** (Chapter 20).

In conclusion, a hierarchic principle has emerged: genes controlling fundamental decisions in development act by controlling batteries of subordinate and subsidiary genes.

10.5   A Master Gene Programming an Entire Organ: *eyeless* and the Monster Fly with Fourteen Eyes

The analysis of genes needed for the correct development of eyes has provided one of the most spectacular results in all of developmental biology: the genetically managed induction of supernumerary eyes in *Drosophila*.

In *Drosophila* several loss-of-function mutations are known that lead to eye size reduction or to the complete loss of the eyes. Among these mutations is *eyeless* (*ey*). The gene is homologous (orthologous) to the gene *small eye* (*Sey* = *Pax-6*) in the mouse and to *Aniridia* in humans. Humans heterozygous for a defective *Pax-6* gene lack irises. The involvement of a common gene in the development of highly different eyes is astonishing enough. Apparently, a highly conserved master gene is involved in the development of eyes as fundamentally different as the compound eyes of insects and the camera eyes of vertebrates. One of the few common attributes of these eyes is that they share the same light-sensitive pigment, **rhodopsin.** The genes coding for the diverse opsins are probably among the targets of the selector gene in question. The *eyeless* gene possesses two motifs coding for DNA-binding protein domains: a homeobox and a paired box. The gene must be active during the

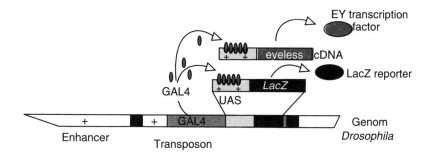

Figure 10–1   Ectopic eyes in *Drosophila* I. Genetic manipulation leading to supernumerary eye formation in novel places. As background for the experiment, a transposon containing the yeast gene for the transcription factor GAL4 was inserted randomly into the *Drosophila* genome. Strains of *Drosophila* were selected where the transposon inserted downstream of, but close to, a promoter (upstream activating sequence, UAS) that becomes activated in an imaginal disc. To select appropriate strains, a *UAS + LacZ* reporter construct was injected into eggs. When the GAL4 gene was activated in an imaginal disc, the GAL4 transcription factor stimulated the expression of *LacZ;* the presence of LacZ protein (bacterial β-galactosidase) is made visible by a blue stain. Flies having blue spots in any of their imaginal discs were selected and inbred. In the ultimate experiment a *UAS + eyeless* construct was injected into the eggs. Now GAL4 could activate not only the *LacZ* gene but, in addition, also the *eyeless* gene. The result is shown in Fig. 10–2.

entire process of eye development. Walter Gehring (1995) and his coworkers succeeded in producing a cDNA of this gene and introducing it into eggs of *Drosophila.*

Moreover, Gehring and his coworkers introduced the *eyeless* cDNA into eggs of transgenic flies that had been prepared by a clever manipulation. In metamorphosis, when many genes become expressed in the imaginal discs, these transgenic flies switch on alien genes simultaneously with their own disc-specific genes (Fig. 10–1). A disc otherwise destined to become a leg but that has incorporated *eyeless* cDNA, switches on the *eyeless* gene—and develops a supernumerary eye, provided the incorporated *eyeless* gene is the wild type and not the mutated allele. Up to 14 supernumerary eyes have been developed by a single transgenic fly! Additional eyes have been found on legs, wings, and antennae (Fig. 10–2). Of course, these do not confer the ability to see, as proper innervation is still lacking. Who knows what curiosities imaginative scientists will make in the future?

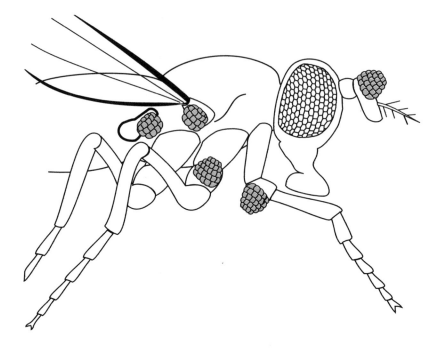

Figure 10–2   Ectopic eyes in *Drosophila* II. A fruit fly expressing the *eyeless* gene in imaginal discs of wings, legs, and antennae. The ectopic expression of intact *eyeless*⁺ can lead to additional (small) compound eyes.

10.6   Homeotic Genes Specify the Attributes of Body Regions

The existence of a particular class of homeobox-containing genes was revealed by spectacular mutants. A famous example is the *Antennapedia* mutation in *Drosophila* in which the antennae are transformed into legs (Fig. 3–20). A "homeotic gene" is not the same as a homeobox-containing gene. The classical homeotic genes, such as *Antennapedia*, do have a homeobox, but not all homeobox-containing genes are homeotic genes. Canonic homeotic genes are not involved in either the construction of the basic body plan or the elaboration of segments and appendages as such. Rather, they specify the particular quality and attributes of a segment or appendage—whether a segment will become the wingless prothorax or the winged mesothorax, for example, or whether an imaginal disc of an appendage will acquire the characteristics of an antenna, a foreleg, a middle leg, or a hindleg. Homeotic genes are now known from all metazoans, including even cnidarians such as *Hydra*.

Recent research has assigned a pivotal role in the evolution of metazoan phyla to the homeotic genes. This view is supported strongly by an astonishing conservatism not only in the nucleotide sequence within the homeoboxes of such genes but also in their arrangement within the genome.

10.7   Homeotic Genes Are Positioned Along the Chromosome in a Sequence Corresponding to the Spatiotemporal Pattern of Their Expression in the Body

In *Drosophila* eight closely related (paralogous) homeotic master genes are clustered on the third chromosome. The whole group is designated as the **HOM-complex (HOM-C)**. It is subdivided into two subgroups, the **Antennapedia complex (Antp-C)** and the **bithorax-complex (BX-C)** (Fig. 10–3). Remarkably, the sequence in which these genes are ordered along the chromosome corresponds to: (1) the temporal order in which these genes become activated and, even more intriguingly; (2) to the spatial order they are expressed along the length of the body. There is a **colinearity** in the positions on the chromosome and the pattern of expression in the body. Only minor inversions do not fit into this general rule.

The order along the chromosome may be an accident of evolution but now represents a temporal program. The genes appear to be activated by some process that spreads along the chromosome. The distance this process propagates roughly correlates with the body region. In the most anterior head region only the first gene is expressed, and the last gene of the cluster is expressed in the most posterior body part. However, the number of genes does not correspond to the number of segments, and there is some overlap in the so-called expression domains. An **expression domain** is the region of the body in which a gene is transcribed. Such domains usually start with a sharp anterior boundary but fade out posteriorly.

Remarkably, not only are the same genes found in *Drosophila*, other invertebrates, and even in mammals, but even the physical arrangement on the chromosomes and the spatiotemporal order of expression are essentially the same.

Terminology among the homeoboxes and homeotic genes can be confusing.

In *Drosophila* the genes are called **HOM** genes.

In vertebrates they are called **Hox** or **HOX** genes.

In mouse, *hox*.

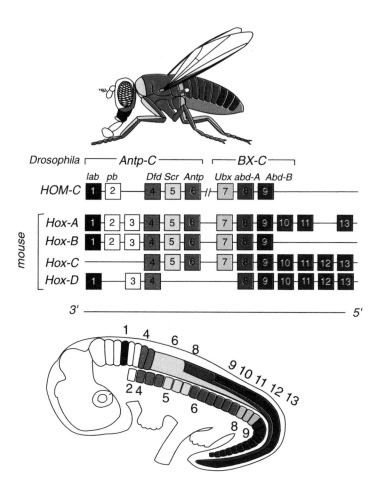

Figure 10–3 Evolutionary conservation of homeotic gene organization and spatiotemporal expression patterns in *Drosophila* and the mouse. *HOM-C* represents *Antennapedia complex* (*Antp-C*) and *bithorax-complex* (*BX-C*) clusters of the homeotic genes on *Drosophila* chromosome 3. The corresponding *Hox* genes in the mouse are present in four clusters (*Hox-A, Hox-B, Hox-C, Hox-D*). Matching numbers or colors in columns indicate particularly strong structural similarities across species. All mouse homeobox genes are transcribed in the same direction, starting at the 5′ end. In the animal itself the genes expressed more anteriorly are transcribed earlier, whereas those expressed in the posterior body regions are transcribed later. In the mouse the expression patterns in the central nervous system and in the somites are shown (bottom). Some expression domains overlap. Frequently, only the anterior boundary is demarcated sharply, while in the posterior direction the expression domains fade out.

In the human, *HOX*.

In *Xenopus, Xhox*.

Genes are written in italics, for example, *Hox-A4*. The corresponding protein is written Hox-A4, **Hox-A4,** or Hox-A4.

Due to duplication in evolution, the entire complex is present four times in mammals: *Hox-A, Hox-B, Hox-C, Hox-D* (formerly: *Hox-1, Hox-2, Hox-3, Hox-4*).

These four clusters are distributed among four chromosomes:

HOX/Hox-A1 to *A13:* Human chromosome 7, mouse chromosome 6

HOX/Hox-B1 to *B13:* Human chromosome 17, mouse chromosome 11

HOX/Hox-C1 to *C13:* Human chromosome 12, mouse chromosome 15

HOX/Hox-D1 to *D13:* Human chromosome 2, mouse chromosome 2

(Not all groups contain all 13 genes. Some genes have been deleted, and some positions remain open for future discoveries.)

Of particular significance is the finding that each *Hox* group in mammals contains genes displaying a particularly high sequence similarity with the corresponding genes in *Drosophila* (vertical columns in Fig. 10–3). In addition, a colinearity also exists in mammals between the physical order along the chromosome and the spatiotemporal order of expression. As a rule, the near-3′ genes are expressed early and in the anterior body, the near-5′ genes are expressed late and in the posterior body. It is as though the expression pattern of homeotic genes defines body regions. Actually, however, the activity pattern of sets of homeotic genes reflects a temporal developmental program that is encoded in the physical position of these genes along the chromosomes. Moreover, the expression domains of the homeotic genes define borders, as indicated by the following findings.

1. **Longitudinal body axis.** All four *Hox* clusters are expressed along the body axes with sharp anterior boundaries. However, the anterior boundaries of, for example, *Hox-A2, Hox-B2, Hox-C2,* and *Hox-D2* are different in the spinal cord and in the series of somites (compare Fig. 14–1). *Hox* genes appear to act only to demarcate relative positions rather than to specify any particular structure.

It is worth noting that activity pattern of homeotic genes does not specify a certain cell type (in contrast to a gene like *MyoD1*) but covers several diverse tissues. For example, the *Hox-D* group is expressed in the spinal cord, the somites, and the limb buds (Fig. 10–4) and these structures include the future epidermis (in the limb buds), nerve cells (in the spinal cord), cartilage, and muscle tissue (in the somites and the limb buds).

Just as in *Drosophila*, irregular *Hox* expression leads to homeotic transformations in mice. These can be read out in the particular quality of the vertebrae. As a rule, knockout mutations cause transformation toward more anterior qualities, whereas overexpression causes transformation toward more posterior fates. For instance, the atlas may be transformed into an additional proatlas or into an additional axis.

The example of the *Hox-D* group also shows that the expression pattern not only covers diverse tissues but also diverse body parts. For instance, *Hox-D13* is expressed not only along the main body axis in the tail but also in the outgrowing limb buds. Unfortunately, it is not possible to assign one finished structure to one homeotic gene. All genes identified so far are not only expressed in structures considered homologous on morphological criteria but also in several nonhomologous organs. A body part is characterized not by the activity of a single master gene but by particular and unique combinations of master gene activities.

2. **Limb bud, anteroposterior axis.** At the posterior margin of the mouse limb bud, a center is established from which waves of *Hox-D* expression originate. The center is coincident with or adjacent to the ZPA in the chick embryo—that is, to the center of the zone of the polarizing activity that organizes the order of digits. It is, however, not the HOX-D proteins that disperse. Rather, the HOX-D proteins are retained in the nuclei of the cells where they fulfill the task of transcription factors. It is the spatial expression domain as such that extends. First, a wave of HOX-D9 emerges from a point and spreads to the anterior margin. Next, a wave of HOX-D10 appears. This wave, however, spreads a shorter distance and does not reach the anterior margin. The expression domains of HOX-D-11 and HOX-D-12 are even shorter, and HOX-D-13 expression finally remains restricted to the spot of origin (Fig. 10–4). An implanted bead that releases retinoic acid at the anterior margin induces the establishment of a second *Hox-D* center where waves of expression start.

3. **Limb bud, proximal-distal axis.** Here the outgrowth is accompanied by a similar spatiotemporal expression pattern, but it is the expression domains of HOX-A genes that spread shorter and shorter distances (Fig. 10–4).

HOMEOTIC GENE EXPRESSION

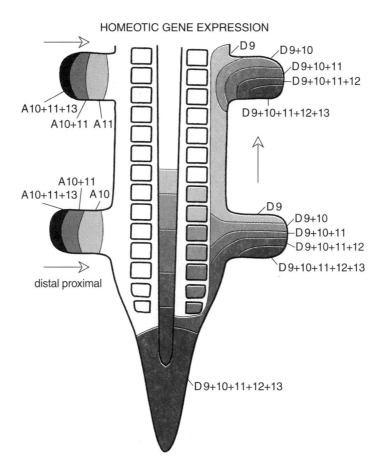

Figure 10–4   Expression pattern of *Hox-A* and *Hox-D* genes in the mouse embryo. The gene with the lowest number is expressed first, the gene with the highest number is expressed last. All expression waves of the *Hox-A* series start at the distal margin and extend more or less in the proximal direction. Likewise, the expression waves of *Hox-D* genes start in the most posterior areas and extend more or less in the anterior direction.

10.8 Emission of, and Response to Inductive Signals Are Reflected in the Expression Pattern of Homeobox Genes

The homeobox gene *goosecoid* is expressed in the area of the Spemann organizer in the amphibian gastrula as well as in Hensen's node on the avian and mammalian blastodisc. Ectopic expression of this homeobox gene, for instance by injection of *goosecoid* mRNA into the wrong places, enables recipient cells to adopt the function of a blastopore lip. In an amphibian gastrula the cells expressing *goosecoid* involute and organize the development of a secondary body axis.

While *goosecoid* confers inducing power to the cells expressing it, induced structures in turn express other homeobox genes, for instance, the *Hox-A*, *Hox-B*, *Hox-C*, and *Hox-D* complexes mentioned in Section 10.7 (and shown in Fig. 10–4 and Fig. 14–1), or *Brachyury*, which becomes expressed in the dorsal midline cells before they segregate to form the notochord.

10.9 Genes Controlling Development Are Integrated into Interacting Networks

Some genes increase and stabilize their state of activity through **positive feedback loops (autocatalysis).** Examples are the muscle-cell-determining genes *MyoD1* and *myf-5*, and the *Drosophila* segmentation gene *fushi tarazu*. Other genes switch off their own activity by **negative feedback.** There are cases in which master genes switch on other genes by a *cis*-activating **heterocatalysis,** while other genes are switched off by a type of intracellular **lateral inhibition.**

Moreover, the promoter and enhancer regions of genes to be controlled are often accessible not only to one transcription factor but to several. For instance, the hormone-responsive element of some genes can be occupied by heterodimers jointly formed by one receptor for thyroxine and one receptor for cortisol. Various transcription factors that aggregate in steps along the control region of genes may synergistically enhance the activity of the regulated gene, or they may cancel each other's effect and diminish the activity of that gene.

In different body regions distinct arrays of selector genes become effective in a complex spatiotemporal pattern. This hierarchical organization of gene regulation during development explains why finite numbers of genes can generate a remarkable diversity of body parts, tissues, and cell types in such a precise and consistent manner. For a combinatorial analogy, consider that the piano only has 88 keys. Nonetheless, innumerable pieces of music have been composed by varying the spatiotemporal activity of the keys. With a complement of 5,000 genes (*Caenorhabditis elegans*), 50,000 genes (*Drosophila*), or 100,000 genes (human), an inexhaustible diversity of combinations is possible.

10.10   DNA Methylation and Heterochromatization Can Silence Genes

Several mechanisms can lead to a permanent functional silencing of genes. In addition, the inactive condition can be transmitted to daughter cells through mitotic cell division.

10.10.1 DNA Methylation Leads to Paternal and Maternal Genomic Imprinting

In the DNA of eukaryotes cytosine can be converted to **5-methyl-cytosine** (indicated by **C***) through the addition of a methyl group. Such methylation takes place only in **CG** nucleotide pairs. The following sequence:

- - C * - G - -

- - G - - C * - -

enables methyltransferase enzymes to copy the methylation pattern so that both daughter cells exhibit the same pattern. However, while this mechanism makes the methylation pattern heritable, any general theory of the role of methylation in, say, differentiation, must take into consideration the fact that DNA methylation has not been found in all organisms. For example, *Drosophila* DNA is not methylated at all.

In mammals methylation is extensive—about 70% of the CG sequences are methylated. The consequences of DNA methylation cannot be predicted for each single CG group or each single gene. In general, however, a high degree of methylation results in lower levels of transcription of the methylated gene.

In the course of mammalian germ cell development the methylation pattern is (largely) erased, and the DNA becomes remethylated. Sperm DNA is more highly methylated than the DNA of egg cells, perhaps in preparation of the subsequent condensation of sperm DNA. However, because the methylation pattern persists past fertilization, the accessibility of the maternally and paternally inherited genes to transcription is not identical ("*maternal and paternal genomic imprinting,*" see Chapter 5).

10.10.2 In Female Mammals One X Chromosome Is Silenced

In a process called **heterochromatization** large regions of chromosomes or even entire chromosomes are brought into a state of increased condensation. This conversion of euchromatin into heterochromatin blocks or impedes transcription, but permits DNA replication and duplication of the chromosomes. Daughter cells can inherit the pattern of heterochromatization.

Heterochromatin contains proteins that possess a domain called the **chromatin-organizing modifier** or **chromo-domain.** With this domain the proteins attach to repetitive DNA sequences that recur in regular distances along the chromosome. By aggregation the proteins are brought closer together and the DNA becomes folded and compressed. A similar DNA-condensing protein is the POLYCOMB protein encoded by the *Drosophila* gene *polycomb.* POLYCOMB is used widely to silence genes in development. For example, it suppresses the expression of abdomen-specific homeotic genes in the anterior body. Defects in the *polycomb* gene may result in transformation of thoracic segments into abdominal segments.

In female mammals all cells are equipped with two **X chromosomes,** while males have one X and one Y chromosome. Consequently, X chromosome-resident genes are present in females twice, in males only once, including those many genes that are needed in both sexes (for instance, genes for certain blood-clotting factors). To compensate for this inbalance in X chromosome copy number, the mechanism of **dosage compensation** evolved. Through heterochromatization, one of the two X chromosomes becomes inactivated and changes its microscopic appearance. In the interphase nucleus this condensed X chromosome is observed as **sex chromatin** or the **Barr body** (Fig. 10–5).

Heterochromatization occurs after the egg cell has already undergone some divisions. It is a random event: chance determines whether in a particular blastomere the inactivated X chromosome is the "paternal" X chromosome provided by the sperm, or the "maternal" X chromosome contributed by the egg cell. Once heterochromatization has taken place for the first time, in all descendants of a blastomere the same X chromosome will undergo heterochromatization. Every human female, like all mammalian females, comes to be a mosaic of cell clones, in which the paternal X chromosome is silent, and clones in which the maternal chromosome has been silenced and converted into a Barr body.

Red-black spotted female cats (the very few red-black spotted tomcats have the aberrant XXY constitution!) display the result visibly on their coat. One of the two X chromosomes bears a gene enabling the synthesis of com-

Figure 10–5   Silencing of X chromosomes in female mammals (dosage compensation). In the blastomeres of the early embryo (upper figure) one of the two X chromosomes is inactivated through heterochromatization and is visible as the condensed Barr body. Because the selection of which X is condensed is a random event, in one blastomere it may be the "paternal," in the other blastomere the "maternal" X that is inactivated (upper figure). After this random event, the same X chromosome is always converted completely into heterochromatin after DNA replication in the daughter cells of the blastomeres. Therefore, in some clusters of cells (cell clones) it is the maternal X chromosome that is silenced; in other clusters it is the paternal X. Red-black spotted female cats (tomcats are red-black only if they show the XXY Klinefelter's syndrome) display the result on their coat. The X chromosome carries genes needed for the synthesis of melanin. In heterozygous cats if the one allele is active, complete black melanin is synthesized; if the other allele is active, only incomplete, red melanin can be produced.

plete, highly polymerized black melanin. The other X chromosome bears an allele with which only incomplete red melanin can be synthesized (Fig. 10–5). The synthesis of melanin takes place in chromatophores that derive from neural crest cells and immigrate into the hair follicles (Fig. 13–1).

Because the process generating heterochromatin is based on the aggregation of DNA-binding proteins, it has a tendency to spread beyond the nominal region to be silenced. Such extended spreading results in **position effect variegation:** genes close to a silenced region suffer a reduction in their accessibility to transcription. Position effect variegation suggests the participation of an autocatalytic feedback loop in heterochromatization. Gene silencing by heterochromatization is considered to contribute to persistent and heritable determination and differentiation.

10.11   Programming of Determination and Inheritance of the Determined State May Include Unknown Mechanisms

When cells are fated to take a distinct developmental path, certain genes remain accessible and ready to deliver their information on demand, while other genes become silenced permanently. There is some evidence that the programming (called **determination** or **commitment**) often occurs shortly before or during DNA replication in the S phase of the cell cycle. In other cases, the programming is independent of the cell cycle, occurring gradually over a prolonged period. Frequently, the committed -*blasts* (myoblasts, neuroblasts, erythroblasts) still undergo several rounds of replication until terminal differentiation begins and the cell-type-specific program of protein synthesis is realized completely.

In committed cells still capable of dividing, the state of determination is transmitted to daughter cells. The term **"cell heredity"** refers to this phenomenon that gives rise to clones of identically committed cells. Textbook examples are the imaginal discs of *Drosophila*, clones of which have been propagated artificially by fragmentation and subsequent regeneration of founder discs (Chapter 3, Section 3.6; Fig. 3–25). However, a variety of cell lines maintained in culture testify to the general heritability of the determined state. From certain myoblasts, for example, myriads of offspring have been grown. If growth factors are withdrawn, the myoblasts fuse and become muscle fibers.

The molecular basis of cell heredity is not well understood. The following mechanisms have been proposed:

1. **Autocatalytic self-activation,** according to the model of the *Myoblast-Determining gene 1 (MyoD1)* (Section 10.3). Several master genes code for proteins that immigrate into the nucleus and activate not only other subordinate genes but also their own gene. Because such transcription factors are also present in the cytoplasm where they are produced on ribosomes, they end up in both daughter cells. By positive feedback the

factors allotted to the daughter cells boost their own production; thus the dilution caused by cell division is compensated. Examples of such persistent self-amplification are, besides *MyoD1*, the *fushi tarazu* (*ftz*) and *Ultrabithorax* (*Ubx*) subsystem of the *HOM complex* in *Drosophila*. These two systems, among others, may contribute to the heritability of the determined state in growing imaginal discs.

The sex-determining gene *sry* (*Sry*) of mammals (Chapter 20) also maintains its own expression.

2. **Heterochromatization.** This process also includes a self-enhancing subprocess, but instead of activation, it results in long-term silencing of genes (Section 10.10).

3. **Methylation.** If an organism's DNA is methylated, this process can also contribute to a lasting change in the accessibility of genes for transcription.

The biochemical disparity in these mechanisms for heritability of determination suggests that more mechanisms may be found.

11

Cell Differentiation Frequently Is Irreversible and Causes Cell Death; Early Cell Death Can Be Programmed

---

11.1   A Reversible State of Determination and
Differentiation Facilitates Regeneration

W hen the channeling of cells into diverse developmental pathways is finished, the cells have acquired a state of stable determination. This state can be "inherited" over many rounds of cell division. Occasionally, however, the state of determination can be suspended and replaced by another condition. Two such phenomena are distinguished: trans-determination and transdifferentiation (metaplasia). **Transdetermination** designates an event by which committed but not yet terminally determined cells suddenly change their developmental program. In **transdifferentiation (metaplasia)** cells that have already undergone terminal differentiation undergo dedifferentiation and take new developmental pathways. Each of these phenomena are observed in the course of regenerative processes and will be discussed with examples in Chapter 21.

11.2   Irreversible Somatic Recombination Takes Place during Lymphocyte Development

Recombination—the rearrangement of genes—is characteristic of sexual reproduction and takes place during meiosis. Surprisingly, a similar event occurs in the development of lymphocytes, when lymphoblasts combine their individual genetic programs for preparing the future production of antigen receptors or antibodies (which are liberated antigen receptors). This process, called **somatic recombination,** occurs at the "birth place" of these cells: the bone marrow, the thymus, or in the lymph nodes (Fig. 11–1).

Lymphoblasts are genetic "gamblers." They try their luck when they combine various short DNA sequences at random to program the variable region of their receptors or antibodies, increasing the chance of creating a

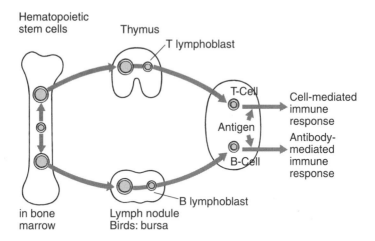

Figure 11–1   Pathways of the lymphocytes. All lymphoblasts are born and committed to take the lymphoid pathway in the bone marrow. The lymphoblasts leave the bone marrow via the bloodstream and colonize in the thymus or the lymph nodules. Those taking the route via the thymus become T cells, whereas those taking the route via lymph nodules become B cells. The mature T and B cells again enter the bloodstream and colonize in other lymph nodules, or the spleen, or migrate through the body in the search for foreign antigens.

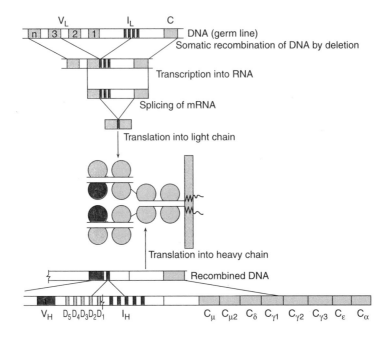

Figure 11–2 Somatic recombination in the generation of antibodies. Three DNA segments (partial genes) are selected from a series of V, J, and C segments, and are joined to a mosaic gene coding for the light chain of the antibody. Likewise, V, D, J, and C segments of another cluster are selected and joined together to a mosaic gene coding for the heavy chain of the antibody. In each lymphoblast the V, J, and D elements are selected randomly to construct a distinctive antigen-binding region (combinatorial diversity). Once an individual combination has been selected at the level of the DNA by somatic recombination, this individual combination is retained in the particular lymphoblast and its descendants. Some variability is left in the production of antibodies because splicing sites are not exactly defined. However, the highest degree of diversity is based on the original, individual somatic recombination.

combination appropriate to recognize and bind a future foreign antigen. In each B cell or T cell a different random combination is tried (Fig. 11–2). Subsequently, the T lymphoblasts emigrate and settle in the thymus; the B lymphoblasts remain in the bone marrow for some time. Wherever lymphoblasts stay, they learn to discriminate between self and nonself. A strict negative selection eliminates all those lymphoblasts whose receptors would bind molecules belonging to the self. The surviving cells are those that can bind only nonself molecules and are thus capable potentially of recognizing foreign substances. The surviving lymphocytes migrate through the body, colonize in other places such as in the spleen and in the lymph nodes, and eventually serve as T cells or antibody-producing B cells.

Somatic recombination, to date found only in lymphocytes, leads to an irreversible alteration in the genetic constitution of the cell.

11.3   Differentiation Is Often Accompanied by Quantitative Changes in the Inventory of Genes: Gene Amplification, Genome Amplification, and Elimination of Chromosomes

1. **Selective gene amplification.** Some differentiated tissue or cell types—for example, gland cells—may require large amounts of gene product. Quantitative changes in the genome of cells and cell lines may occur in the course of differentiation. The selective replication of particular genes, although a rare event, is one of several possibilities that can lead to the non-equivalence of cell genomes that were previously identical. A textbook example is the selective amplification of the ribosomal 18S, 5.8S, and 28S rRNA genes in the nuclei of many oocytes (Chapter 5.3; Fig. 3–45). This amplification is manifested by the appearance of numerous nucleoli (centers of ribosome production). Selective gene amplification is also known from follicle cells in the ovaries of *Drosophila*, where genes encoding the secreted proteins of the egg envelope (chorion) are amplified.

2. **Genome amplification.** Amplification of the entire genome by polyploidy or polyteny is relatively common. **Polyploidy** refers to multiplication of the chromosomes. Instead of the two sets of chromosomes observed in diploid cells, four or more sets of chromosomes are found. In mammalian development the cells of the trophoblast and the liver amplify their genomes through polyploidy. These cells are highly active metabolically and polyploidy may facilitate the continuous production of large amounts of enzymes. **Polyteny** refers to the enlargement of chromosomes through endoreplication: the DNA is replicated repeatedly, but the newly generated strands are not separated but stick together forming cables of DNA strands. During *Drosophila* development the nurse cells in the ovaries become polyploid, and the cells of the salivary glands and the Malpighian tubules become polytene.

3. **Elimination of genetic material.** Instead of amplifying genetic material, some cells lose chromatin or entire chromosomes. A textbook example is the horse roundworm (*Ascaris lumbricoides = Parascaris equorum*), whose early embryos taught Theodor Boveri much about the significance of chromosomes in directing development. Boveri studied a bizarre phenomenon: the

haploid egg has only two chromosomes, and two more are contributed by the sperm. However, even these few four chromosomes appear to be too much for ordinary somatic cells. During the cleavages the four chromosomes are allocated only to the cells of the germ line in full length and complement. In the worm's somatic cells the chromosomes are fragmented and part of the chromatin gets lost, a phenomenon known as **chromatin diminution.** It is not known whether genes that are only needed for gametogenesis are lost, or merely noncoding DNA.

Some cases are known where entire chromosomes or the entire nucleus are eliminated. In the midge, *Wachtiella persicariae*, many nuclei lose 38 of their original 40 chromosomes. Well-known examples of cells devoid of the entire nucleus are the mammalian red blood cells (except those of camels!) and the keratinocytes of skin, feathers, and hair.

11.4   Programmed Cell Death Is Part of Normal Development, Even in the Nervous and Immune Systems

In all multicellular organisms—even the small nematode, *Caenorhabditis elegans*, known for its invariant number of cells—programmed cell death (**apoptosis**) is part of normal development. Some cells die early in embryo development, particularly during the development of the nervous system. Apoptosis in *C. elegans* is initiated by two genes (*ced-3*, *ced-4*). Loss-of-function mutations in either of these two genes allows survival of cells normally programmed to die.

In vertebrates more than 70% of putative neuroblasts die during certain stages of nervous system development, especially those that are not connected correctly with their targets (Chapter 14). Many lymphoblasts and germ cells also die. In the development of the hand and the foot, programmed cell death causes erosion of interdigital tissues and thus the separation of fingers and toes (Fig. 11–3).

Apoptosis is a suicide prepared by the active synthesis of proteases and some other special proteins. The cell takes itself to pieces, blebbing small morsels that can be eaten easily by neighboring cells (later in life, by macrophages). No necrotic complications take place (**necrosis** is nonprogrammed, catastrophic cell death due to damage).

Apoptosis can be initiated by external cues and can be prevented by specific survival factors. For example, the nerve growth factor **NGF** ensures the survival of sympathetic neurons (Chapter 14).

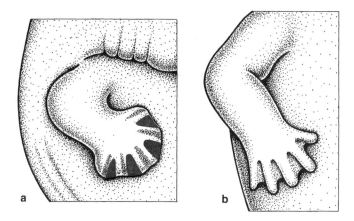

Figure 11–3 Programmed cell death (apoptosis) in the separation of the palm in a human embryo. (a) A 41-day-old embryo; beginning cell death in colored areas. (b) A 51-day-old embryo; separation of the fingers is still incomplete. The separation is finished by Day 56.

Programmed cell death of lymphoblasts is an essential part of the learning process that enables the immune system to discriminate between self and nonself. It may seem strange, but there is good reason for nature to use a strategy based on cell death. Millions of lymphoblasts, the future B cells and T cells of the immune system, rearrange DNA segments at random to produce a huge variety of antigen receptors or antibodies (antibodies are released antigen receptors). This strategy will inevitably not only generate lymphocytes whose receptors fit onto foreign antigens, but will also generate self-reactive lymphocytes with receptors that bind self-molecules. This would be fatal if such lymphocytes were not eliminated quickly because the immune system is extremely aggressive. Most of the self-reactive lymphocytes are eliminated in humans around the time of birth. Due to this process of recombination followed by elimination, the immune system recognizes foreign substances rather reliably. How all these learning processes are accomplished in detail and controlled is largely unknown. Knowledge about these control mechanisms will be very important in understanding **autoimmune diseases.** Presumably, intolerance to self-antigens in later life is due to continued production of self-reactive lymphocytes from stem cells.

Steroid hormones of the adrenal cortex—in particular **cortisol** and cortisone—promote the collective suicide of T lymphocytes in the thymus. Birth and death in the immune system support the life of the entire cell community of our body.

12

Animal Morphogenesis Is Shaped Actively by Adhesion and Cell Migration

A s a result of cell division and cell differentiation, a large variety of cells with different shapes and divergent molecular constitutions and makeups appear. In turn, these cells create associations of cells serving a common function: tissues and organs. In cell associations the number, the size, and the shape of the individual cells eventually will determine the shape of the whole association. In contrast to the development of plants, in animals active shaping through intracellular contractile filaments and forces of cohesion and adhesion are observed, and extensive displacement and migration of cells take place.

12.1   Cell Migration à la Amoebae

In sea urchins gastrulation begins at the vegetative pole with the immigration of cells into the cavity of the blastocoel. Descendants of the micromeres untie

the bonds to their neighbors, pull out of the society of the epithelial blastoderm (dissociate), and immigrate into the blastocoel. The descendants of the
large micromeres reaggregate at certain places and collectively manufacture
the larval skeleton (Fig. 2–1b).

In the gastrula of sea urchins and in the amphibian gastrula at the tip of
the archenteron there are cells that retain their epithelial association but
extend pseudopodia and take on the function of pathfinders (Fig. 2–1b, 12–1).
In the course of gastrulation the roof of the archenteron slides along the overlying ectoderm (inducing it to form the central nervous system). A layer of
fibronectin covering the inner surface of the ectoderm serves as a lubricant.

In favorable, translucent embryos many freely wandering cells can be
seen, such as the deep cells seen in fish embryos (Fig. 3–38). This amoeboid
behavior is also displayed by cells in the blastodisc of birds or mammals that
invade through the primitive groove into the depths and spread in the blastocoel (subgerminal cavity) to form the axial mesoderm (chordamesoderm).

The **neural crest cells** (Chapter 13, Section 13.4; Fig. 13–1) of vertebrates display especially extensive migratory activity. Primordial germ cells
also go on extensive tours (Fig. 5–1).

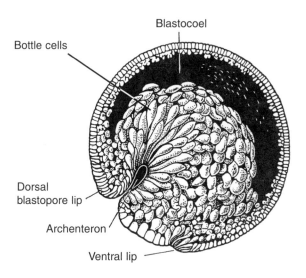

Figure 12–1   Bottle cells in the amphibian gastrula. (After Holtfreter 1946.)

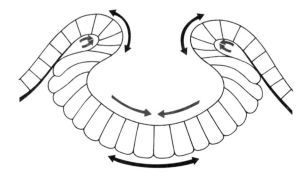

Figure 12–2 **Bending moments in neurulation,** generated by the expansion of the cells on one side and the contraction on the opposite side. In addition, sliding movements participate in morphogenesis.

12.2 Folding and Invagination: Cells in Epithelial Associations Develop Coordinated Bending Forces

During invagination of the archenteron, neurulation, and similar processes that create shapes from epithelial sheets, cells enlarge their surface on one side (e.g., the basal side), while on the opposite side the surface area is diminished (Fig. 12–2). Through this coordinated process an epithelial sheet creates bending moments leading to folding or invagination. Animal cells can be made narrower at their apical surface by the contraction of bundles of actin and myosin filaments that run beneath the upper cell surface, where the cells are connected strongly by adhesion belts.

12.3 Forces of Adhesion Can Shift the Relative Positions of Cells

A drop of oil contracts on a hydrophilic substrate to form a sphere, while it spreads on a hydrophobic (lipophilic) surface to form an extended film. A drop of water displays the opposite behavior. Similarly, adhesion or repulsion of cells is influenced by physical forces, such as tension of wetting, which is the result of several surface forces. In addition, electrostatic and ion exchange properties influence the strength of adhesion or repulsion. Cells can alter the degree of hydrophobic or hydrophilic interactions with neighboring cells, and the strength of electrostatic attraction or repulsion. They can do this, for example, by inserting membrane proteins and by glycosylating or deglycosylating surface proteins. Physical forces acting at interfaces are of significance when in the course of gastrulation sheets of endoderm, mesoderm, and ectoderm slide along each other. Even if cells do not move actively, cell sheets can slide and spread over other sheets by sheer physical forces of adhesion and cohesion, thereby minimizing the strength of surface tensions. Minimizing surface energies also accounts for, or contributes to, the process of sorting out cells of different origin in aggregates (Fig. 12–3).

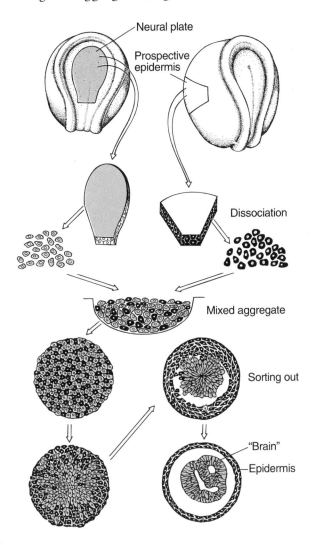

Figure 12–3   Sorting out cells of different origin and fate. After segregation, prospective epidermal cells form an outer epithelial layer, and the cells derived from the neural plate form hollow bodies reminiscent of the neural tube or the brain, respectively. (After Townes and Holtfreter 1955.)

During embryonic development of the vertebrates, coherent groups of cells stream to their destinations by liquidlike spreading movements. According to the "**differential adhesion hypothesis**" (Steinberg 1970), these movements are driven and guided by cell-adhesion-generated tissue **surface tensions,** operating in the same manner as surface tensions do in the mutual spreading behavior of immiscible liquids, among which the liquid of lower surface tension is always the one that spreads over its partner. Miniature devices have been designed to measure surface tensions of cell groups. The measured values correlate with the behavior of the cells in aggregates. Cells displaying high tensions are enveloped by cells displaying lower surface tensions. For example, the limb bud mesoderm (tension 20.1 dyne/cm) is enveloped by the liver (4.6 dyne/cm) that, in turn, is enveloped by the neural retina (1.6 dyne/cm). In animal embryos the differential spreading of one cell population over the surface of another is a common means of embryonic morphogenesis.

12.4   Specific Adhesion Molecules Also Mediate Cell Recognition

The cell membranes of animal cells are equipped with proteins and glycoproteins. Several of them mediate the physical cohesion of cells in cell associations by forming noncovalent bonds with corresponding surface molecules of neighboring cells. In addition, such cell adhesion molecules serve in cell recognition (Box 6). Specific **cell adhesion molecules (CAMs)** bind to others. Binding can be **homotypic** between identical CAMs or **heterotypic** between different CAMs. At least five classes of CAMs have been distinguished.

1. **Cadherins (calcium-dependent adherins).** Cadherins enter into homotypic binding only in the presence of calcium ions. Removal of calcium in the surrounding medium breaks off the mutual adhesion of the cells, which become dissociated. The class of cadherins includes, among others, epithelial **E-cadherin** (called **uvomorulin** in mouse embryos), the liver-specific **L-cadherin,** and the neural **N-cadherin.** The adhesiveness of diverse cadherins causes or facilitates sorting out cells in cell aggregates (Fig. 12–3).

2. **CAMs of the immunoglobulin superfamily.** This superfamily includes the subfamily of the **N-CAMs (neural cell adhesion molecules).** One of the N-CAMs was the first CAM to be detected (by Edelman, review 1983). N-CAMs are found on the surface of neuronal cells, including glia

cells, but also appear on the surfaces of other cells, in particular, during embryo development. N-CAMs are glycoproteins containing sialic acid residues. They are anchored in the cell membrane by a transmembrane domain or are coupled to membrane-resident inositol phospholipids via a glycan bridge. All N-CAMs display some sequence similarity to the immunoglobulins (Fig. 12–4), that is, to **antibodies,** to the **antigen receptors of B and T lymphocytes,** and to the molecules of the **major histocompatibility complex (MHC).** As a rule, binding is homotypic: the CAMs of one cell bind to identical CAMs of adjacent cells. Calcium ions are not needed.

3. **Integrins.** Integrins are heterodimers whose two subunits are integrated into the cell membrane. By means of integrins, cells not only establish contact to other cells but also to molecules of the extracellular matrix (ECM) (Box 6).

4. **Selectins and lectins.** Selectins are found on the surfaces of blood cells and endothelial cells of the vascular system. Selectins are also lectins, meaning that they are capable of binding certain oligosaccharides or carbohydrate moieties of glycoproteins. Because lectins, in general, and selectins, in particular, are proteins but they recognize and bind carbohy-

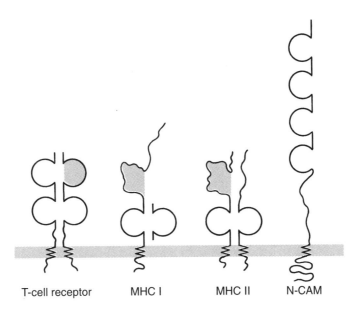

T-cell receptor   MHC I   MHC II   N-CAM

Figure 12–4   Some membrane-anchored members of the immunoglobulin super-family.

drates, they are heterotypic. Through heterotypic interaction between blood cells and the endothelial cells of blood capillaries, lymphocytes and macrophages can establish contact with the walls of blood vessels, and subsequently leave the vessels.

5. **Glycosyltransferases.** These are monosaccharide-transferring enzymes located on the exterior surface of the cells and are considered to mediate reversible cell contact. For instance, a galactosyltransferase anchored in the plasma membrane of one cell can attach galactose to an acceptor molecule located on the surface of an adjacent cell. As long as no galactose is present in the surrounding medium, the enzyme-acceptor bridge mediates cell adhesion. In the presence of galactose the sugar is transferred to the acceptor and the binding is released. The interaction of the mouse sperm with the zona pellucida of the egg envelope is thought to involve such a mechanism.

Cell adhesion molecules mediate many important events in animal development. For instance, they:

- Mediate permanent attachment of cells to each other or to the extracellular matrix, or permit detachment
- Generate boundaries between tissues and facilitate segmentation, such as the subdivision of cell associations into repetitive units
- Prepare the formation of cell junctions
- Act as signal molecules (as an example see the NOTCH/DELTA system described in Chapter 9)
- Guide the migration of neuroblasts and the outgrowth of the nervous processes (dendrites or axons; see Chapter 14)

13

Cell Journeys: Even Germ Cells and Cells of the Peripheral Nervous System Originate from Emigrant Precursors

———◆———

13.1 Extensive Cell Migration Takes Place in the Development of Animals

Animal embryos, especially those of vertebrates, are like cities full of tourists. In a translucent fish embryo migratory cells can be seen crawling and swarming (Fig. 3–38). In an avian and a mammalian blastodisc the cells of the meso "derm" do not form a "skin" (Greek: *derma* = skin) or a coherent germ "layer"; instead, the mesodermal cells creep around like amoebae to colonize the spaces between the epiblast and the hypoblast (Fig. 3–39, 3–48). When the somites split up, the cells of the sclerotome and the dermatome emigrate (Fig. 3–32, 9–5). Primordial germ cells, blood cells, and neural crest cells travel particularly long distances.

13.2   Primordial Germ Cells: Epic Travelers

As a rule, primordial germ cells are not generated in the gonads. In *Drosophila* the stem cells of the future germ cells are the first cells to be produced at all. There is no trace of a gonad present at this early stage. The stem cells, called **pole cells,** are located at the posterior pole of the egg. During gastrulation, the pole cells are drawn into the interior of the embryo with the invaginating hindgut (Fig. 3–15). Later, they actively leave the gut and immigrate into the developing ovarioles.

In *Xenopus* eggs a localized "**germ plasm**" has been discovered. It contains RNA-rich granules and can be made visible with certain fluorescent dyes. The granules are tethered to the precursors of primordial germ cells. Fluorescent labels have made it possible to follow the entire tour of the wandering primordial germ cells (Fig. 5–1). They are seen first in the ventral archenteron; from here they migrate in an amoeba-like fashion along the mesenteries (the epithelial ligaments by which the gut is suspended in the coelomic cavity). Eventually, the primordial germ cells creep into the **genital ridges,** which will give rise to the gonads.

The RNA-containing granules that facilitate fluorescent tracking of germ cell precursors in *Xenopus* are missing from many animal groups, including newts, birds, and mammals. To trace the path of germ cell precursors in these animals, other markers are used, such as monoclonal antibodies that attach to germ-line-specific surface molecules.

In the mouse embryo primordial germ cells are observed first in the extraembryonic mesoderm; later, they appear in the region of the allantois (Fig. 3–43). Subsequently, they creep along the allantois and the gut toward the genital ridges (Fig. 5–1).

In birds primordial germ cells are detected during neurulation along the border of the blastodisc in a crescent-shaped endodermal zone (the germinal crescent). To reach the embryo, they use the bloodstream for transportation over large distances. In this respect, they behave like lymphocytes and macrophages. However, do they know where to leave the blood vessels?

13.3   Blood Cells: Island Ancestry

Blood cells originate from pluripotent stem cells, called **hematopoietic stem cells** (Chapter 16). In adult mammals these are found in the bone marrow (in the mouse, in the spleen as well). However, the stem cells do not originate in these tissues; in the embryo they arise anywhere in **blood islands.** In birds blood islands are observed first outside the embryo proper in the "area opaca," where clusters of angiogenic cells form the first blood vessels. The peripheral cells of the clusters form the endothelial linings, whereas the central cells come to be blood cells. Many blood islands merge to form a capillary network.

In mammals blood islands are detected anywhere in the extraembryonic mesoderm, especially in the mesodermal coatings of the yolk sac and in the region of the allantois. These locations coincide with places where associations of angiogenic cells form the first blood vessels (Fig. 3–43). Later, the stem cells of the blood cells colonize in the liver, the spleen, the thymus, and the bone marrow.

13.4 Neural Crest Cells with Wide Developmental Potential Migrate into Many Target Areas

The story of neural crest cells is among the most amazing tales in all of developmental biology. When the neural plate involutes to form the neural tube during neurulation, and the neural tube detaches from the ectoderm, a series of cells is left over on either side. The cells are neither integrated into the neural tube nor taken up into the ectoderm. Instead, they set off on travels; colonize in various regions of the body; and give origin to a bewildering variety of cell types, tissues, and organs (Fig. 13–1). These include the following:

1. The **pigment cells** in skin, feathers, and hairs. These include the black or brown melanophores of our skin.
2. The **nerve cells of the spinal ganglia (dorsal root ganglia).** Once fully developed, these feed information to the spinal cord coming from sensory organs of the skin, from muscle spindles, or from other somatic senses.
3. The **nerve cells of the autonomic vegetative system,** as far as the perikarya (cell bodies with the nucleus) of these cells are located outside the central nervous system. Classified among the vegetative system are the **parasympathetic ganglia of the head,** such as the ciliary ganglion; the **paravertebral sympathetic chains of ganglia;** the celiac ganglion; and the **nerve nets around the digestive tract (myenteric plexus).**
4. The hormone-producing derivatives of neuroblasts, such as the producers of epinephrine in the **adrenal medulla,** and the **neuroendocrine cells of the digestive tract,** such as those producing cholecystokinin.

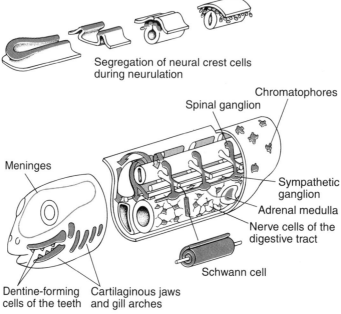

Migration routes and descendants of neural crest cells

Segregation of neural crest cells during neurulation

Chromatophores

Spinal ganglion

Meninges

Sympathetic ganglion

Adrenal medulla

Nerve cells of the digestive tract

Schwann cell

Dentine-forming cells of the teeth

Cartilaginous jaws and gill arches

Figure 13–1 Neural crest cells. Top: Segregation of the neural crest cells (red) during neurulation. Bottom: Migration routes and cells/tissues formed by the neural crest descendants that have arrived at their destination.

5. The **peripheral glia cells** and the glia cells that accompany nervous processes, such as the Schwann cells that envelope and isolate long peripheral axons.

6. The **meningeal coatings** of the brain.

7. The **dorsal fin** in fish.

8. The **cartilage elements of the jaws and pharyngeal arches.** For instance, the branchial arches in the region of the pharynx and the foregut (Fig. 4–2, 4–4) and their derivatives, the sound-transmitting elements of the inner ear: **incus, malleus, stapes,** and the **laryngeal and tracheal skeleton** (Fig. 4–5).

9. The **dental papilla** (producing bonelike material) of the tooth germs.

10. In the head region descendants of neural crest cells appear to contribute to the **bones of the face.** However, some authors prefer to classify the respective **ectomesenchymal cells** forming bones as a separate cell type.

Xenoplastic transplantation experiments, in which neural crest material is inserted into foreign species, have facilitated analysis of the origin, wander routes, and destination of neural crest cells. Classic transplantations between the newts *Triturus torosus* and *Triturus rivularis* have been extended by transplanting neural crest cells of the quail (*Coturnix coturnix*) into embryos of the chick (*Gallus gallus*). Quail cells integrate themselves easily into the chick and can be identified easily by the unusually large amount of heterochromatin in their nuclei.

In their movements during embryo development, neural crest cells of the trunk follow two major routes (Fig. 13–1):

1. The **ventral route.** The majority of cells travel down the anterior portion of the somites through clefts between the myotome and the sclerotome. The cells settle and aggregate at various places. Some aggregate in clusters close to the neural tube and form the **spinal ganglia.** Others take longer pathways and assemble to form the **sympathetic ganglia** and the **adrenal medulla.** Other neural crest cells migrate into the muscle layers of the gut, forming the nerve net of the **myenteric plexus.**
2. The **dorsal route.** Cells traveling and spreading in the posterior region of the somites between the dermatome and the ectoderm differentiate preferentially into **pigment cells.**

From experiments to stimulate and guide the migration of cells in petri dishes it has been inferred that neural crest cells orient themselves in at least three ways:

1. By following paths labeled with components of the extracellular matrix (fibronectin, laminins, collagen IV, or hyaluronic acid).
2. By feeling out characteristic surface components (such as cell adhesion molecules) of the cells bordering the pathways.

The phenomenon of orienting along threads and cables composed of extracellular matrix and along other cells has been called **contact guidance** (Fig. 13–2; Box 6).

3. By sensing long-range chemoattractants. The existence of such long-range signals has been presumed but actual signal molecules have not been identified yet.

Perhaps the most fascinating aspect of the extensive migration, besides the question of how the cells are guided to their various destinations, concerns the programming of the specialized cell types that arise from the neural crest cells.

Figure 13–2   Contact guidance. A neuroblast in the central nervous system migrating from its birthplace close to the central canal of the spinal cord, or close to the ventricle in the brain, into the periphery along glia fibers. (Redrawn after P. Rakic, *J. Comp. Neurobiol.* 145:61–84, 1972.)

How can these roving, unspecialized cells be induced to settle down and differentiate? Two extreme hypotheses have been put forward. One holds that neural crest cells already have been educated and trained for their future job where they were born. Those who miss the target area predetermined by their programming will undergo suicide (apoptosis); those who find their destination will be multiplied through cell division. The second view maintains that neural crest cells are initially pluripotent and are not committed to a distinct job. Only when they cease migrating will they become committed to a distinct task.

Both of these hypotheses have evidence in their favor. For example, it is known that even at the places of origin not all neural crest cells are identical. Only neural crest cells of the head and neck region have the potential to form cartilage and bone. No head-specific skeletal elements are formed if the neural crest of the head region is removed experimentally or replaced by neural crest material of the trunk. On the other hand, when cells of the anterior neural crest, normally destined to form cartilage or bone, are shifted into a thoracic region, they give rise to spinal and sympathetic ganglia in accordance with their new position. Moreover, the careful tracking of many individually labeled cells has shown that in the trunk region neural crest cells are indeed pluripotent. Apparently, their developmental potential is narrowed gradually toward their eventual fates while they migrate toward their target area, and is determined definitively only after they have arrived at their final destination (for examples, see Chapter 14).

14

Development of the Nervous System: Cell Migration, Pathfinding, and Self-Organization

———⊰•⊱———

14.1 A Glance Back to the Foundation of the Nervous System

Recall that in vertebrate embryos, the central nervous system forms along the dorsal midline of the body axis. The formation of the nervous system begins during gastrulation, when the dorsal mesoderm (chordamesoderm) moves into contact with the overlying ectoderm. The dorsal mesoderm emits signals, presumably NOGGIN and CHORDIN protein, which induce the overlying ectoderm to adopt a neural fate. The induced neural plate forms neural folds, the folds form the neural tube, and the neural tube forms the brain and the spinal cord. Still missing are the spinal ganglia and the essential parts of the autonomous nervous system, in particular, the sympathetic system and the peripheral components of the parasympathetic system, such as the neuronal network of the gut. These parts derive from the **neural crest cells.**

14.2   The Basic Architecture of the Central Nervous System, and How the Main Sensory Organs of the Head Arise

It seems curious that most of our brain is initially a hollow cavity. In its anterior part the neural tube balloons into three main vesicles: the forebrain (**prosencephalon**), the midbrain (**mesencephalon**), and the hindbrain (**rhombencephalon**). In a second step the forebrain and the hindbrain are subdivided into two parts each and we see the classic five parts of the textbook brain (Fig. 14–1, 14–2):

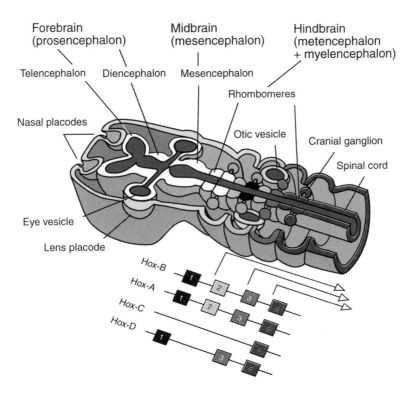

Figure 14–1   Development of the brain and adjacent sensory organs in a generalized mammal. The posterior part of the brain, the rhombencephalon, is subdivided into repeating units called **rhombomeres** or **neuromeres**. Cranial ganglia are associated with the even-numbered rhombomeres (r2 to r8). While the eye vesicle arises from the posterior forebrain (diencephalon) by evagination, the eye lens, the nasal epithelium, and the inner ear arise from placodes, thickenings of the epidermis that are invaginated. In addition, the expression pattern of some *Hox* genes is shown. This pattern helps to identify and demarcate rhombomeres.

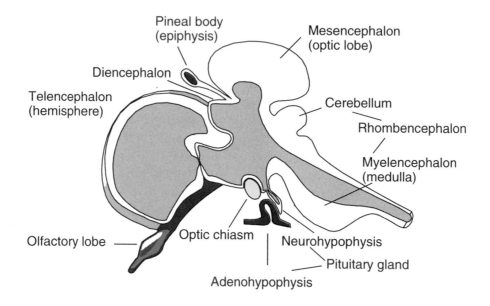

Figure 14–2 Advanced vertebrate brain. The pineal body (epiphysis) is formed as an evagination of the dorsal midbrain. The pituitary gland is composed of two different parts: (1) the neurohypophysis (posterior lobe) arises from an evagination of the ventral diencephalon (hypothalamus); and (2) the adenohypophysis arises from an evagination of the roof of the pharynx.

- **Prosencephalon:**
    1. Telencephalon (cerebral hemispheres, neocortex)
    2. Diencephalon
- **Mesencephalon:**
    3. Mesencephalon
- **Rhombencephalon:**
    4. Metencephalon (cerebellum)
    5. Myelencephalon (medulla oblongata)

In humans the following principal functions are assigned to these regions:

- **Telencephalon:** Smell, evaluation of visual and acoustic information (in lower vertebrates performed by the mesencephalon).
- **Diencephalon:**
    Thalamus: Relay center for optic and acoustic tracts; gateway for sensory input passing from the spinal cord to the cerebellar hemispheres.

Epithalamus and **hypothalamus:** Supreme commander of basic vegetative functions, control center of inner homeostasis, sleep, alertness. Connects the nervous and hormonal systems. To prepare for the latter task, the dorsal epithalamus forms the **pineal gland (epiphysis)** and the ventral hypothalamus extends the **infundibulum,** which will form the posterior lobe of the pituitary, the **neurohypophysis.**

The **pituitary gland** is completed by a structure that emerges not from the brain but from the roof of the pharyngeal cavity. By a local evagination, the roof forms **Rathke's pouch,** which transforms into the anterior lobe of the pituitary gland, the **adenohypophysis.**

- **Mesencephalon:** Forms in nonmammalian vertebrates the **tectum opticum** (optic lobe), the main center of visual data processing. In mammals it is merely a relay station for visual and acoustic reflexes.
- **Cerebellum:** Coordination of complex movements.
- **Myelencephalon:** Reflex and control center for vegetative functions.

The hindbrain is patterned initially into repetitive compartments, called **rhombomeres** (Fig. 14–1). These are connected with the **cranial ganglia,** which arise from neural crest cells. The first ganglia are attached to the even-numbered rhombomeres, whereas the odd-numbered rhombomeres are connected with ganglia only later (Fig. 14–1).

The brain is surrounded and supplemented by the main **sensory organs** needed for long-distance orientation and for complex behavioral control.

- The **eye** is essentially a derivative of the brain itself. At the border of the diencephalon and the mesencephalon, the **optic vesicles** bulge outwards and extend laterally. The lens of the eye is supplemented by the **lens placode,** an ectodermal thickening formed under the inductive action of neighboring cells, in particular, the eyeball. This thickening detaches and forms the lens, while the overlying ectoderm forms the translucent cornea.
- **Ectodermal placodes** that invaginate to contact the brain also give rise to two important sensory systems and to an important hormonal gland. The **olfactory epithelia** of the **nose** derive from paired **nasal placodes.** The **vestibular apparatus** of the **inner ear** includes the membranous **labyrinth** and the **cochlea,** and derives from the **otic placode.** On each side of the brain an otic placode invaginates, detaches, and forms an **otic vesicle** (Fig. 14–1). Each otic vesicle undergoes complex developmental transformations until they are suited to mediate the senses of angular and linear accelerations, the sense of equilibrium, and the sense of hearing.

14.3   The Peripheral Nervous System Is Formed from Emigrated Neural Crest Cells, whose Fate Is Determined by Their Path and Destination

The fate of the migrating neural crest cells is believed to be specified by the following: (1) the origin of the cells (lineage and place of origin), (2) influences acting on them while they are traveling, or (3) the conditions they encounter at the target area. Determination is accomplished by a series of alternative choices where their paths bifurcate as follows:

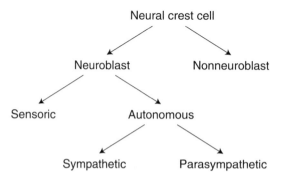

The ultimate determination of the type of neurotransmitter to be produced is only made at the destination. An experiment to address this issue took advantage of the fact that cervical (neck) neural crest cells normally give rise to parasympathetic, cholinergic nerve cells producing the transmitter acetylcholine, while thoracic (chest) cells give rise to adrenergic sympathetic ganglia producing epinephrine. If the cells are exchanged before they leave home, the original cervical cells colonize the chest and produce epinephrine. The original thoracic cells arrive in areas where parasympathetic innervation is wanted and produce acetylcholine. Such experiments suggest that the final fate and function of neural crest cells are place dependent.

14.4   Even in the Construction of the Central Nervous System, Extensive Cell Migration Occurs

When the neural tube grows to form the brain in the head region and the spinal cord in the trunk, the existing neuroblasts first must be multiplied. Mitoses take place close to the central spinal canal and along the borders of the brain ventricles where the supply of nutrients is best. However, these are not the sites where most neurons are required. The immature neurons move from their birthplace into the periphery. In the brain they construct the peripheral **cortex** and in the spinal cord they move to peripheral ventrolateral sites. It is the task of these cells to become **motor neurons** and to bundle their outgrowing axons to form the ventral roots of the spinal nerves. Neuroblasts move actively along glial fibers that stretch from the central canal or ventricles, respectively, to the periphery, and serve as guiding threads (Fig. 13–2).

14.5 Patterning Nerve Connections Is a Process of Epigenetic Self-Organization

Neuroblasts undergo terminal differentiation at their destination. One cell pole becomes the input region and forms dendrites for receiving information. The opposite pole is specified to become the output and forms an axon for transmitting information. Dendrites and axons must grow in the direction of their targets (such as sensory cells, other nerve cells, muscle cells, or other effector cells) and have to establish synaptic contacts—a task of unimaginable complexity!

The human brain grows during embryo development at a rate of 250,000 nerve cells per minute to reach a population of 1 trillion (10^{12}) nerve cells. Every nerve cell is connected synaptically with about 100 other nerve cells. It is inconceivable that the DNA encodes an exact plan for all these connections; the storage capacity of the entire genome is far too small.

The patterning of the neural connections is based on **self-organization,** the phenomenon where properties of higher-order structures emerge from the interaction of lower-order components. Thus, brain development depends upon the interaction of neurons with other neurons, with glia cells, and with supporting structures of the extracellular matrix. Genetic information enables the interacting cells to produce signal molecules, to construct receptors for picking up signals, and to install signal-transducing systems for initiating adequate responses.

14.6 The Growth Cone Is a Sensory Antenna Located on the Tips of Growing Dendrites and Axons

Growing dendrites and axons terminate in a peculiar structure that looks like a hand with many fingers: the **growth cone.** The fingers are filopodia

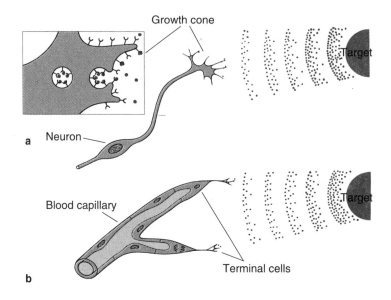

Figure 14–3 Guidance by diffusible substances of growing nerve fibers and blood capillaries. (a) The elongating axon of a sympathetic neuron finds its way toward the target by means of sensors located on its terminal growth cone. The target emits the chemoattractant NGF (nerve growth factor). (b) The terminal cell at the tip of a blood capillary is also equipped with filopodia and sensors to search for its target areas. The targets emit angiogenic factors.

enabling the cone to crawl actively over a substrate, to force its way through surrounding tissue, and to advance towards its destination. The filopodia are equipped with membrane-associated receptors serving as molecular antennae. The filopodia are extended and retracted, bent to the left and right, and thus explore the environment. They have the function of pathfinders (Fig. 14–3).

14.7 "Nerve Growth Factors" Can Guide Growth Cones Chemotactically and Serve as Survival Factors for Correctly Connected Nerve Cells

The so-called **nerve growth factor NGF** (discovered by Rita Levi-Montalcini 1987) was the first identified long-range signal molecule guiding growth cones to their distant target. This glycoprotein is emitted by target areas (Fig. 14–3) and acts on immature spinal and sympathetic neurons. However, it does not act as a mitogenic agent stimulating cell proliferation, as initially assumed and indicated by its name. Rather, NGF guides the direction of growth cone movement and thus determines the direction in which a neurite (dendrite or axon) is extended. The directional cue is provided by the concentration gradient of NGF. If a micropipette releasing NGF is placed close to a growth cone, its direction of advance can be deflected to the artificial source of NGF.

NGF binds to membrane-bound receptors on the growth cone. Bound NGF releases a cascade of signal transduction events (like those shown in Box 3) and subsequently is internalized by endocytosis. Internalized NGF is carried in vesicles by retrograde transport into the main body of the neuron, the perikaryon. By unknown means, internalized NGF ensures the **survival** of those neurons that succeeded in arriving at the correct destination and therefore are "fed" with NGF.

Several subtypes of NGF and NGF-like factors are known. They enable the survival of diverse subpopulations of nerve cells and are known collectively as **neurotrophins.** Simultaneously and confusingly, the known neurotrophins are neurotropic (*tropic* = guiding; *trophic* = nourishing). However, the term "trophic" is not very appropriate because neurotrophins are signal molecules effective in very minute concentrations and are not used for true nourishment.

14.8 Multiple Molecular Cues and Glues Mark the Route for Pioneer Fibers and Tie Up Fibers into Nerves

14.8.1 Substrate- and Cell-Bound Road Signs

Diffusible long-range signal molecules, such as NGF, cannot mediate precise long-distance guidance, if billions of neurites are to be connected with specific target cells.

Long-term observation of nerve cells growing in petri dishes on various substrates reveals a behavior similar to that known from migrating neural crest cells. Growth cones at the tips of elongating neuronal processes orient themselves using cues provided by the physical and chemical qualities of the extracellular matrix covering the bottom of the dish, and by sensing the molecular surface structure of neighboring cells. Growth cones show preference for substrates to which they can adhere firmly and avoid areas apparently displaying a repulsive quality.

Analysis of the behavior of neuronal cells in vitro reveals several sources of information that might also be used in vivo by neuronal processes to navigate in the direction of their target areas:

- Channels, folds, and grooves can physically favor or restrict the path of neurite growth.
- The path is coated by macromolecules that delineate the way past obstacles. For example, many of the roads upon which axons travel are paved with laminins.
- Some signposts encountered by growing axons have a repellent label.
- Some components of the extracellular matrix are more adhesive than others and bias axon movement. In addition, adhesive gradients may indicate the correct direction.

Of particular significance are **cell adhesion molecules (CAMs,** Box 6). Cell surface and adhesion molecules are highly diverse and suited to mediate precise **contact guidance.**

In living organisms the coherent growth of bundles of nervous fibers is facilitated by **pioneer fibers** sent out to create a track that later fibers can follow. In paving the way proteases that are secreted by the growth cones may be helpful. Such enzymes remove barriers and digest channels through the extracellular matrix.

14.8.2 From the Pioneer Fibers to a Cable

After the pioneer fiber has laid the guiding thread, the growth cones of the following fibers cling to it, or to other fibers that already adhere to it. However, only the growth cones move freely; the older parts of the fibers are immobile. This situation, in which many immobile fibers attach to one another, is used to weld together all of the fibers. CAMs serve as glue. One of the CAMs is known as **fascilin.** Fascicles of parallel bundles are known as **nerves.**

14.9 How Is the Ultimate Destination Recognized?

If one tracks the cable of the **opticus** connecting the eye with the brain (for simplicity, do this in fishes or amphibians), we see the cable fraying at its end. The bundled axons detach from each other, diverging over a broad area. The axons terminate at defined neurons in the **optic tectum** of the midbrain. In embryo development axons grow out from the **retina** and find their way into the brain along glial pioneer fibers (like that shown in Fig. 13–2). The pioneer fibers pilot the axons to the area of destination but not to the particular spot within this area where they will find their target cells. On reaching the tectum, the retinal axons must choose which particular region they have to seek, and they must find their ultimate destination according to their provenance in the retina, so that each part of the retina will be connected with a distinct corresponding part of the tectum. The term for this correspondence is "**retino-tectal projection**" (Fig. 14–4, 14–5).

To analyze the behavior of retinal neurons, retinal precursor cells can now be placed on chips that have been covered with stripes of small paving stones. The paving stones are membrane vesicles prepared from various areas of the optic tectum; likewise, the neuronal precursor cells are taken from defined parts of the retina. It has been shown that membranes of the target cells are attractive, while membranes of nontarget cells can be repulsive. However, how can one particular axon find one particular small area in the tectum?

The **theory of neuronal specificity** (also called the **theory of chemoaffinity** or the **theory of neurotropism**) assumes that the target cells are marked by specific surface molecules. The tectum would contain a map that can be read and interpreted by the growth cones. However, because there

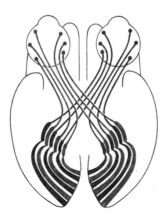

Figure 14–4   Retinotectal projection in humans. The major visual pathways bring together the input from the left and right eye. All the information supplied by the left halves of both eyes is brought together in the left side of the brain, and information supplied by the right halves of the retinas is relayed to the right side of the brain. The pathways (interrupted by synapses in the lateral geniculate nuclei) terminate in the visual cortex, where information arrives in alternating stripes of cortical neurons, called ocular dominance columns. One column is assigned to information coming from the left eye, the next is devoted to information arriving from the corresponding point of the right eye, and so forth. In the end, each spot of the visual field is seen by both eyes, and the two corresponding streams of information arrive in neighboring collectives of cortical neurons.

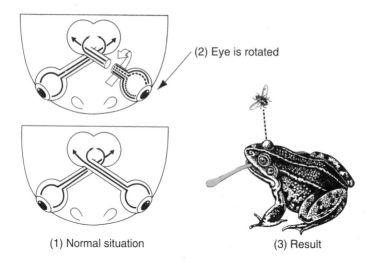

(2) Eye is rotated

(1) Normal situation   (3) Result

Figure 14–5   Experimentally rotated eyes in the frog. Eyes are removed, rotated in their sockets, and reinserted. The optic nerves regenerate back into the brain and reestablish connection to the optic tectum. If the eyes are rotated by 180°, the frog behaves as though it sees the world upside down. The frog projects its tongue downward even though the fly is above its head.

are billions of nerve cells, not every cell can have its own unique molecular identity. Instead, combinatorial cues are used to label territories, subterritories, and ultimate goals. Such cues could guide the growth cones to their ultimate destination step by step and thus reduce the necessity for long searches.

Much of the required molecular diversity can be coded—in addition to the amino acid sequence of proteins—in the combinations of carbohydrate units of glycoproteins and glycolipids (sialoglycolipids, gangliosides). There is evidence that nature indeed uses the intricacy and diversity of such surface molecules to label paths and targets. However, positional information coded on the surface of the cells is probably not very precise. Subsequent corrections are possible (see Section 14.11).

In mammals axons growing to the brain first reach the optic chiasm. Here they must choose a direction in which to continue growing. The fibers coming from the nasal half of the retina must cross to the opposite side and join the contralateral optic tract, and the fibers arriving from the temporal side must continue straight and join the ipsilateral tract (Fig. 14–5). If all of these decisions are made correctly, the axons are allowed to come to rest in the **lateral geniculate bodies.** From here, other neurons have to find the way to the

visual cortex. However, the problem remains: how to find the right way and recognize the ultimate target?

14.10   Interrupted Neurons Can Find Their Targets Again

A famous experiment conducted with frogs explored the issue of neuron connections and how the brain interprets action potentials arriving from the eyes. The optic nerve was cut, the eyeball rotated by an angle of, say, 180°, and reimplanted. Because neurons that have only lost the distal part of their axons can regenerate, the retinal axons grow back to the optic tectum. However, they make the connections appropriate to their original position, not to their new position. Thus, the retinal neurons reconnect with that part of the tectum to which they had been linked before the eye was rotated. Because

- The original connections are reestablished,
- The brain principally evaluates visual information according to its origin in the retina, and
- The eye has been rotated without "instructing" the brain of the altered situation,

the frog "sees" all things upside down. To demonstrate this result, the frog sees a fly flying above it as being instead in a position below it, and snaps downward (Fig. 14–4).

In humans a much simpler experiment can be performed: wear glasses made of prisms that show the world upside down. The high capacity of the human brain for learning by experience enables us to mentally "rotate" the seen world back into the original position, but learning to do so takes weeks.

14.11   Extra and Incorrect Neuronal Connections Are Eliminated

In young animals the retinotectal projections initially are quite fuzzy and imprecise. In amphibians, fishes, and birds retinal axons first branch widely and form synapses even in territories destined for neighboring axons. The territories innervated by the axons overlap initially. Later, the territories are trimmed back by elimination of superfluous and wrong connections.

The phenomenon of subsequent corrections can be studied more easily in the innervation of the skeletal muscles by the motor neurons of the spinal cord. It is common during development for too many axons to arrive at a muscle fiber and form synapses (Fig. 14–6). In the embryo the central nervous system sends action potentials to test the muscles. Axons and axonal branches

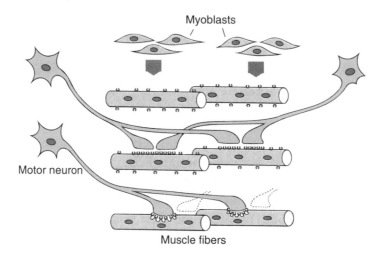

Figure 14–6 Correction of synaptic connections. Myoblasts fuse to form multinucleated, cross-striated muscle fibers. Initially, the fibers are innervated by several motor neurons. Incorrect connections are eliminated by a process of competition. Infrequently used synapses lose out and die; only those synapses survive that are correct and therefore are frequently used.

only survive if they are connected correctly and therefore are used frequently and regularly to transmit action potentials. In the struggle for survival, competition between synapses might be significant: circumstantial evidence suggests that transmitters released by synapses suppress unused synapses nearby. On the other hand, muscle fibers appear to secrete **survival factors** at the synapse. By analogy to the "trophic" function of NGF (Section 14.7), survival factors secreted by active muscle fibers might be taken up by the presynaptic membrane of the motor neuron and delivered to the cell body by retrograde transport. Only correctly connected neurons survive; neurons that fail to establish or maintain contacts are eliminated by programmed cell death (Chapter 11, Section 11.4).

14.12 Even after Birth New Links Are Established and the Pattern of Connection Is Molded by Experience

No bird, mammal, or human is born with a brain whose connections are finished completely in every detail and fixed permanently. On the contrary, in the cortical layer of the mammalian brain most synaptic connections are established only after birth under the influence of visual input and tactile experience. Basically, the main tracts are established. When a kitten is born, axons from the retina have reached the geniculate bodies and axons from the geniculate neurons have reached the visual cortex. The cortical areas of either half of the brain receive inputs from both eyes. However, the cortical architecture itself is still under construction and completed only under the influence of visual input.

The course of the optical tracts and the experiential changes in the cortical input areas after birth can be observed by an astounding method: dyes or otherwise-labeled low molecular weight substances are simply injected into the eyeball. The labels are taken up by the retinal neurons and transported into the brain. Surprisingly, the labels even cross synapses in the geniculate bodies and finally arrive in the visual cortex.

In the visual cortex of humans and several other mammals the axonal imputs coming from the nasal part of the left retina alternate with the inputs arriving from the nasal part of the right eye, and vice versa (Fig. 14–3). Thus, inputs that will convey information about the same part of the seen world are brought in close juxtaposition. At birth, the various inputs must be fed into networks of neurons to evaluate the information. It is only after birth that most of the neurons that fulfill this task are generated and organized into networks. Initially, adjacent inputs project into overlapping cortical areas. During the first few weeks after birth, the areas allocated to the left and right eye become sharpened and largely segregated from each other. In some mammals, **ocular dominance columns** can be identified in the cortex that are allocated alternately to the left and right eye.

Among the most intriguing experiments ever done in developmental neurobiology were those carried out by Hubel and Wiesel (1963). When the eyelids of the right eyes of newborn kittens were sewn shut and left closed for 3 months, the right eyes of the kittens became functionally blind. The corresponding ocular dominance columns were reduced for the benefit of those allocated to the open left eyes. Nervous connections are not directly encoded by the genome, nor are all connections the result of autonomous and independent processes of self-organization. Experience can lead to the strengthening and consolidation of active connections, and to the elimination of silent synapses.

A popular current hypothesis assigns **imprinting** in birds to similar mechanisms. During the short postnatal periods when key optic or acoustic stimuli are learned for long time periods and even for life, synapses are thought to be newly established or consolidated, while others are eliminated.

14.13 Long-Term Memory May Be Based on Continuing Neuronal Differentiation

In birds and mammals existing neurons send out and withdraw neurites for life; regular and frequent use strengthens and consolidates existing synapses. The refinement, retuning, and remodeling of synaptic connections are thought to be related to associative learning (including imprinting) and to long-term memory.

15

Heart and Blood Vessels: Divergent Developmental Roads but One System in the End

The heart arises from migrating cells that convene to form two parallel tubelike heart rudiments which eventually fuse. Blood capillaries arise independently, far away from the heart. Yet from these divergent and seemingly chaotic origins, a common vascular system emerges. In fact, the vascular system faithfully repeats evolutionarily ancient structures in a highly conservative manner.

15.1 Aristotle's Jumping Point: The Heart

Aristotle, when opening incubated eggs, observed a "jumping point" in the blastodisc. Even without magnifying glasses he identified the jumping point as a beating heart.

The avian embryo lends itself well to studies of heart development. The events described as follows occur in a similar way in other vertebrates as well.

The circulatory system is established early in development during neurulation when diffusion is no longer sufficient to supply the embryo with nutrients and oxygen. It is the first functional system and the heart is the first functional organ.

Mesodermal cells, having entered the interior of the embryo through the blastopore or primitive groove, crawl like amoebae anteriorly along the endoderm and assemble on either side of the developing foregut forming two separate aggregates. Both aggregates organize themselves to form a double-layered length of tube. The inner wall of the tube becomes the **endocard,** the outer wall becomes the muscular **myocard.** As the endoderm acquires the form of the foregut canal, the two cardiac tubes come in direct contact with each other. The two tubes fuse, and rhythmic contractions and expansions initiate the pumping activity that continues throughout life. In spite of the uninterrupted beating, the heart tube is folded and subdivided into separate functional units (**sinus venosus, atrium, ventricle, truncus arteriosus,** and **aorta**). This simple heart, consisting of one pump, satisfies the needs of a fish all its life. The embryos of the land vertebrates also require only a single pump, because a separate circulatory system enclosing the lung is not yet necessary. Even before birth, however, land vertebrates—especially birds and mammals—must be prepared for a life in air.

Without missing a beat, the mammalian heart is folded and reconstructed to yield two functional pumps with two chambers each; one pump to propel the volume of blood through the body (large circulation), the other pump in anticipation of the needs after birth when the lungs must function. The process by which the folded tube is transformed into the final heart is described in more extensive embryology textbooks (see Bibliography).

15.2   How Do Blood Vessels Grow and Find Their Target?

15.2.1 Vasculogenesis, the Formation of Large Vessels

The large blood vessels originate as elongations of the heart tube. However, while the vessels emanating from the heart grow and branch, additional blood vessels are being formed independently in many locations. In the extraembryonic mesoderm that covers the yolk sac and allantois, double-walled tubes like the heart tube arise. However, these are thinner and the outer wall forms smooth muscle fibers rather than cross-striated fibers, as in the heart. With time, the growing and branching vessels encounter each other and fuse. How is the overall vascular system designed and how is it all regulated so that in the end a species-specific circulatory system with a huge number of identifiable vessels results?

15.2.2 Angiogenesis, the Formation of the Capillaries

The development of the capillary network is extraordinary. Capillaries are generated independently from the heart and the large blood vessels at many sites, known as **blood islands.** The capillaries extend, branch, and find their way into their target areas that require blood. Arterioles find their way to venoles; both find the correct pathway to the large vessels.

Expansion of the network takes place from endothelial cells of existing capillaries. Endothelial cells retain the characteristics of stem cells and remain capable of dividing. The expansion of the vascular system must be adjusted to the needs of a growing organism. Target tissues without sufficient blood supply emit **angiogenic factors** that stimulate the proliferation of endothelial cells and guide the advancing capillaries toward the respective target area (Fig. 14–3).

There are astonishing parallels in the ways by which targets are sought out in the vascular and nervous systems. Moving **terminal cells** are located at the tips of capillaries. They are equipped with sensory filopodia and fulfill the function of pathfinders (Fig. 14–3). Further terminal cells arising along the flanks of extending capillaries give rise to **branches.** The angiogenic factors emitted from the targets promote the generation of branches. Thus, a wide area can be provided with blood. Among the known angiogenic factors are **prostaglandins** and the **basic fibroblast growth factor (bFGF)**, also known as **mesoderm-inducing factor** in the blastula stage embryo.

15.3   Before and After Birth: How to Adapt to Changing Conditions

The circulatory system of the embryo is not constructed in such a way as to meet the demands of an independent animal. The embryonic circulatory system is adapted to the needs of the embryo, but also displays "unnecessary" structural reminders of the evolutionary past. For instance, the four to six branchial arches of the embryo in terrestrial vertebrates are considered to be such unneeded echoes, as they are no longer required to supply gills and must be modified substantially (Fig. 4–4, 4–6). It is possible, of course, that the arches have retained a function. However, if they have, their role—presumably in the organization of the development in the visceral head and neck region—must still be defined.

The reconstruction of the embryo's circulatory system is performed in preparation for the moment when the embryo or fetus suddenly has to switch to air breathing. The embryo, whether reptile, bird, or mammal, receives both nutrients and oxygen via the umbilical vein, and in exchange the embryo clears off carbon dioxide through the umbilical arteries. In birds the location of gas exchange is the allantois and in mammals it is the placenta (Fig. 3–50).

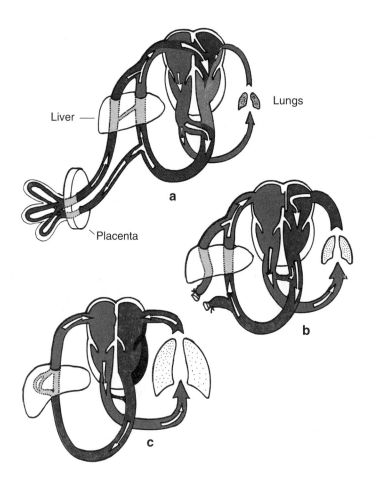

Figure 15–1   Switch in the blood circulation from a monocircular to a double-circular system in birth. (a) In the unborn fetus the lung circulation is practically nonfunctional; the wall between the two atria is perforated and the heart functions as though only one atrium and one ventricle are present; blood is pumped to the placenta, which has the function of a gill, and flows back to the heart. The body, represented by the liver, receives only partially oxygenated blood. (b, c) After birth, pulmonary circulation is established and connected crosswise with the greater circulation.

The umbilical vein leads into the right heart, where blood is pumped to the lungs after birth. However, the lungs are still collapsed. It would not be useful, and because of hydrodynamic resistance it would be almost impossible, to propel the entire blood supply through the nonfunctional, collapsed lungs. Because the embryo needs to direct the oxygen-rich blood from the umbilical vein toward the brain and body, the bloodstream must be guided from the right into the left heart. The passage between the two atria is known as the **foramen ovale.** (Fig. 15–1a)

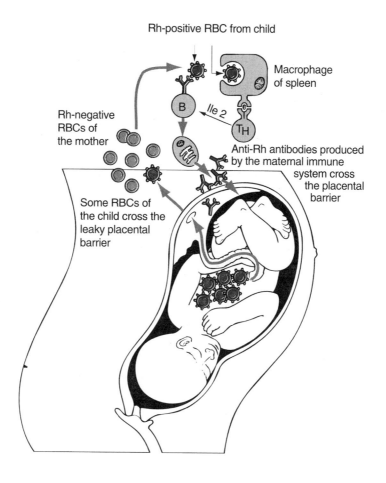

Figure 15–2   Rhesus factor. Some Rh-positive erythrocytes of the child pass the placental barrier and enter the maternal blood. The red blood cells (RBCs) of the mother are Rh-negative. Therefore, the Rh-positive erythrocytes of the child stimulate the immune system of the mother. The maternal anti-Rh antibodies infiltrate the blood of the child and induce the destruction of the fetal erythrocytes.

After birth, the lungs must be inflated immediately, the blood's path to the lungs opened, and the passage between the two hearts closed (Fig. 15–1b, c). The first breath is a perilous event.

15.4   The Placenta: Closeness to the Mother Can Be Dangerous

The embryo itself develops the placenta, enlarging and improving the chorionic villi in the part of the trophoblast that comes into intimate contact with the uterine wall. The contact is made even more intimate by fingerlike outgrowths of the villi that incorporate ramifications of the umbilical blood vessels and penetrate into depressions in the wall of the uterus. The details of placenta formation, and the degree of connection between fetal and maternal tissues, are rather different among the mammals. The mother assists her growing and demanding child by removing barriers between her blood and the child's blood. The child demands nutrients and oxygen, and tries to get rid of carbon dioxide and urea. For those purposes, a complete destruction of barriers would be helpful. On the other hand, with the gradual removal of barriers, the child comes increasingly into contact with the immune system of the mother.

In humans almost all barriers are demolished and the degree of connection between the fetal and the maternal circulatory systems becomes unusually intimate. On the mother's side all tissue layers are removed, including even the endothelia of the blood vessels. The placental villi of the fetus dip directly and deeply into the lacunae of maternal blood (Fig. 3–50). How the child evades the maternal immune system is not known. In principle the fetus is foreign tissue as it carries not only maternal but also paternal genes. Therefore, the immune system would be expected to destroy the child. In fact, sometimes fatal incompatibilities do arise as exemplified by the serious complications associated with the rhesus factor (Rh) (Fig. 15–2).

16

Stem Cells Enable Continuous Growth and Renewal

As a rule, cells, tissues, and organs generated in early embryogenesis must be enlarged in the course of larval or juvenile growth. However, even after obvious growth ceases, processes of proliferation persist in most organisms. For example, many cells live for only a short time and must be replaced repeatedly by new ones.

Occasionally, even fully differentiated cells, such as the **hepatocytes** in our liver, retain the ability to divide. Normally, hepatocytes are long-lived and divide slowly. However, in response to injury or poisoning they divide rapidly to replace damaged neighbors.

Most kinds of cells lose their mitotic potential in the course of terminal differentiation. Some even lose their nucleus, such as the keratinocytes of our skin or the erythrocytes of our blood. In all such cases a population of cells must be present that remains able to divide and to terminate differentiation only after being multiplied. Such cells are called **stem cells.** When stem cells retain their "embryonic" characteristics permanently, and can even be multiplied without limit in culture, they are designated **immortal stem cells.**

16.1 The Principle of the Stem Cell Was Invented Early in Evolution

Even sponges, considered to be the most primitive metazoans, have multipotent stem cells called **archaeocytes.** Hydrozoans have pluripotent **I–cells** (**interstitial cells**) from which primordial germ cells, stinging cells, and even nerve cells can be recruited throughout the life of the organism—which, in *Hydra,* may last for centuries (Chapter 3, Section 3.3; Fig. 3–6). Turbellarians have pluripotent **neoblasts.** In contrast, in nematodes and arthropods no stem cells appear to persist after the last molt, apart from the primordial germ cells and possibly some hemolymph cells.

16.2 Unipotent Stem Cells: Renewal of the Skin and the Muscle

During fetal and juvenile stages, skin must be expanded; throughout the whole life, skin must be renewed. The skin is exposed to stress caused by mechanical strain, chemical irritation, UV irradiation, and microbial attack. Unceasing growth and renewal are accomplished through the mitotic activity of unipotent stem cells. These lie upon the basal lamina; their descendants are displaced outwards, and on their outward journey they differentiate to keratinocytes, die, and finally are sloughed off after 2 to 4 weeks (Fig. 16–1). Stem cells are able to continue dividing in contact with laminin in the basal lamina. Stimulating growth factors, such as KGF (keratinocyte growth factor), EGF (epidermal growth factor), and TGFb (transforming growth factor beta), and inhibitory signal substances, such as the stress hormone **epinephrine,** participate in the control of skin proliferation.

Skeletal muscles are renewed only in the event of particular need, as when the muscle is injured. Residual myoblasts, called **satellite cells,** are left over from embryo development attached to bundles of muscle fibers. If needed, the quiescent satellite myoblasts are reactivated and multiplied. They fuse to myotubes and stop dividing. The myotubes fill their interior with bundles of actin and myosin, enlarge, and take over the function of damaged muscle fibers.

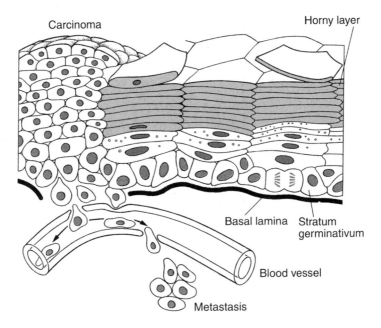

Figure 16–1   Stem cells and carcinogenesis in the skin. The stem cells of the stratum germinativum produce descendants that differentiate into keratinocytes and are eventually sloughed off. In carcinomas the differentiation is incomplete and cells derived from the stem cells continue to divide, thus forming a tumor. Transformed cells may dissociate from the tumor and enter the blood-stream to establish colonies (metastases) in other places.

16.3   Pluripotent Stem Cells: Renewal of the Gut

Numerous protuberances in the small intestine, called **villi,** are composed of cells that move to the top of the villi, where they eventually die. During their displacement, they serve the function of absorptive brush-border cells or mucus-secreting goblet cells. When they arrive at the top of the villi, the cells sacrifice themselves to set free digestive enzymes; the residuals are shed into the lumen of the gut. New cells are supplied by stem cells that lie in a protected position in the depths of crypts that descend into the intestinal wall in between the villi (Fig. 16–2). Because villi consist of several cell types, their renewal depends on pluripotent stem cells.

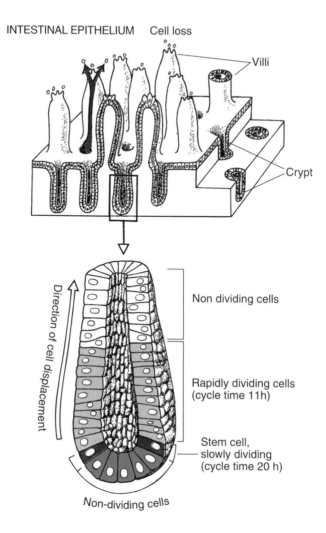

Figure 16–2   Renewal of the villi from stem cells in the small intestine. The multipotent stem cells are found not in the villi but at the bottom of adjacent crypts. The descendants of the stem cells are displaced in a distal direction, enter the villi, and eventually are sloughed off at the tip of the villi. Sloughed off cells release digestive enzymes.

16.4   Pluripotent Stem Cells: Formation and Renewal of Blood Cells

In humans, 6 million erythroblasts are born to replace 6 million dying red blood cells each second. The bloodstream carries many cell types with different functions ranging from the storage of oxygen to the combat of infections and the production of antibodies. An astonishing finding is that all these diverse cells are the offspring of a single type of stem cell. The production and differentiation of blood cells is known as **hematopoiesis** (Greek: *haimat* = blood, *poiein* = to make).

16.4.1 Origin and Significance of Stem Cells

As previously introduced, production of blood cells takes place during embryo development in mesodermal **blood islands,** where the first blood vessels are also formed from angiogenic founder cells. The first blood-forming islands are found in the chick outside the blastodisc in the extraembryonic mesoderm covering the yellow yolk ball, and in humans in a corresponding position—the mesodermal layer covering the yolk sac (Fig. 3–48). Later, the stem cells migrate into the liver, the spleen, and possibly the thymus. Ultimately, they populate the bone marrow. In the postnatal mammal significant numbers of **hematopoietic stem** cells are found only in the bone marrow. A surprisingly small number of pluripotent stem cells (1 per 10,000 cells in the bone marrow) provide all of the blood cells.

Modern hematopoiesis research began with the search for a means to protect against the disastrous effects of whole body X-irradiation. Mice, whose blood-forming system had been destroyed by high doses of X rays, could be rescued by injection of suspensions of bone marrow cells into the circulating blood. However, stem cells only represent 0.01% of the total mass of bone marrow, and much effort had to be invested to isolate and identify stem cells. Monoclonal antibodies were key tools, because they could be selected for their ability to recognize cell-type- or lineage-specific surface antigens.

Without bone marrow stem cells we would soon die because the life span of blood cells is restricted to days or weeks. In humans the half-life of erythrocytes is 120 days. From this value, a turnover of 6 million erythrocytes per second has been estimated. Aged and dying blood cells are sorted out in the lymphatic organs, notably the spleen, and are removed by macrophages through phagocytosis. The process of sorting out is called **sequestration.**

Most leukocytes are also short-lived. Among the immune cells, only "memory cells" (B-memory cells and T-memory cells) are long-lived.

16.4.2 Stem Cells, Blasts, and Terminal Differentiation

The rare pluripotent stem cells of the bone marrow are endowed with the capability of dividing unceasingly. Part of their offspring remain as pluripotent stem cells (**self-renewal**), while the remainder becomes committed to be

the progenitors of more specialized cells with restricted developmental potential. The committed progenitor cells, designated by the suffix -blast, are still
able to divide, but the number of rounds of divisions they can undergo is
restricted. The limited reproduction of the progenitor cells (**amplification
divisions**) serves to regulate the number of specialized blood cells. After their
last division, the cells undergo terminal differentiation.

16.5 In Mammals a Decision Tree Leads to Eight Categories of Blood Cells

The **pedigrees** of blood cells can be traced using engineered retroviral vectors carrying a selectable marker gene (neo^R). The marker virus, like other
retroviruses, inserts its own DNA into the genome of the host cells and is
multiplied during cell divisions, together with the host genome. The virus was
engineered to remove genes that would enable it to spread from cell to cell.
Bone marrow cells were first removed and infected in vitro with the retroviral
vector, then they were transferred into lethally X-irradiated mice. Offspring
of the retrovirus-marked donor cells were found predominantly in the spleen,
but also in the bone marrow. Clusters of blood progenitor cells in the spleen
are known as **colony-forming units (CFU)**. Such units were derived, statisti-
cally, from one founder cell, but could give rise to one or several types of
blood cells, depending on the position of the founder cell in the treelike pedigree.

Blood progenitor cells can now be cultured in vitro, where various factors
can be added, and single cells and their offspring can be monitored. A few cells
behave like pluripotent stem cells, forming colonies of various types of granulocytes, macrophages, and other cell types classified as follows as the "myeloid
lineage." Other isolated and seeded cells reveal more restricted potencies, generating only neutrophil granulocytes and macrophages. Still other seeded cells
give rise to a single cell type, such as neutrophil granulocytes or macrophages.
From such studies, the whole pedigree has been reconstructed (Fig. 16–3).

Starting from **one basic multipotent stem cell**, at least eight types of
blood cells are generated (Fig. 16–3).

The **myeloid lineage** supplies the following:

1. **Erythrocytes**, which carry hemoglobin and, in mammals, lose their
   nucleus;
2. **Neutrophil granulocytes;** and

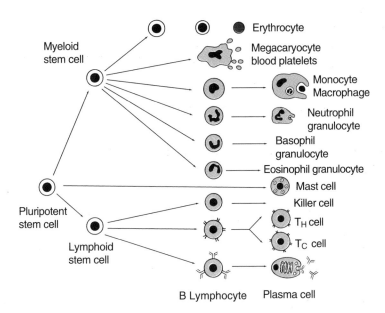

Figure 16–3 Pedigree of the blood cells. **Hematopoiesis,** the formation of blood cells, starts in the bone marrow from multipotent stem cells.

3. **Macrophages.** Together, neutrophil granulocytes and macrophages devour a large variety of infectious invaders and much waste produced in the body. **Monocytes** circulating in the blood are considered to be modifications of the macrophages found in the interstitial spaces of tissues and in the lymphatic organs (lymph nodes, tonsils, and spleen). In the spleen the macrophages eat infectious bacteria as well as aged erythrocytes.

4. **Basophil granulocytes,** which emit alarm substances such as histamine, leukotrienes, and prostaglandins during inflammatory responses.

5. **Eosinophil granulocytes,** which also fulfill tasks in inflammatory and allergic responses.

6. **Megacaryocytes.** Unlike the other blood cells, these "large nucleus cells" remain in the bone marrow, become very large and polyploid, and bud off numerous **platelets** into adjacent blood capillaries.

The **lymphoid lineage** supplies the following:

7. **B lymphocytes.** In the form of plasma cells, the B cells produce and release antibodies.

8. **T lymphocytes** are subdivided into several subtypes (T4 = T-helper cells, T5 = T-suppressor cells, T8 = cytotoxic T cells), and together

constitute a system to discriminate between self and nonself, and to adjust the mechanisms of defense to the current needs.

9. **Natural killer cells** presumably also derive from the lymphoid lineage and fulfill some tasks in the immune response.

The various nonerythrocytes are often summarized as leukocytes (white blood cells).

16.6   Blood Cell Manufacturing Is Controlled by Many Cytokines and Hormones

Many soluble, extracellular polypeptides influence the types and the numbers of blood cells manufactured. Many of these proteins have been isolated, identified, and their genes cloned. A term designating such factors in general is "**cytokine.**" An example of a cytokine is as follows:

• **Stem cell factor (SCF),** which stimulates the stem cells of the myeloid pedigree to proliferate. Other cytokines do not manure the root of the tree but stimulate branching and growth of special branches. Such secondary cytokines are the following:
• **GM-CSF,** the **G**ranulocyte and **M**acrophage **C**olony-**S**timulating **F**actor
• **G-CSF,** the **G**ranulocyte-**S**timulating **F**actor
• **M-CSF,** the **M**acrophage-**S**timulating **F**actor

Cytokines that stimulate progenitors of lymphocytes are known as **interleukins.** The name indicates that sources of such cytokines are other leukocytes. Interleukins are messengers that are exchanged between leukocytes, for instance between lymphocytes and macrophages, or between B and T lymphocytes.

Probably the best-known cytokine is the hormone **erythropoietin** (Greek: "making red blood cells"). Erythropoietin is a glycoprotein that is synthesized in the kidney under oxygen-deficient conditions and is transported by the bloodstream. It stimulates the proerythroblasts in the bone marrow to divide more rapidly. In this way the manufacturing of red blood cells is increased and the oxygen-binding capacity of the blood is improved.

Such factors, of which a considerable number have been identified, are often designated as **hormones** if they have been found in the blood, following the tradition of using the term "hormone" to designate substances active in trace amounts and transported to their site of activity by the blood.

Box 6   How Cells Communicate and Interact   271

─────────────── Box 6 ───────────────

HOW CELLS COMMUNICATE AND INTERACT

Complex processes, such as spatially and temporally ordered cell differentiation, directed cell movement, and shaping of structures composed of several cell types, all require cell-to-cell communication and coordinated cell interaction. Many molecules are used to convey signals.

Ions and small polar (hydrophilic) molecules, such as Ca^{2+}, IP_3 (inositol triphosphate), cAMP, and cGMP, are able to diffuse from cell to cell if the cell membranes of adjacent cells are joined through **gap junctions** (tubular transmembrane canals).

Small molecules with low polarity may directly diffuse across cell membranes. This applies to molecules such as CO_2 and NH_3, which are not considered generally to serve in communication, but also to several well-established signal molecules. For example, the function of a transcellular signal molecule is assigned to **arachidonic acid,** produced in extended signal transduction systems by cleavage of diacylglycerol or phosphatidylcholine (see Box 3). Many metabolites of arachidonic acid, such as prostaglandins, leukotrienes, and hydroxy fatty acids (collectively referred to as **eicosanoids**), are liberated into the extracellular spaces and may have the function of specific, locally acting signal molecules. From the extracellular space, such molecules may infiltrate and cross the membrane of neighboring cells, or they may be picked up by neighboring cells with membrane-associated receptors.

The significance of low molecular weight, nonpeptide molecules for embryogenesis is largely unexplored, because such molecules are frequently very unstable and are extremely difficult to trace.

Larger, apolar (lipophilic) molecules, such as retinoic acid, thyroxine, and steroid hormones (Box 5), are commonly said to freely cross cell membranes and to bind receptors only within the cytoplasm. However, while their lipophilic nature enables them to infiltrate membranes, it impedes their detachment from the membrane and spread into the aqueous cytosol. Therefore, the cytoplasmic receptors probably collect lipophilic signal molecules from the cell membrane.

STATIONARY SIGNAL MOLECULES

The counterparts of diffusible signal molecules are glycoproteins that act as signal molecules but remain anchored in the membrane of the signal-presenting cell. Only adjacent target cells can sense the presence of such a signal exposed on the cell surface. Glycoproteins that enable the mutual recognition of neighboring cells typically have the additional function of glue, mediating the physical coherence of the cells. They are thus called **CAM** (Cell Adhesion Molecules, Chapter 12). CAMs bind to other CAMs of the same type (**homotypic** or **homophilic binding**) or of a different type (**heterotypic** or **heterophilic binding**).

Occasionally, neighboring cells interact via membrane-associated enzymes. For example, membrane-anchored glycosyltransferases may transfer monosaccharides present in the extracellular fluid onto acceptor molecules in the membrane of adja-

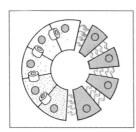
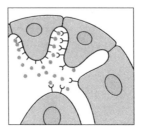
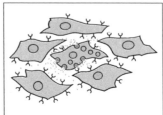
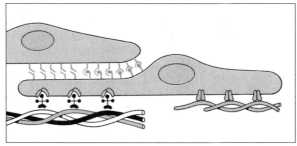
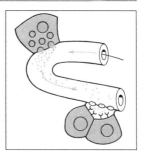

cent cells. The addition of a sugar moiety may result in a conformational change in the acceptor molecule, in turn activating a signal transduction pathway.

Sometimes, membrane-anchored polypeptides are cleaved off by enzymes exposed on the exterior surface of neighboring cells. The peptide cleaved from the precursor becomes a diffusible, extracellular signal molecule. For example, the **epidermal growth factor (EGF)** is split off from a membrane-bound precursor polypeptide (Chapter 16, Section 16.2, and Chapter 17, Section 17.1).

SIGNAL MOLECULES SPREADING IN INTERCELLULAR SPACES

Diverse molecules with signaling functions are released into the extracellular fluid that fills the spaces between cells. Depending on their biological function, the location of their release, or tradition, such substances are called **morphogens, inducers, growth factors, differentiation factors, tissue hormones** or **paracrine hormones, modulators, mediators, elicitors,** or **transmitters** (Chapter 17). The distances of their spread range from a few nanometers to several millimeters. Some of these factors may also be chemoattractants, directing the movement or growth of cells. An example is the **nerve growth factor NGF,** which guides the growth of axons and dendritic fibers in sympathetic neurons (Chapter 14).

HORMONES

If signal molecules are delivered into blood or lymph vessels, they have the potential to be distributed throughout the body and thus to coordinate developmental processes over large distances in diverse and remote parts of the body. Such hormonal signal molecules are well suited to coordinate extensive developmental processes, such as metamorphosis and sexual development (Chapters 19 and 20), but cannot mediate positional information or act as local inducers.

Box 6   How Cells Communicate and Interact   273

SIGNALS MEDIATED BY THE EXTRACELLULAR MATRIX

Many macromolecular substances are released into extracellular spaces, to fill them, and to confer certain physical properties to the tissue. The proteins, glycoproteins, and proteoglycans that constitute the **extracellular matrix (ECM)** serve as filling and hardening materials, but also can serve as signposts and guides, directing the movement of wandering cells, and the growth of nerve fibers and blood capillaries (Chapters 13, 14, and 15). Fibronectin, laminin, and hyaluronic acid are known components of the ECM having signal function.

17

Signal Molecules Control
Development and Growth

17.1 Categories of Signal Systems

Every cell of a multicellular animal is programmed during development to respond to a set of signals that act in various combinations to decide which of many possible pathways a cell will enter, whether it will proliferate or stay quiescent, and whether a cell will continue to live or die.

The wide range of controlling signal systems already introduced in this book—including those that mediate induction processes in the amphibian embryo, modify pattern formation in the avian limb, foster branching and connection in the nervous or vascular systems, or stimulate the supply of blood cells—all make it clear that in the construction of tissues and organs an impressive array of molecules controls the various events of development (overview: Box 6).

Because biological signal molecules are effective in extremely minute amounts and are present only in trace quantities, their characterization is difficult. The number of identified molecules is growing slowly but steadily.

Rich sources of soluble signal molecules are cells that can be grown in cell culture and release "growth factors" into the medium (called "**conditioned media**"). Most signal molecules identified to date are polypeptides consisting of 100 to 500 amino acids. Their isolation, identification, and amplification has been facilitated greatly by the tools of the molecular biologist (Box 7). In addition, with new biophysical techniques, such as nuclear magnetic resonance spectroscopy (NMR) of fragments obtained after atomic bombardment of target molecules, a large variety of peptides and lipids has been identified.

When the amino acid sequence of polypeptides is analyzed, and the sequence is compared with those of known polypeptides stored in diverse databases, partial or complete sequence identity between established and the new sequence is observed frequently. A common observation from such comparisons is that the same factor, or closely related members of a family of polypeptides, can be used by nature for very different purposes.

For instance, **insulin**—known as the hormone regulating the level of glucose in the blood—and the **insulin-like growth factor** (**IGF**) are both found in the embryo even before the islets of Langerhans are formed in the pancreas (see Section 17.4). Many factors isolated from adult organisms may participate in the control of organ development.

At least three classification systems have been developed to classify biological signal molecules. They are based on biological function, mode and range of signaling, and the chemical nature of the signaling molecule, respectively.

1. **Biological function**

- **Determination factors** specify the destiny and commitment of a cell or group of cells. Among the determining factors are inducers and morphogens.
- **Morphogens** participate in biological pattern formation. They act in morphogenetic fields, organizing there the spatial order of cell differentiation (Box 4).
- **Inducers** are produced by emitter cells and act upon neighboring cells.
- **Growth factors** and **cytokines** mediate proliferation.
- **Differentiation factors** initiate terminal differentiation.

From a practical standpoint, this scheme has shortcomings. The biological role of a given signal molecule may depend on its local concentration, the time it is released, and the type and sensitivity of cells it encounters. Even the traditions and assumptions of the scientists studying a factor determine which glossary is used. For example, signal molecules identified in amphibian embryos traditionally are called "inducers," whatever their function is. Researchers who focus their interest on biological pattern formation generally speak of "morphogens." Botanists classify any signal molecule as "hormones."

2. **Mode and range of signaling**

- **Endocrine factors (hormones)** are transported into the bloodstream. Examples include erythropoietin (discussed in Chapter 16). Insulin and IGF are locally acting factors in the early embryo but they are hormones released into the bloodstream once the pancreatic islands and the liver have begun to function in the fetus or the newly born child.
- **Paracrine factors** diffuse in the clefts and spaces between cells (interstitial spaces). An example is the **platelet-derived growth factor** (**PDGF**), which plays a role in the regeneration of injured blood vessels and damaged tissue.
- **Autocrine factors** feed back upon the emitting cells, which are equipped with receptors for the signals they produce. The signals can act in either a stimulatory or inhibitory way, causing the emitting cell to divide or to continue signaling. Autocrine stimulation by **TGFs** (**transforming growth factors**) can foster the development of tumors.

3. **Chemical nature of the signaling molecule.** Because the majority of signal molecules identified to date are polypeptides, comparison of the amino acid sequences can reveal protein families whose members display nonrandom sequence identity or similarity.

In embryos, for instance, members of the following families have been found:

- The **EGF family,** including the **epidermal growth factor.** Other members of this family are **KGF (keratinocyte growth factor)** and **TGFα (transforming growth factor alpha)**.

   EGF stimulates the proliferation and expansion of epthelial tissues. The EGF molecule is a peptide consisting of 53 amino acids. It is cleaved off of a precursor molecule that is exposed at the exterior surface of EGF-producing cells and contains several copies of the EGF domain. Proteases split off EGF peptides one by one from the precursor, like slices of bread.

   In invertebrates several signal molecules of developmental significance belong to the EGF family too. Examples include the products of the genes *Notch* and *delta* in *Drosophila*. The proteins encoded by these genes remain attached to the cell surface, serve as cell adhesion molecules, and play roles in the segregation of epidermal and neuronal precursor cells (see Chapter 3, Section 3.8, and Chapter 9).
- The **insulin family,** including IGF. IGF continues to be produced after the embryo development has finished. After birth, the family members are manufactured by the liver and are known as **somatomedins.** Their release

from the liver is controlled by the pituitary **growth hormone.** Growth hormone and somatomedins stimulate the growth of the skeletal elements in the juvenile body.

- The **FGF family,** including **bFGF (basic fibroblast growth factor**), a weak mesoderm-inducing inducer in the amphibian blastula (see Chapter 9), and **aFGF (acidic fibroblast growth factor),** also known as **ECGF (endothelial cell growth factor).** The FGF family members are **heparin-binding growth factors.** This designation refers to the fact that these polypeptides readily attach to heparin-containing extracellular matrices.

- The **PDGF family (platelet-derived growth factor).** PDGFs are so named because platelets release them to induce blood clotting. However, PDGFs also have effects in embryogenesis—fostering, for example, the growth of smooth muscle cells and glia cells. The PDGFs occur as homo-dimers or heterodimers composed of A and B chains (AA, AB, or BB). After birth, they mediate regeneration of injured vessels.

- The **TGFβ family (transforming growth factor beta).** These dimeric factors include the **activins** and **Vg1,** known as strong mesoderm-inducing factors in the amphibian embryo (Chapter 3, Section 3.8, and Chapter 9). In addition, several members of this family are involved in setting up polarity axes and pattern formation. Examples are **activins** in the *Xenopus* embryo and **dpp (decapentaplegic)** in *Drosophila,* as shown in Fig. 9–6.

There are also "negative" growth factors, causing the destruction of tissues and organs. An example from sexual development (Chapter 20) is **AMDF (anti-müllerian duct factor),** also known as **MIS (müllerian duct inhibiting substance).** AMDF belongs to the TGFβ family. This list is by no means complete.

17.2 Hormones Synchronize Reorganization and Remodeling in Metamorphosis and Sexual Development

When and where do hormones play their roles? In medical and animal physiology hormones are defined as signal molecules that are produced by special emitters, the **hormonal glands,** and are distributed in the body via circulating fluids, called blood, lymph, or hemolymph. Only cells equipped with matching hormone receptors respond to a particular hormonal signal. A commonly used analogy is the broadcasting system: messages are propagated everywhere but picked up only by receivers that are tuned appropriately. However, this parallel has its limits. Unlike broadcasts, the meaning of the hormonal message is not so much dependent on what the emitter says but on what the receiver thinks. How a cell responds to a hormone is a function of its previous programming. That is, its response is a function of the cell type and its individual history. Thus, various cells respond differently to the same hormonal signal. Moreover, combinations of hormone receptors are specific to different cell types. Thus, a wide array of responses can be elicited even if only a few different hormones are sent out: hormones have multiple effects.

Classically defined hormones can come into play only after hormonal glands have been developed and the target cells are equipped with receptors (exceptions are discussed in Section 17.4). As hormones circulate and their concentration evens out, they cannot convey positional information (unlike paracrine factors) but only act as **temporal trigger signals.** On the other hand, unlike paracrine factors, hormones potentially can reach many distant targets. Therefore, hormones are well suited to control the timing of developmental processes in which many tissues and organs must be reorganized and remodeled. In particular, **metamorphosis** (Chapter 19) and **sexual development** (Chapter 20) are initiated and coordinated by hormones. However, this does not mean that substances listed as hormones in textbooks of biology and medicine never influence early embryonic development.

17.3 In Early Development the Mother May Have the Say in the Matter

Ecdysone, which acts as the molt-inducing hormone in arthropods (Chapter 19), is synthesized in *Drosophila* in adult females, directed into the oocytes, associated with yolk platelets, and eventually liberated from the platelets in the course of embryogenesis. Presumably, embryonic ecdysone initiates and synchronizes synthesis of the first larval cuticle.

The hormonal control of **vitellogenesis** (yolk formation) in the growing egg cell is very similar in insects and vertebrates. In both groups reproduction is dependent on the seasonal behavior of the parental animals. Environmental cues are integrated in the brain of females and stimulate the release of **gonadotropic hormones** that set off a cascade of events. In most vertebrates, including humans, maturation of the egg cells is initiated by gonadotropic polypeptide hormones that are liberated by the pituitary gland at the beginning of the—often annual—reproductive periods. In frogs the pituitary hormones prompt follicle cells in the ovary to release the steroid hormone **progesterone.** In turn, progesterone acts upon the oocyte, inducing the formation of active **maturation pro-**

moting factor (**MPF**, Chapter 7) and **germinal vesicle breakdown** (the start of meiosis), the formation of the polar bodies, and ultimately ovulation.

As a mammalian embryo has implanted into the wall of the maternal uterus and the placental contacts are established, the fetus comes under the influence of maternal hormones that can pass into the blood of the embryo through the leaky placental barrier. The effects of maternal hormones upon the human embryo and the fetus are largely unknown but may well be of great significance. Maternal stress hormones, such as epinephrine and cortisol, presumably reach the embryo, as do sexual hormones. In certain human diseases the mother produces male sexual hormones. These can redirect the somatic and mental development of a girl in the male direction, although the fetus's genetic gender does not change (see Chapter 20, Section 20.3).

17.4 Even before Hormonal Glands Are Formed, an Embryo Produces Some Classic Hormones

Even before gland development, an embryo produces several "classic" hormones. As early as the blastocyst stage, the trophoblast of the implanting embryo produces **progesterone** to influence the surrounding maternal tissue and to prevent, at least in its immediate surroundings, menstrual bleedings that would be disastrous for the embryo. In addition, the embryo produces several polypeptide **gonadotropins** that direct the maternal organism to entirely stop menstruation and prepare for pregnancy instead.

Sensitive analytical methods have demonstrated the presence of additional hormones and hormone receptors in various regions of the embryo. For example, **insulin** and **insulin-like growth factors** (**IGFs**) have been found even before the pancreas and the islets of Langerhans are formed. These hormones are thought to act as growth factors in embryos.

17.5 Physiologically, Hormone Systems Are Structured Hierarchically

In all animals morphological, physiological, and biochemical research has disclosed a remarkably similar hierarchical organization to the hormone system, regardless of whether this organization is based on homology or has evolved convergently. A few parallels between insects and vertebrates may be pointed out (see also Fig. 19–3 and 19–4).

The brain plays a vital role, because in the control of metamorphosis and sexual development, environmental cues must be evaluated and information must be integrated about the duration of the daily light phase, the current temperature, and the presence of sexual partners.

The bridge between the nervous and hormonal systems is erected by **neurosecretory (neuroendocrine) cells.** These cells receive orders via synaptic inputs from the nervous system and in turn emit endocrine signals. Neuroendocrine cells are found along the hypothalamus–pituitary axis in vertebrates, and along the pars intercebralis–corpora allata–corpora cardiaca axis in insects. The **peptidergic** (peptide hormone producing) neurosecretory cells release their contents at axonal terminals within structures known as **neurohemal organs** into the circulating body fluid (blood or hemolymph). Thus, neurohemal organs connect the nervous and circulatory systems. They are known as the **neurohypophysis (posterior lobe of the pituitary gland)** in vertebrates and the **corpora cardiaca** in insects.

Alternatively, neurosecretory cells in the brain direct their signals towards adjacent **first-order hormonal glands.** First-order glands are the **adenohypophysis (anterior lobe of the pituitary gland)** in vertebrates or the **corpora allata** in insects. These first-order hormonal glands emit controlling (Greek: *tropic*) hormones to govern the activity of **second-order hormonal glands,** such as the **thyroid gland** in vertebrates and the **prothoracic gland** in insects. The second-order glands then liberate those hormones (e.g., thyroxine in vertebrates or ecdysone in insects), which ultimately reach the final targets in the body.

17.6 However Signal Molecules Are Classified, Frequently the Same Signal Transduction Systems Are Switched On

17.6.1 Peptide Hormones Trigger Membrane-Associated Signal Transducing Cascades

Polypeptide signals are picked up by **membrane-associated receptors.** Through signal transduction pathways, the signals are amplified and guided by **second messengers** into the diverse compartments of the cell. The following classes of signal transduction systems predominate:

1. **Transmembrane receptors with intracellular tyrosine kinase activity.** Systems of this type are turned on by signal molecules, such as **EGF, TGFα, PDGF, insulin,** and **IGF** (introduced in Section 17.1). When such receptors are occupied by complementary ligands (extracellular sig-

nal molecules), two receptors join to form a dimer. The two joined partners phosphorylate and activate each other by coupling phosphate taken from ATP onto Tyr residues positioned along the intracellular domain of the receptors, a process called **autophosphorylation.** In addition, the activated receptor may in turn phosphorylate a wide array of intracellular target molecules. Moreover, enzymes and regulatory factors, such as phospholipase-gamma, may attach onto the phosphorylated domains of the dimerized receptor. In turn, phospholipase-gamma generates second messengers, such as diacylglycerol, by cleaving membrane-associated phospholipids.

2. **Transmembrane receptors with intracellular serine/threonine kinase function.** Functional receptors occupied with ligands form heterodimers. Ligands for these receptors are members of the TGFβ family, including induction factors, such as the **activins** and **decapentaplegic.**

3. **The cAMP signal transducing system.** This classic system is described in many textbooks.

4. **The PI-PKC (phosphatidylinositol-protein kinase C) signal transducing system.** This system, described in Box 3, is involved in a variety of responses, including activation of the egg upon fertilization, stimulation of B cells by antigens, induction of the central nervous system in vertebrates by neuralizing signals emanating from the roof of the archenteron, and stimulation of the thyroid gland by thyroid-stimulating hormone (TSH) during amphibian metamorphosis.

Because cells usually are equipped with a number of receptors and their corresponding signal transduction systems, network interactions between these systems multiply the variety of possible responses. For example, transmembrane tyrosine kinase receptors with attached phospholipase-gamma can generate diacylglycerol and thus activate a branch of the PI signal transducing pathway.

17.6.2 Lipophilic Signal Carriers Attach to Intracellular Receptors

Retinoic acid, steroid hormones, and **thyroxine** penetrate into the cell membrane where they may immediately activate membrane-associated enzymes and ion channels. Eventually, the signal molecules are picked out of the membrane by cytosolic receptors. Occupied by ligands, the receptors enter the nucleus and bind to specific binding sites in the regulatory region of genes that are under their control. Ligand-occupied receptors form dimers (homodimers or heterodimers) with other receptors, and function as transcription factors controlling gene activity (Box 5).

18

Cancer Comes from Disturbed Growth and Differentiation Control

Of all afflictions, cancer is perhaps the most relevant to the study of developmental biology because it represents alterations in otherwise normal processes of growth. **Growth** is defined as an increase in mass. Growth can result from more than one process. For example, enlargement of individual cells is a cause of growth. Muscle fibers grow by expanding in length and diameter. Also, the deposition of extracellular matrix material contributes to the enlargment of the growing juvenile. However, most increase in mass is based on cell division followed by the enlargement of the daughter cells to the size characteristic of the cell type. However, an increase in mass can also result from the activity of certain "oncogenes" (from the Greek: *oncos* = mass and *genein* = to generate), and such an increase can be dangerous because it can result in the formation of a tumor.

18.1   How Does Cancer Arise?

In multicellular organisms the individual cells are subject to social control. Unlike free-living unicellular organisms, the members of a multicellular community must reduce or cease their multiplication at the right time and place.

The process of terminal differentiation alone is associated with a delay in the cell cycle and thus causes a slowdown of cell multiplication. Frequently, terminal differentiation results in a complete cessation of growth. In mammals, for example, mature blood cells and nerve cells completely lose the ability to divide. Termination of growth in these cells is part of their normal developmental program. In contrast, stem cells and several types of progenitor cells must continue to divide.

Cancer arises when basic rules of the cell-specific or social control of proliferation are violated. Excessive multiplication may occur in the following situations:

1. Progenitor cells multiply too rapidly or frequently. Terminal cell differentiation cannot cope with this excessive cell number and remove enough of the descendants from the pool of cells capable of dividing.
2. Even if the cell cycle is not accelerated, uncontrolled growth can arise:
   - When both daughters of a dividing stem cell retain the traits of the stem cells. Normally, on average only one of two daughter cells retains the characteristics of a stem cell, while the other becomes committed to cell differentiation.
   - When differentiating cells do not stop dividing. Cell divisions continue, even when the program of differentiation is finished largely. For example, **melanomas** develop from nearly mature derivatives of neural crest cells that synthesize black melanin but nevertheless do not stop dividing.

18.2   Glossary of Cancer Terms

The deterioration of a civilized cell into an asocial cancer cell is called **neoplastic transformation** (Greek: *neo* = new; *plastein* = to form). Agents causing **carcinogenesis** are called **carcinogenic** (Latin/Greek: cancer generating) or **oncogenic** (Greek: *oncos* = mass; *genein* = generating). However, the common use of the term "oncogene" may be confusing. While the adjective "**oncogenic**" means "cancer generating," the noun "**oncogene**" refers to a gene causing cancer.

A **tumor** (Latin: swelling) is an association of cancerous cells, generally deriving from one transformed founder cell. A tumor may be **benign** (Latin: mild, good natured) or **malignant** (Latin: malicious, ill natured). A tumor becomes malignant when cells migrate from the primary tumor to invade other tissues and establish colonies of secondary tumors, called **metastases** (Greek: *meta* = subsequent, *stasis* = location).

Depending on its origin, a tumor may be one of the following:

- **Carcinoma:** a malignant tumor derived from epithelial tissue (Fig. 16–1)
- **Adenoma:** a benign tumor of epithelial origin and glandular appearance
- **Adenocarcinoma:** a malignant tumor of glandular appearance
- **Sarcoma:** derived from connective tissue
- **Melanoma:** derived from melanocytes (pigment cells of the skin)
- **Neuroblastoma:** derived from neuroblasts
- **Glioma:** derived from glia cells (most brain tumors are gliomas)
- **Myoma:** derived from myoblasts
- **Myeloma:** derived from blood progenitor cells
- **Lymphoma:** derived from lymphoblasts
- **Leukemias:** derived from various, undefined blood progenitor cells

Strikingly, tissues and cell types capable of self-renewal from stem cells suffer neoplastic transformation more frequently than others. Among them, carcinomas are the most frequent tumors. Epithelia contain many dividing stem cells in which mutations may disturb the mechanisms of control. In addition, epithelia are exposed to carcinogenic agents, such as UV solar radiation, to a particularly high degree.

18.3   Carcinogens and Oncogenes

18.3.1 A Variety of Agents Trigger Cancer Development

Cancer is caused by at least four classes of agents:

1. **A variety of chemicals,** most of which are also DNA-damaging mutagens. Significant exceptions to the general rule that carcinogens are also mutagenic are small fibers of asbestos and glass, which frequently injure cells and can cause lung cancer if inhaled.
2. **Ionizing and UV irradiation,** and
3. **Viruses**—in particular, retroviruses. Such viruses frequently carry eukaryotic oncogenes, picked up from their eukaryotic host cell in the course of their evolution.

4. **Cancer-causing genes** are inherited in the germ line. An example is *Rb-1:* when inherited as two defective alleles the gene causes **retinoblastomas** (lethal tumors of the eye) in babies and children.

18.3.2 Two Classes of Cancer-Causing Genes Are Known

Genes causing cells to escape from their normal growth control may be inherited. More frequently, such dangerous genes enter the genome along with retroviruses. Remarkably, when oncogenes are intact, located at their correct, chromosomal position, and expressed in adequate amounts, they are not dangerous. In fact, they are required for normal development and growth. Only when overexpressed, mutated, or translocated to inappropriate chromosomal sites do such genes cause cancer.

Two classes of genes causing cancer are known:

1. **Oncogenes proper.** They derive from normal (noncancerous) **proto-oncogenes,** when mutations give rise to dominant *gain-of-function* alleles. Because of the dominant nature of the mutation, it only requires one of the two proto-oncogene alleles to be mutated for growth and differentiation to be deregulated. Most tumor genes of viruses belong to this class.

As a rule, tumor-inducing genes are designated by a three-letter code. For example, the first viral oncogene to become identified was *src* (from sarcoma). The normal gene that does not cause cancer is designated *c-src,* or *src-c* (*c* = cellular).

In general, the proteins encoded by proto-oncogenes are involved in the control of the cell cycle. Many participate in the generation of mitogenic signals. The mutant oncogenes cause inappropriate multiplication of cells. Frequently, the products of proto-oncogenes are as follows:

- Growth factors
- Receptors for growth factors
- Elements of the signal transduction pathway that conveys the mitogenic signal to the nucleus
- Elements of the machinery in the nucleus that starts cell division

2. **Tumor suppressor genes.** Tumors appear when mutated *loss-of-function* alleles of these genes are present in two copies. People who inherit only one defective allele are often predisposed to cancer, but at a lower frequency than individuals with homozygous defective alleles.

Among the tumor suppressor genes are the *Rb-1* gene mentioned above and the *p53* gene, involved in the recognition of damaged DNA and in apoptosis (more about *p53* in Section 18.5).

As a rule, tumor suppressor genes are involved in initiating and carrying out terminal cell differentiation, and in the simultaneous termination of cell division. Loss-of-function mutations relieve cells from inhibitory influences that normally hold their numbers in check.

18.4   Putative Mechanisms of Cancerogenesis

Although in no case has the complete chain of causal events been elucidated, a general framework for explaining the origin of cancers can be outlined. Cancer originates from disturbed control of growth or an insufficient change from proliferation to terminal differentiation. Defects have been found at all levels and in all members of the controlling system:

- **Growth factors:** too much or not inactivated quickly enough (example: *sis*).
- **Signal reception disturbed:** Some oncogenes code for defective receptors that are not deactivated quickly enough, or are constitutively active, even in the absence of growth factors (examples: *erbB*, *trk*).
- **Signal transduction disturbed:** A variety of known oncogenes code for subsequent elements of signal transduction pathways (examples: *src*, *ras*— often involved in human colon carcinomas).
- **Signal response in the nucleus disturbed:** The ultimate members of such signal transduction pathways are transcription factors (examples of oncogenic transcription factors: *myc*, *myb*, *fos*, *jun*). The JUN and FOS proteins form heterodimers known as **AP-1** protein, which prompts initiation of DNA replication.

Interestingly, inverse mechanisms have been proposed for the mechanisms of the less-known tumor suppressor genes: Production, reception, or transduction of signals that normally slow down cell division and foster cell differentiation are disturbed. In all these cases the common ultimate defect is that the cell cycle starts too frequently.

18.5   The Initial Event Is Frequently (Always?) a Mutation, but Additional Tumor-Promoting Influences Usually Are Required to Make the Defect Disastrous

Most carcinogenic agents are known as **mutagens.** Conversely, most mutagens can initiate cancer formation when proto-oncogenes are affected. Fortunately, not all of these primary events do cause the development of tumors. Frequently, more than one mutation must occur in a single cell, or additional **tumor promoters** must subsequently foster growth (**two-stage carcinogenesis**). Examples of known tumor promoters are **phorbol esters,** heterocyclic molecules produced by plants in the family Euphorbiaceae (the milkweeds). These tumor-promoting phorbol esters disturb signal transduction in the PI-PKC (phosphatidylinositol-protein kinase C) system by long-term activation of the key enzyme protein kinase C (see Box 3).

In every round of DNA replication mistakes are made. DNA repair mechanisms check newly synthesized DNA strands and correct faults with astonishing efficiency. Damaged DNA is also subject to guardian mechanisms that abort the cells. Checkpoints in the cell cycle decide whether a cell is permitted to enter a new cell cycle or is withdrawn from service and prompted to undergo apoptosis. A pivotal role in this process has been assigned to the protein coded by the tumor suppressor gene *p53*. Strangely, although p53 protein is a transcription factor that can activate many genes, mice that are unable to express any functional p53 undergo normal embryonic and fetal development, but they develop cancers early in their juvenile life. The *p53* gene is mutated frequently in many types of human cancers.

18.6   Cancer Cells Display an Unusual and Selfish Behavior

Cancer cells neglect and ignore many social rules. For instance, many cancer cells do not obey the rule of **contact inhibition.** Most cell types maintained in cell culture stop moving when they encounter other cells and stop dividing when they are surrounded completely by adjacent cells. Adhesiveness links the cells in a **confluent monolayer.** Cancer cells are different. They show persistent locomotion, crawl over their neighbors, and continue to divide, producing **foci** (small cell clusters). Transformed cells are rounder in cross section, lack adhesion plaques and stress fibers, and are less covered with fibronectin.

The ability to perform amoeboid movements is another characteristic of many cancer cells. This behavior enables them to colonize other body regions and to form **metastases.** Crossing the walls of blood vessels and invading solid tissues is facilitated by the ability of many malignant cells to produce enzymes that digest holes into the basal lamina of blood vessels and channels through connective tissues. Cancer cells share these properties with some normal cell types, notably granulocytes, macrophages, and lymphocytes. Thus, several types of cancer cells can spread through the body like white blood cells, but unlike white blood cells, cancer cells continue dividing.

Solid tumors can promote their own growth by another trick. Like embryonic tissues, they produce and emit **angiogenic factors** that stimulate blood capillaries to grow toward and into the tumor. Thus, nutrient supply is improved.

Altogether, cancer cells resume and reactivate many faculties that are characteristic of embryonic cells. Thus, besides being a serious medical problem, cancer is a matter of developmental biology.

19

Metamorphosis:
A "Second Embryogenesis"
Creates a Second Phenotype

19.1 Most Amphibians and Invertebrates Undergo
Major Phenotypic Transformations

Metamorphosis refers to fundamental remodeling of the whole body, and is associated with a fundamental change in the mode of life. The new phenotype occupies a new ecological niche and colonizes a different habitat. Larvae and adults utilize different nutritional resources and often live in different environments.

Most benthic marine invertebrates produce planktonic larvae. These free-living stages enable the species to first make use of planktonic prey and subsequently to exploit a scattered or transient ecological niche suited for the adult phase. Planktonic invertebrate larvae include those of echinoderms, such as the pluteus of sea urchins (Fig. 2–1); the trochophore of the polychaete annelids (sandworms, tube worms; Fig. 3–11); the veliger larvae of mollusks;

the actinotrocha of the phoronids (slender wormlike sedentary creatures); the nauplius of the crustaceans (lobsters, crabs, barnacles); and finally the tadpole-like larvae of the tunicates. Planktonic larvae enter the adult population through metamorphosis and settlement (see Section 19.5).

In insects and amphibians the acquisition of a new phenotype in metamorphosis implies the transition to a different habitat or at least a change in the nutritional resources that are exploited. Though the amphibians were the first vertebrates to spend a significant portion of their lives in terrestrial habitats, most present-day amphibians still return to the water to reproduce. The larvae live in water and feed preferentially on algae and plants. The terrestrial adults are carnivorous. Likewise, no adult insect (imago) competes with its larval stage for food.

Regardless of the species in which it is taking place, the process of metamorphosis implies several universal events:

- Destruction of specific larval structures (abdominal legs of caterpillars, gills and tail of tadpoles)
- Adaptive remodeling of tissues that persist in the adult stage (nervous system, excretory organs)
- Development of structures unique to adults (wings in insects, lungs in amphibians)

In **insects** and other arthropods **molts** occur during development and growth. Arthropods are covered with a **cuticle:** a hardened extracellular shell that cannot expand. The cuticle must be shed and replaced periodically to allow growth and changes in body form. At each molt, the epidermis withdraws from the old, partially dissolved cuticle and subsequently expands, forming folds and secreting beneath the old cuticle a new, soft one. When the old cuticle is sloughed off, the epidermis and the new, still elastic cuticle expand. Once the individual has hatched, it supports expansion and unfolding of the epidermis and the cuticle by generating hydrostatic pressure with compressed body fluid.

Metamorphosis in insects is the transformation of the larval into the **imaginal** (adult) phenotype. Among the various insect orders, metamorphosis takes two broadly different forms:

1. **Hemimetabolous development** (Greek: *hemi* = half; *metabolous* = changing).

In its basic organization the larva already resembles the imago in hemimetabolous species and even has compound eyes, although the early larval body is very small and lacks structures such as wings and genitalia. With each molt, the size and appearance of the larva (often called a **nymph** in hemimetabolous insects) approaches the size and appearance of the final imago (Fig. 19–1). Sometimes, only minor remodeling of the larval body is necessary (**paurometabolic development**); in other species conspicuous lar-

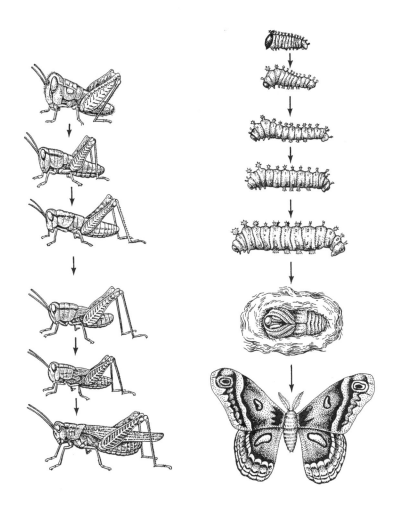

Figure 19–1   Hemimetabolous development of a grasshopper (from Saunders 1970) and holometabolous development of a butterfly, *Attacus cecropia*. (Redrawn after Gilbert.)

val organs must be destroyed (such as gills in mayflies or the extensible labium of dragonflies, with its hooklike jaws). Frequently, the nymph enlarges the wing pads rather suddenly in the last two molts, but a pupa does not occur.

Examples of hemimetabolous development include wingless insects (Apterygota), grasshoppers (Orthoptera), cockroaches (Blattidae), termites (Isoptera), earwigs (Dermaptera), mayflies (Ephemeroptera), dragonflies (Odonata), bugs (Hemiptera), sucking lice (Anoplura), and cicadas (Homoptera).

2. **Holometabolous development** (Greek: *holos* = whole, complete).

From the egg, the young hatches with a wormlike segmented body (called the caterpillar, grub, or maggot). The larva may have short legs; it lacks wings and compound eyes. The successive larval instars increase in size through several molts but do not abandon their wormlike appearance until **complete metamorphosis** takes place in two major steps. In the first step the last instar larva transforms into an immobile **pupa** that is encased in a protective cuticle with a form and color characteristic of this stage. In the second step the imago is formed within the pupal cuticle. Beneath the cuticle, many larval organs break down and are absorbed by phagocytic cells, while adult structures arise concurrently. At the end of pupation, the finished imago hatches. Some of these dramatic events are described for *Drosophila* in more detail as follows.

Holometabolous development is characteristic of goldeneyes and ant lions (Neuroptera); caddis flies (Trichoptera); moths and butterflies (Lepidoptera); true flies and mosquitos (Diptera); fleas (Siphonoptera); beetles (Coleoptera); and wasps, ants, and bees (Homoptera).

19.2 In Metamorphosis Dramatic Remodelings Occur Down to the Molecular Level

Metamorphosis entails reshaping and restructuring at every level of the organism, from the body's morphology and anatomy to its physiology and cellular machinery. Such multiple changes are initiated and synchronized by hormonal signals, which are discussed in Section 19.3. However, as in all hormonally controlled events, various tissues and organs respond quite differently to changing levels of hormones, depending on their programming in previous stages of development. To exemplify the remarkable range of metamorphic events, we will compare the metamorphoses of an amphibian (the frog) and an insect (the fly).

19.2.1 Frog

Destructive and constructive metamorphic processes proceed gradually and take weeks. There is no quiescent stage; the tadpole has to manage a smooth changeover to the adult organization while it is moving and feeding.

- Because the type of locomotion will change from smooth swimming to jumping with legs, hindlimbs emerge in **prometamorphosis.** During the **climax of metamorphosis,** the tadpole extends its previously prepared but hidden forelimbs out of the branchial cavity, and gradually resorbs its tail (Fig. 3–28, 19–4).
- When the lungs are formed, the gills vanish. The transition in the mode of breathing is accompanied (and made feasible) by reconstruction of the circulatory system. The aortic arches and several large body vessels are restructured. In the erythroblasts a new isoform of hemoglobin, that binds oxygen less avidly, is synthesized.
- To avoid desiccation, the skin is made more tight with adult keratins and is interspersed with mucous glands; the eye becomes protected by an eyelid.
- The horny teeth for tearing plants disappear; a long tongue develops and the intestinal tract is adjusted to a carnivorous diet.
- While the tadpole can readily release ammonia into the surrounding water by diffusion (**ammoniotelic excretion**), in metamorphosis the liver and the kidney are retooled with new sets of enzymes, enabling the frog to convert ammonia into urea (**ureotelic excretion**).
- The lateral line sensory system used by the larva to detect slow wave or current movements in water is reduced. The first branchial pocket is remodeled and becomes the ear tube (pharyngotympanic tube), closed by the tympanic membrane (Fig. 4–2) and incorporating the sound-transmitting **columella.** In the retina of the eye the visual pigments are exchanged: the fish-type **porphyropsin** (opsin + retinal A2 molecule) is replaced by **rhodopsin** (opsin + retinal A1), characteristic of terrestrial vertebrates.

19.2.2 *Drosophila*

The larva undergoes a holometabolous (complete) metamorphosis. When entering the prepupal stage, the third and last instar wanders around seeking a place in which to transform into a pupa. Having found a suitable location, it cements itself onto the substrate with a glue produced in its salivary glands. Now the larva softens its last larval cuticle and inflates into a barrel-like shape. The expanded cuticle is rehardened and is now called the **puparium** (Fig. 3–12). Beneath the protective envelope most larval tissues are destroyed. The fly is constructed like a mosaic, from residual **imaginal discs** and **imaginal cells.**

The exterior epidermal structures that become covered with a cuticle are manufactured from imaginal discs. In *Drosophila* most of the adult is built from 10 pairs of such discs (Fig. 3–24). Unlike caterpillars of butterflies, in dipteran larvae these discs are not seen as epidermal thickenings but lie hidden inside the body. However, the discs do not originate inside the body.

As early as the embryonic blastoderm stage small patches of 20 to 70 blastoderm cells are segregated out. As discussed in Chapter 9 (see Fig. 9–6), imaginal disc fields are specified at the intersection between stripes of cells secreting certain signal molecules (encoded by the genes *hedgehog, decapentaplegic,* and *wingless*). These clusters of cells remain diploid, become invaginated, are wrapped in thin epithelial sheets, and remain stored inside the larval body, where they grow but do not develop further until they are hormonally stimulated in the pupa. Within the puparium, the discs are everted and elongated. The cells in the center of the disc telescope out to become the most distal portion of an antenna, leg, or wing (Fig. 19–2). Shieldlike extensions at the base of the discs expand and assemble to form the capsule of the head and the thoracic segments.

The abdominal epidermis and many internal structures are generated by **imaginal cells** (also called **histoblasts**) that lie interspersed among larval

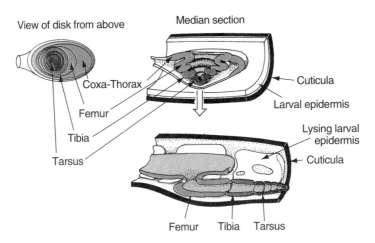

Dipteran leg imaginal disc

Figure 19–2 Development of a leg imaginal disc in the pupa of a fly. The leg is everted like a telescope.

cells. In its basic structure the nervous system is taken over from the larva but becomes modified substantially.

19.3 Hormonal Control of Metamorphosis Shows Many Analogies between Insects and Amphibians

In both insects and amphibians the principle of dual control is realized. There are hormones that foster growth but curb the transformation of the body into the adult phenotype; there are also hormones promoting this transformation. Both systems are governed by neurosecretory cells in the brain. The neuro-hormones released by the neurosecretory cells control the liberation of development-controlling hormones from subordinate hormonal glands (Fig. 19–3, 19–4).

19.3.1 Releasing the Brake

In the larvae of insects and in amphibian tadpoles metamorphosis is made possible by lowering the "titer" (level or concentration) of a hormone that allows growth but inhibits the differentiation of adult structures.

In insects this is the **juvenile hormone.** The hormone is a terpenoid (sesquiterpene) and as such has a structure reminiscent of many secondary metabolites in plants. The hormone is produced by a pair of glands called the **corpora allata.** As long as the juvenile hormone is present in sufficient concentration, molts result in another larval instar. In the last larval instar a neural input from the brain inhibits the corpus allatum from releasing the juvenile hormone. The lowered hormone level allows metamorphosis to proceed. Timely experimental removal of the corpora allata causes larvae to undergo premature metamorphosis from which small adults arise, while implantation of additional glands results in additional larval molts, delayed metamorphosis, and giant adults. In the adult production of the juvenile hormone is resumed, but now the hormone acts as a gonadotropic hormone, stimulating the development of the gonads.

In amphibians a hormone from the anterior pituitary, called **prolactin,** has a comparable "juvenilizing" action by preventing premature metamorphosis. This hormone is also used in other regulatory circuits. Its name points to its role in stimulating the production of milk in mammals. In other mammals it acts as **luteotropic hormone (LTH;** not identical with the luteinizing hormone, LH) stimulating the conversion of an ovarian follicle into a corpus luteum (yellow body).

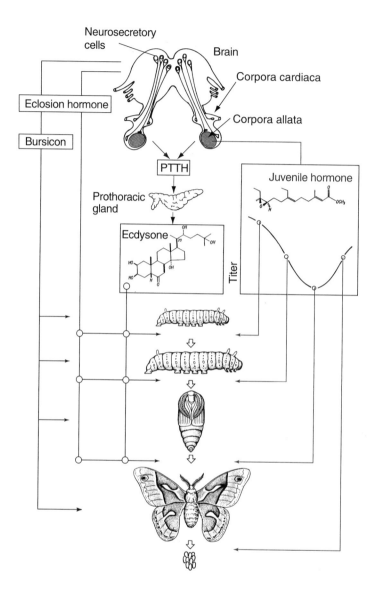

Figure 19–3   Hormonal control of metamorphosis in insects. The source of bursicon and eclosion hormone are ganglia of the ventral cord. (Redrawn after Wehner and Gehring, Zoologie, Thieme-Verlag, Stuttgart, 1994.)

19.3.2 Stimulating Developmental Progress.

The timely switching on of genes needed to redirect development toward the adult phenotype is triggered in both insects and amphibians by neurohormones delivered by the brain.

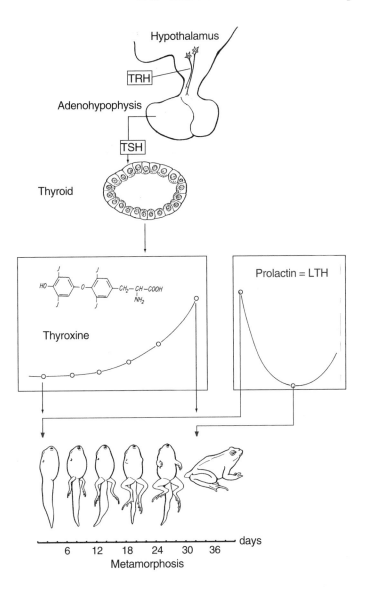

Figure 19–4 Hormonal control of metamorphosis in amphibians. (Redrawn and modified after Eckert, Tierphysiologie, Thieme-Verlag, Stuttgart, 1993.)

In amphibians the chain of signaling starts with a neurohormone produced in the hypothalamus. It functions as a releasing factor, stimulating the adenohypophysis (anterior pituitary gland) to liberate **TSH (thyroid-stimulating hormone)**. This peptide hormone conveys the message to the **thyroid**

gland, which in turn emits the final, metamorphosis-inducing hormone, **thyroxine T3** (T3 = triiodothyronine). Administering T3 to tadpoles elicits precocious metamorphosis, resulting in small frogs. Blocking the thyroid gland by thyreostatica (drugs blocking the thyroid gland) delays metamorphosis and results in giant axolotl-like tadpoles.

In insects the chain of signals starts with the neurohormone **PTTH (prothoracico-tropic hormone).** This peptide is released via the corpora allata and functions as a "glandotropic" hormone, stimulating the **prothoracic gland** to liberate **ecdysone.** Ecdysone is then converted in the target tissues into the ultimately effective **ecdysterone** (20-hydroxyecdysone). For simplicity, we will use only the term "ecdysone."

The role of ecdysone in metamorphosis has been investigated intensively. Ecdysone is present prior to the beginning of metamorphosis: in the larval stage each molt is triggered by pulses of ecdysone, but the presence of juvenile hormone prevents the larva from forming a pupa.

Thyroxine and ecdysone are both indispensable for metamorphosis. The molecular structure of these hormones is very different—thyroxine is a derivative of the amino acid tyrosine, whereas ecdysone belongs to the family of steroid hormones. Therefore, a common mechanism of action was not expected. However, **ecdysone, thyroxine,** and the chemically unrelated **retinoic acid** all bind to the same type of homologous intracellular receptors. Occupied by their respective ligand, the receptors become transcription factors controlling gene activity. The receptors contain zinc fingers that enable them to bind to DNA (Box 5).

The transcription-stimulating activity of ecdysone is visible as the pattern of puffs in the polytene chromosomes (see Chapter 10). A close look at this pattern reveals that each stage of metamorphosis and each type of tissue is characterized by a specific combination of puffs.

Various tissues respond differently to hormonal signals. For example, when a limb bud is transplanted experimentally into the tail region of a tadpole, the bud does not grow unless stimulated by thyroxine at the beginning of metamorphosis. In metamorphosis both structures—tail and leg bud—are exposed to the same titer of hormones, but the tail is reduced, while the leg bud persists and grows out.

When comparing amphibians and insects, differences should be noted in addition to the parallels described previously. In insects the shedding of the old cuticle and the acquisition of a new one are controlled not only by ecdysone and the juvenile hormone but also by several additional neurohormones. For example, **eclosion hormone** triggers the behavior associated with molts, and **bursicon** induces the hardening (sclerotization) of the new cuticle after the molt.

19.4   False Hormones and Plant Defenses ·

A remarkable story involving plant biochemistry and insect development is exemplified by the plant bug *Pyrrhocoris apterus*. When brought from Europe into an American laboratory, these bugs failed to undergo metamorphosis. Instead, they underwent additional larval molts, and finally died as giant nymphs. The systematic search for possible causes eventually took into account the paper lining of the rearing dishes. It was discovered that larvae reared on European paper, including pieces of the distinguished British journal *Nature*, underwent metamorphosis. However, bugs reared on American paper, including *Nature's* competitor *Science*, refused to metamorphose. The American paper was made from balsam fir, a North American tree that produces compounds with juvenile hormone-like activity.

Numerous plants have been found to produce **natural pesticides** that act by interfering with the hormonal system of insects. Some members of the plant family Asteraceae produce **precocenes,** which cause the corpora allata of several hemimetabolous insect species to wither. As a result, the level of juvenile hormone drops too early, metamorphosis is initiated precociously, and the result is dwarf imagoes that are unable to reproduce. The Indian **neem tree,** *Azadirachta indica,* and related African trees synthesize a large variety of terpenoid compounds, collectively named **azadirachtin,** that kill a number of insect pests by interfering with the molting process, disrupting development into adults, or reducing fecundity and egg hatching. The compounds include a large complement of terpenoids, including sterol-like tetranor-triterpenoids, pentanor-triterpenoids, and diterpenoids, as well as a number of nonterpenoidal ingredients. The terpenoids are thought to act by interfering with ecdysone-mediated processes. In addition, many of the compounds depress feeding activities in insects.

19.5   Metamorphosis Is Often Released by External Cues

Changes in the ecological roles of larvae and adults can reflect prevailing environmental conditions. The free-living larvae of sedentary marine animals must locate a suitable substrate on which to settle, because a poor choice cannot be corrected after metamorphosis. Frequently, specific chemical cues promote settling and induce metamorphosis. The larvae of epiphytic animals may search for plant-derived chemical cues. Thus, the larvae of the hydroid *Coryne uchidai* settle on *Sargassum* algae that produce metamorphosis-inducing heterocyclic compounds. With surprising frequency, the source of the specific metamorphosis-inducing substances is bacteria that cover the substrate on which the animal settles. For example, certain bacteria isolated from the natural habitat of the hydrozoan *Hydractinia* (Fig. 3–7) induce metamorphosis in its larvae. Likewise, metamorphosis in sea urchins (Fig. 2–1b) is stimulated by bacteria that cover the natural substrate on which the adult feeds.

In insects and amphibians metamorphosis is part of the developmental program and is triggered mainly by internal factors. However, the period when metamorphosis occurs and even the precise moment (think of mayflies!) must be adjusted to environmental conditions, such as temperature and day length.

In many insects a stage of quiescence (**diapause**) is inserted into the life cycle. Diapause facilitates survival during adverse environmental conditions, such as cold or extreme heat. To synchronize the internal developmental program with external cues, the hormonal system is placed under the control of the nervous system. Sensory input can thus be used as trigger signals.

Insects have provided evidence for intriguing mechanisms coupling environmental cues, the hormonal system, and development. For example:

- Adults of the marine chironomid midge, *Clunio marinus*, hatch from the pupal envelope on distinct days of the lunar month. These days are correlated with the spring-neap tide cycles. Ecdysone is involved in programming the time point for synchronized hatching in a local population.
- In some butterfly genera (*Araschnia*, *Precis*, *Bicyclus*, and *Polygonia*) two very differently colored forms ("morphs") appear each year (**seasonal polyphenism**). In *Araschnia levana* butterflies with red pigmented wings hatch from diapause pupae in the spring, whereas butterflies with black and white patterned wings hatch from nondiapause pupae in the summer. The color pattern is modified in response to ecdysone signals.

20

Sex and the Single Gene

20.1 Most Organisms Have a Bisexual Potential

Sex is devoted to the exchange and recombination of genetic information. Most eukaryotic organisms exist in two complementary forms, females and males, who invest genetic material in a mutual game of chance. The random rearrangement of genetic material is associated with the production of offspring and therefore is called **sexual reproduction.** The offspring literally embody the new combination of genes and subject it to the test of natural selection.

Unlike asexual reproduction that is based exclusively on mitotic division, sexual reproduction incorporates two characteristic events enabling novel genetic combinations: (1) fusion of two genetically different cells (gametes) in **fertilization;** and (2) **meiosis,** a process that usually takes place in preparation of fertilization (in several "lower" eukaryotes meiosis occurs at some other point of the life cycle) and enables the rearrangement of individual chromosomes and genes.

Gametes occur as two types of highly specialized cells, the egg and the sperm. In a number of plants and "lower" animals, each individual produces both types of gametes. This condition is called **hermaphroditism,** combining *Hermes* and *Aphrodite;* it is exhibited by **monoicous species** (Greek: *mono* = one, *oikos* = house). In the vast majority of animals, however, the role of producing eggs and sperm is allotted to two distinct types of individuals, and the species are **dioicous** (Greek: *di* = two, twofold; *oikos* = house) or **heterosexual,** with females and males.

As a rule, the differences between the sexes are not restricted to the reproductive organs but extend to other traits, such as size, ornaments, and weaponry, and always include sex-specific, reproduction-associated behavior. Thus, the typical animal species displays **sexual dimorphism.**

To understand aberrant but also normal sexual development in humans, it is important to appreciate that each individual is endowed basically with **bisexual potential** and possesses almost all of the genes required to develop both female and male traits. Otherwise, a mother could not transmit "male" genes from her father to her sons, and a father could not transmit "female" genes from his mother to his daughters.

20.2   Environmental Conditions or Chance Determine the Distribution of Sex-Determining Genes

In all organisms, beyond the sets of genes required to develop both types of gonads and sexual organs, additional **selector** or **master genes** determine which of the two sets of genes actually is being used in developing an individual's sex. Typically, one decisive master gene is present but allotted to only half of the offspring. This is referred to as **genotypic sex determination.** Alternatively, master genes are activated differentially by environmental influences. Such cases are subsumed under the heading **environmental sex determination.**

What is the ultimate decisive event? As indicated, in the animal realm (and likewise in plants) two modes of ultimate decisions are realized:

1. In **environmental sex determination** the decision is made only *after* fertilization, and **external cues,** such as temperature or presence of a sexual partner, make the choice, presumably by influencing the activity of master genes. For instance, in the polychaete annelid *Ophryotrocha puerilis* two individuals that encounter each other by chance influence each other by means of chemical signals (pheromones) and eventually one individual plays the role of the male and one plays the role of the female. Environmental sex determination has been discovered in several invertebrate species but also is known to

occur in fish and reptiles (all crocodilians, many turtles, and some lizards). Environmental sex determination is indicated, but not proven, by sex ratios deviating significantly from 1:1.

2. In **genotypic sex determination** the decision is left to **chance in fertilization.** Traditional belief, offered by ancient writers such as Aristotle, ascribes sex determination, even in humans, to nongenetic determinants such as nutrition or the heat of passion during intercourse. How such factors ensure a sex ratio of 1:1 usually is not the subject of much attention in superstitious explanations. However, how can chance yield a ratio consistently close to 1:1 (in humans, 105 male births per 100 female births)?

The principle may be explained using humans as the example. All humans carry $2 \times 23 = 46$ chromosomes in their diploid cells. Of these, 2×22 chromosomes are indistinguishable in males and females, and are called **autosomes.** In females the 44 autosomes are complemented by two more homologous chromosomes, the X chromosomes. In males two morphologically distinct chromosomes are found: one is an X, like those found in females, while the other is a modified X, termed the Y chromosome. X and Y chromosomes are often called "**sex chromosomes,**" but this designation is highly misleading. A better, but still misleading, expression is "**sex-determining chromosomes.**" As discussed in Section 20.3, ultimately only one gene located on the Y chromosome is of significance. With respect to the two nonautosomes, in humans and other mammals, females are **homogametic** (XX), while males are **heterogametic** (XY). The suffix "gametic" suggests consequences in gametogenesis. In birds, incidentally, females usually are heterogametic and the males are homogametic.

When meiosis takes place during spermatogenesis and a haploid set of 23 chromosomes is parceled out to each future sperm cell, this set includes, besides 22 autosomes, either the X or the Y. As a consequence, 50% of the sperm cells obtain the X chromosome by chance, while the other 50% obtain the Y chromosome. In oogenesis all future egg cells obtain, besides 22 autosomes, an X chromosome.

In conception it is (largely) a matter of chance whether a sperm cell carrying X or a sperm cell carrying Y fertilizes an egg cell. On the average, Y-carrying sperm appear to be slightly faster in the race to the egg than X-carrying sperm, hence the ratio of 105 males to 100 females.

Unexpectedly, intensive research into the nature of genotypic sex determination has shown that the role of chance in distributing chromosomes or genes is the only common denominator among the various types of genotypic sex determination identified in the animal kingdom. Despite nature's tendency to retain established "solutions" to biological problems, and despite the universality of sexual dichotomy, there is no universal, molecular mechanism of genotypic sex determination.

In the simplest case sex is determined by a gene that occurs as two alleles. One sex is heterozygous, the other homozygous (*Mm/mm* system). This mechanism occurs in the common housefly, *Musca domestica.* However, in most animal species the mechanism is more complex and includes cascades of mutually dependent events. Frequently, the sex-determining genes occur on two morphologically different chromosomes, usually called the X and Y. In such cases one of the two sexes is XX and the other is XY. Despite the consistent XY formalism, the roles of these chromosomes vary in different organisms.

In *Drosophila*, for example, the ratios between the number of X chromosomes and the two sets of autosomes (AA) in each individual diploid cell is decisive in determining the sex of each individual cell.

XX/AA (a ratio of 1:1) is female

X/AA (a ratio of 1:2) is male

The Y chromosome in *Drosophila* is of significance in spermatogenesis in the adult male but not in early sex development. Sexual determination is cell autonomous and is not mediated by hormones. If one X is lost due to a faulty mitosis (nondisjunction) in an XX embryo, the resulting X0 cell and all its descendants become male in female surroundings (while the descendants of the corresponding XXX cells are "superfemale"). In *Drosophila* and other insects individuals composed of a mosaic of female and male cells are called **gynanders** (Greek: "wifeman") or **gynandromorphs.** In insects, unlike vertebrates, there are no sex hormones that could moderate differences in X chromosome numbers and ensure a uniform phenotype. An individual with an abnormal ratio of XX/AAA develops an **intersexual phenotype** with genetically uniform cells but an intersexual organismal appearance.

In each somatic cell X-linked genes code for factors aptly called **numerators,** while genes located in the autosomes code for **denominators.** In the presence of two X chromosomes more **numerator proteins** are produced (by genes that apparently escape dosage compensation, or before dosage compensation is accomplished). When numerator proteins predominate, the **master gene *sxl* (*sex lethal*)** is switched on. The SEX LETHAL protein directs a cascade of biochemical events. These include the correct splicing of its own pre-mRNA by which sex lethal makes it possible to maintain its own synthesis. The cascade of events set going by SEX LETHAL terminates in the activation of female-specific genes. In contrast, male development is the **default program,** resulting from insufficient levels of SEX LETHAL protein to trigger female development.

In mammals the situation is the reverse: the Y chromosome contains a male-determining gene, and the female development is the default option when the gene is absent or defective.

20.3   Sexual Development in Mammals Proceeds from a Key Gene to Hormones to Behavioral Imprinting

In mammals, particularly as studied extensively in mice and humans, sexual development is a multistep process. Several stages occur prior to birth.

20.3.1 The Genetic Sex

As all eggs have an X chromosome, the genetic basis of sex depends on the genotype of the sperm cell. If the sperm contributes another X, the embryo will be XX and eventually will become female. If the sperm cell contributes a Y, the embryo gets the constitution XY and will become male. The decisive significance of the Y chromosome arises from the fact that it carries a master gene, *SRY* (*Sex-determining Region of the Y chromosome*). *SRY* activity generates the transcription factor **TDF** (**Testis-Determining Factor**). TDF contains a DNA-binding domain of the **HMG** (**High Mobility Group**) class that enables the factor to direct the activity of subordinate genes (most of which remain unidentified). A 14-kbp DNA fragment containing *SRY* and no other gene was sufficient to cause XX mouse embryos to develop male traits. It appears that a single gene ultimately determines whether an individual will be a man or a woman. This does not mean that other genes are not involved in the realization of the sex; but one particular gene's activity tips the scale.

Abnormal sexual development can result from events that have taken place in the father of the child. During meiosis in spermatogenesis, the *SRY* gene may have been translocated from the Y to the X chromosome. Fortunately, crossover between X and Y is infrequent. However, if a sperm carrying an X-linked *SRY* wins the race and fertilizes an egg cell, the seemingly normal XX embryo will develop male traits. In the few cases known to result from such a crossover, the genitalia did not develop fully normally and the person remained infertile, indicating that additional Y-linked genes are required to support male differentiation.

On the other hand, an XY embryo with a defective *SRY* gene will be a girl, because female development is the default program. The phenomenon is known as **sex reversal.** The somatic cells of such women lack the Barr body (the inactivated second X chromosome, Chapter 10, Section 10.7) that is characteristic of normal female cells.

20.3.2 The Gonadal Sex.

In early embryogenesis the gonads of XY and XX individuals do not exhibit morphological differences. Moreover, the gonad contains both potentially male and potentially female germ cells. Primordial germ cells colonize the gonadal cortex, where they can become oogonia, as well as the central medulla, where they can become spermatogonia (Fig. 20–1).

20.3.2.1 Male Development

In XY embryos the primordial germ cells in the cortex perish under the influence of the TDF, whose synthesis is *SRY* dependent. The gonad becomes a testis.

- The central medulla develops testis cords; at puberty, these cords will hollow out and become the seminiferous tubules in which sperm are produced.

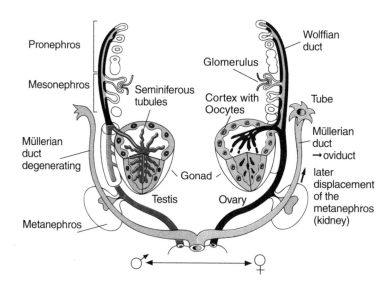

Figure 20–1 Development of the primary sexual organs in humans. In the stage shown female and male development begins to diverge but are still similar. In the male gonad oocytes in the cortex begin to degenerate while the testis cords in the center of the gonad continue to develop into seminiferous tubules. The gonad is converted into a testis. In the female gonad the central testis cords disappear, while the oocytes in the peripheral cortex survive and become surrounded by follicle cells; the gonad is developing into an ovary.

- The supporting cells differentiate into **Sertoli's cells;** these produce the hormonal **anti-müllerian-duct factor (AMDF,** also known as **MIS** or **müllerian duct-inhibiting substance);**
- The steroid precursor cells (interstitial cells) develop into **Leydig's cells** that begin to produce the male sexual hormone (androgen) **testosterone.**

20.3.2.2 Female Development

When the *SRY* gene is not present or is defective, the gonad becomes an ovary:

- The primordial germ cells in the cortex stop mitotic division, become oocytes, and enter meiosis. In contrast to males, where meiosis in the spermatocytes is delayed until the onset of puberty, in females meiosis begins in the ovaries as early as the 12th week of embryonic development.
- The supporting cells assume the function of **follicle cells** and surround the oocytes.
- The steroid precursor cells (interstitial cells) become **theca cells** and produce the female sexual hormone **estradiol** and related estrogens.

Further gonadal development depends on whether or not testosterone is present. **Testosterone** is an essential **androgen.** Its presence triggers male development; its absence allows female development to take over.

20.3.3 The Somatic Sex

The development of the inner and outer sexual organs also starts from an indistinct primordium (Fig. 20–1, 20–2). The embryo has initially two developmental options and its gonads are accompanied by two different kinds of tube-shaped channels. One of these tubes, called the **müllerian duct,** can grow to become an oviduct, the other tube, called the **wolffian duct,** has the potential to become a vas deferens and conduct sperm. Externally as well, genitalia are initially indistinct in their morphology.

In XY individuals the müllerian duct atrophies, mediated by AMDF, while in the presence of testosterone the wolffian duct forms the vas deferens and the pronephros forms the epididymis. The external genitalia form the scrotum and penis under the influence of **dihydrotestosterone,** derived from testosterone by means of a 5α-reductase in the early urogenital primordium (genital ridge). An absence of testosterone or defective testosterone receptors may, in spite of an XY genotype, lead to phenotypic feminization. Other factors can also lead to a female phenotype in the presence of a Y chromosome (e.g., defective *SRY,* as described previously).

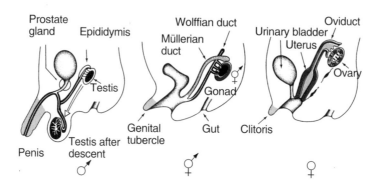

Figure 20–2 Sexual development in humans. Divergent development starts from a common, indistinct stage (middle). In males the müllerian duct regresses; the wolffian duct comes to be the vas deferens the former mesonephros is transformed into the epididymis. In females the müllerian duct is enlarged to become the oviduct and the uterus, while the wolffian tube is reduced. The external genitalia also arise from a common, indistinct stage.

20.3.4 The Psychical Sex

When testosterone is administered to pregnant rats shortly before they give birth, both XY and XX offspring behave like males. To facilitate coitus, a female rodent normally displays a behavior known as lordosis. She arches her back by elevating her head and rump, and moves her tail to one side. Following prenatal exposure to testosterone, females do not show this behavior. Instead, they show the mounting behavior typical of males. Similar observations have been made in other mammals. The phase of hormonal imprinting is generally shortly before birth but in some species may extend into postnatal life. In rats hormonal influence affects a sexually dimorphic nucleus in the optic area. The brain develops differently in males and females, and this different development may determine different behavior in later life.

Transsexual humans have the strong feeling, often from childhood onwards, of having been born the wrong sex. A recent report claims to have found a "female" brain structure in genetically male transsexuals (Zhou et al. 1995). However, sexual behavior is not only determined by genes and hormones. Environmental influences may also have an imprinting effect.

Sex-specific differences in the fine structure of the brain can be investigated most favorably in songbirds. Brain development in birds is not necessarily terminated upon birth. In some species careful examinations reveal minute testosterone-dependent seasonal differences between the sexes, associated with the development of the song in the male bird.

20.4   Puberty Is a Kind of Metamorphosis

During the final sexual maturation, in humans called **puberty,** numerous transformations occur. "Larval" cartilaginous elements become ossified, and many new structures and functions develop, collectively called **secondary sex characteristics.** These changes are also developmental processes. In humans the existence of a hormone has been postulated that, by analogy to prolactin in amphibians and juvenile hormone in insects (Chapter 19), prevents premature sexual maturation. When the production of this putative hormone ceases, the pituitary gland is hypothesized to emit increasing amounts of the gonadotropic hormones **FSH (follicle-stimulating hormone)** and **LH (luteinizing hormone,** also known as interstitial cell-stimulating hormone, ICSH). This takes place in boys as well as in girls. Stimulated by the gonadotropic hormones, the gonads augment the production of testosterone or estrogen, respectively. The main function of these steroid hormones is to trigger spermatogenesis or maturation of the oocytes.

20.5   Periodic Hormonal Cycles Coordinate Cycles
of Sexual Development

Cylic periods of reproduction in invertebrates is a topic so rich in variants that special textbooks and monographs have been written to describe them. Cycles of reproduction are known to be annual (mayflies, many butterflies), lunar, and semilunar (the marine midge, *Clunio,* Chapter 19, Section 19.5). Enthusiastic naturalists can have the pleasure of seeing colorful mating displays in many fishes and amphibians. Sex-related, periodically repeating developmental processes occur in mammals also, as is well known to those who collect horns and antlers. The monthly reproductive (menstrual) cycle of women is described in physiology textbooks.

21

Regeneration and Renewal
versus Loss and Death

21.1 Without Continuous Regeneration Life
Soon Comes to an End

When we think of regeneration, we usually imagine the reconstruction of lost body parts. However, this is only one among several regenerative events that organisms perform.

Re-generation means generation anew. In this general sense regeneration occurs at all levels of life, including the level of macromolecules. Proteins suffer irreversible alterations with time (denaturation) and must be replaced by newly synthesized ones. Without renewal, slow loss of function in essential enzymes eventually would result in the death of the cell. Experimental evidence suggests that renewal of the molecular inventory and resulting **rejuvenation** takes place routinely in the course of cell divisions. Division-mediated rejuvenation confers potential immortality to the single-celled protist. In contrast, terminally differentiated cells, incapable of dividing, will sooner or later die. Rejuvenation by mitotic division can be supplemented or replaced by rejuvenation through sexual reproduction.

Several forms of regeneration may be distinguished:

1. **Physiological regeneration (cell renewal).** Aged or damaged cells are replaced. This type of regeneration is seen in many organisms, but may be absent in small, short-lived animals, such as *Caenorhabditis elegans* (Fig. 3–8). Regeneration in the sense of cell renewal takes place extensively in humans, although usually humans are thought to have a very low capacity to regenerate; after all, humans do not reconstruct amputated limbs. However, without repeated renewal of the inventory of short-lived cells, such as the pool of blood cells (Chapter 16), human life would be restricted to a few weeks.

2. **Reparative regeneration (reconstruction).** Replacing lost body parts is routine in many invertebrates. The regenerative capacity is most spectacular in sponges, coelenterates, and turbellarians, suggesting that the faculty for regeneration reflects the evolutionarily "primitive" position of those organisms. However, this picture is too simplistic. For example, among the turbellarians, *Mesostoma* is unable to regenerate. On the other hand, nematodes display a low capacity to regenerate but are not more complex in their body organization than planarians, which display excellent regenerative capabilities. Together with nematodes, the ascidians among the tunicates are considered to represent the prototype of the mosaic type of development, with low regulative capabilities in embryogenesis. Yet adult ascidians regenerate quite well, in contrast to nematodes. Thus, there is no straightforward correlation between the ability of an organism to repair and supplement body parts, and its phylogenetic ranking.

Arthropods are able to supplement incomplete legs as long as they undergo molts (Fig. 9–11). Among the amphibians, the urodeles (salamanders and newts) preserve an obvious faculty for regeneration, even after metamorphosis. They are capable of regenerating amputated limbs (Fig. 21–2, 21–3), the lost tail, and a removed eye lens (Fig. 21–4).

Nature is not always perfect. Sometimes the wrong organ is formed. This phenomenon—a different organ developing than the one that was removed—is called **heteromorphosis.** For example, in shrimps an antenna may be formed in place of an amputated eye stalk. Heteromorphosis is reminiscent of the phenomenon of homeotic transformation (see Chapter 3, Section 3.6.5).

Autotomy, the self-mutilation and shedding of body parts (tails in lizards, legs in arthropods) to escape an enemy, is, in fact, exceptional and serves to emphasize the rarity of reparative regeneration in land animals.

3. **Asexual reproduction.** Asexual reproduction is also a type of regeneration. Asexual propagation is natural, organismal **cloning.** Self-cloning is accomplished in several ways: through **fission** (various turbellarians and

annelids), through **budding** (*Hydra*, Fig. 3–5), or through multicellular **encysted bodies** (gemmulae in sponges, statoblasts in bryozoans).

4. **Reconstitution** is a special case in that it is induced experimentally. The process of reconstitution documents an astonishing faculty for self-organization of multicellular associations. If embryos (e.g., sea urchin blastulae), excised organs (e.g., the eye of an amphibian tail bud stage larva), or even entire animals (e.g., *Hydra*, Fig. 9–9) are dissociated carefully into single cells and reaggregates are prepared from the resulting cell suspension, the original structures can be more or less rebuilt completely (see Chapter 9, Section 9.9, and Section 21.2).

The processes of regeneration are among the most enigmatic phenomena in developmental biology. Intriguing questions include the following:

- How does an organism recognize that a part is missing, which part is missing, and how much of the part is missing? This raises the question about how regeneration is initiated and how the pattern is controlled.
- Where is the material for the substitute taken from? Is the substitute derived from residual embryonic founder cells, from permanently dividing stem cells, or from differentiated cells undergoing transdifferentiation?
- Is the original structure rebuilt by recruitment and reorganization of cells already present (**morphallaxis**) or restored from a few cells in the wound area that proliferate and grow out to replace the missing structure (**epimorphosis**)? In fact, actual regenerative achievements are based on a combination of transdifferentiation, cell migration, and cell proliferation rather than either extreme of pure recruitment or pure proliferation.

21.2 Multiheaded Hydras, Curious Planarians, and Insects with Broken Legs

When Hercules was sent out to perform his heroic deeds, he was confronted with the frightening hydra water snake, which regenerated two heads for each head that was cut off. Hydras today are perilous only for small animals such as water fleas, and if a forceful animal attaches to its tentacles and the polyp is torn into pieces, every piece of the hydra regenerates the lost part faithfully. Only by tricky experiments—for example, by repeatedly splitting the head lengthwise or treating the animals with tumor promoters—can multiheaded polyps be generated.

21.2.1 Reparative Regeneration, Reconstitution, and Pattern Control

Enormous regenerative abilities are possessed by several groups of animals located near the putative evolutionary trunk of the metazoan bush: sponges, coelenterates, and most turbellarians. A paragon of high regenerative ability is the freshwater polyp *Hydra*. Beginning in 1735, regeneration studies with *Hydra* rang in the era of experimental developmental biology. For example, if a ring is cut out from the middle of the *Hydra* body column comprising only 1/20 of the body length, the excised ring will regenerate a head at its upper end and a foot at its lower end.

Hydra can even be dissociated entirely into single cells. From the milky suspension, amorphous clumps of agglomerated cells can be recovered (Fig. 9–9). Over the course of several days or weeks, these clumps, called **reaggregates,** autonomously reorganize themselves to yield one or more complete polyps, depending on the mass of the reaggregate. Amoeboid movements of the cells, and mutual sorting out of ectodermal and endodermal cells contribute to this amazing process of self-organization, called **reconstitution.**

The high regenerative potential of *Hydra* is based on: (1) the plasticity of the state of cellular differentiation, and (2) a peculiar property of the freshwater polyp: in normal growth the polyp continuously replaces the whole inventory of its cells. Therefore, a system of **pattern control** must operate throughout its life. This system must supervise the replacement of old cells by young cells: the replacement must be quantitatively balanced and occur at the correct places—that is, it must be regulated positionally. An important parameter in this system of control is the **gradient in positional value:** this gradient ensures that the potential to form a head is higher in near-head regions of the body column, whereas the potential to form a foot is higher in the near-foot regions. Thus, the new head will always be made at the end of an excised segment that previously had been closer to the original head, while the foot will be made at the opposite end (see Chapter 9, Section 9.9).

When cells of disparate positional value are brought in close contact with each other by transplantation, conspicuous reactions are observed. At the site of confrontation, heads or feet may be formed (Fig. 9–10); subsequently, missing body parts are intercalated until the continuum of positional values is complete and smooth. For instance, if a fragment with the near-head value 8 is brought in contact with a piece having the near-foot value 2, the regions 7,6,5,4,3 become intercalated.

Positional values can also be assigned to the **imaginal discs** and **legs of insects.** Hemimetabolous insects that will undergo further molts are able to replace an amputated leg. If a middle piece of the leg is excised, and the distal part of the leg is grafted onto the proximal stump, differing positional values are made contiguous. This confrontation stimulates cell proliferation and the missing part is **intercalated** in the correct location. The discontinuity of positional values is smoothed out (Fig. 9–11). Such findings suggest that the molecular makeup of a cell's surface indicates positional value so that adjacent cells can recognize it.

Planarians, the flatworms, have eyes, a brain, and a rich internal organization. In spite of this structural complexity, some planarians (e.g., members of the genera *Planaria, Dugesia,* and *Polycelis*) have a regenerative ability comparable to that of *Hydra*, and pattern reconstitution after cutting follows similar rules.

Planarians may be cut across or lengthwise, and each part of the body will regenerate the missing part (Fig. 21–1). When the cut is made, a regeneration

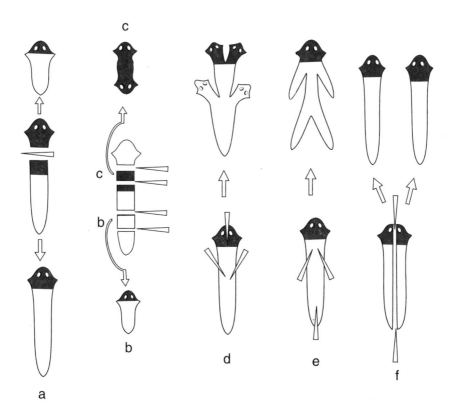

Figure 21–1   Regeneration in planarians. Examples of basic experiments done at the turn of the century by T.H. Morgan (1901).

blastema is formed at the cut surface, and the missing part is developed from the blastema. However, the remaining part is reorganized on a diminished scale. A lateral incision may cause the development of either an additional head or an additional tail. A head forms from a triangular part that faces anteriorly, the tail develops from a part facing posteriorly. Very narrow pieces cut out by two transverse cuts from the anterior body region may form a head on both the anterior and posterior cut surface: the internal gradient in positional values is not sufficiently steep to give the anterior blastema a start. Only a temporal advantage enables it to suppress competitive head formation at the posterior cut in time.

To explain pattern control in formal terms, much could be said about gradients, positional values, and intercalation. Instead of stressing the parallels to hydra and insects, we cast a glance at one of the strangest stories in the history of biosciences.

Once upon a time, in the 1960s and 1970s, many researches were engaged in the search for **memory molecules** that were thought to store all that has been learned. Planarians had to play an involuntary part. Punished by their human trainers, the worms were taught to avoid bright light and painful damage by contracting their bodies. When such planarians were cut, not only the anterior body part with the brain was claimed to have retained the lesson but also the regenerating posterior part. The lesson could be transferred to other worms if pieces of trained animals were fed to untrained ones. The memory molecules, able to store and transfer lessons, were said to be RNA (but encoded by which genes?). When the ghosts of the memory molecules suddenly disappeared, the interest in the regenerative abilities of planarians ceased. Worms (and professors alike) can now put their minds at ease: the risk of being eaten by cannibals who seek to avoid the hardship of learning has diminished.

Returning to well-founded science, we ask, what is the basis of the high regenerative ability of some animals?

21.2.2 Stem Cells or Transdifferentiation?

In the past the high regenerative capacity of "lower" animals has often been attributed to their hypothetical possession of "embryonic reserve cells," sometimes called **neoblasts.** As multipotent stem cells, neoblasts were thought to be capable of giving rise to every sort of cell and thus able to supply every lost cell type. In fact, multipotent (but not omnipotent) stem cells are found in sponges, hydrozoans, turbellarians, tunicates, and vertebrates—including humans! (Chapter 16).

In *Hydra* interstitial stem cells, called **I-cells,** are small cells residing in interstitial spaces between the epithelial cells. Their derivatives are nerve cells, stinging cells, certain gland cells, and the gametes (Fig. 3–6). Remarkably, in contrast to mammals, *Hydra* is able to replace even nerve cells. This unique capacity is the basis for the potential immortality of *Hydra*.

Interstitial stem cells can be eliminated selectively in *Hydra*; as a consequence, eventually nerve cells and stinging cells are lost. Amazingly, nerve-free hydras survive, if force-fed, and they retain the ability to regenerate and to bud new, albeit nerve-free, animals!

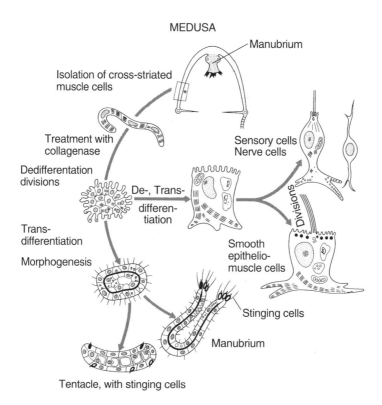

Figure 21–2 Transdifferentiation. This example documents the ability of cross-striated muscle cells of a medusa to dedifferentiate and give rise to other cell types, including germ cells. Cross-striated muscle cells are isolated from the subumbrella and destabilized by treatment with collagenase (to remove the stabilizing extracellular matrix) or by treatment with tumor promoters. The destabilized cells develop into a large variety of other cell types. These can organize themselves to form organs of a medusa, such as tentacles or a manubrium. (Redrawn after Schmid et al., *In Regulatory Mechanisms in Developmental Processes.* Elsevier, 1988.)

Even *Hydra* cannot produce all types of cells from multipotent I-cells. The ectodermal and endodermal epithelial cells, which form the tubelike body column, originate from their own kind. In the gastric body column even differentiated epithelial cells preserve the ability to divide. In the middle of the body column the differentiated state does not exclude the ability to proliferate: only in the head and foot is this ability lost. On the other hand, the state of differentiation can change. Epithelial cells are displaced gradually from the gastric region into the head or foot. Arriving there, the cells convert into battery cells of tentacles or gland cells of the foot. Certainly, this **plasticity of the differentiated state** makes essential contribution to the high regenerative capacity in these animals.

When an epithelial muscle cell of the *Hydra* body column is converted into an epithelial gland cell of the foot, this transformation may be interpreted as a modification of the state of differentiation. However, in hydromedusae (a type of small jellyfish) an amazing gift for **transdifferentiation** has been documented. Cross-striated muscle cells isolated from the underside of the bell-shaped umbrella are capable of undergoing dedifferentiation. Dedifferentiated muscle cells recover the ability to divide and can give rise to a variety of different cell types, including nerve cells and germ cells (Fig. 21–2).

21.3   Entire Animals Cannot (Yet) Be Recovered from Somatic Cells

In horticultural and agricultural breeding plants that cannot be multiplied by natural asexual reproduction or grafting can be cloned from isolated single somatic cells. Small pieces of phloem or leaves are excised and dissociated into single cells by enzymatically removing the cell wall. The resulting **protoplasts** are inoculated into a fluid containing nutrients and growth hormones. In an appropriate medium the cells divide and give rise to a conglomerate of largely undifferentiated and disorganized cells, called **callus.** Surprisingly, within such a callus an embryo can form in the absence of true germ cells or fertilization. The embryos can be seeded onto agar, where they develop into plantlets that grow up, flower, and ultimately produce seed.

Such success has not yet been achieved with isolated somatic cells of animals. The transdifferentiation of striated muscle cells in hydrozoans, mentioned in Section 21.2 (and shown in Fig. 21–2), suggests that in these organisms a method of cloning analogous to the methods used in plant breeding could be developed. However, such research is not yet driven by financial motives.

21.4 Reparative Regeneration and Transdifferentiation in Vertebrates

21.4.1 Regeneration of Limbs in Amphibians

The larvae of urodeles (newts, salamanders, and axolotls) are able to completely reconstruct amputated limbs, a capacity they retain all their lives. Limbs, especially those of the larvae but also of mating adults, are sometimes bitten off by various predators present in the pond. In anurans (frogs and toads) early larvae are capable of supplementing limb buds; with age, however, the potential to regenerate diminishes and disappears completely after metamorphosis.

In the freshly cut limb migrating epidermal cells cover the wound and protect the remaining stump. Beneath this protecting layer, some cells, such as chondrocytes and osteocytes, are lysed or undergo apoptosis. Most cells survive, yet disconnect from each other and undergo dedifferentiation. Irrespective of their origin (dermis, muscle, cartilage, connective tissue), the cells take the uncharacteristic shapes of fibroblasts or mesenchyme cells, and accumulate as a cell mass known as **regeneration blastema.** The dedifferentiated blastema cells proliferate and ultimately give rise to the new extremity. Proliferation is stimulated by growth factors presumably produced beneath the wound epithelium.

In contrast to limb buds appearing in embryogenesis, limb formation from blastemas is also dependent on the presence of residual nerve cells. A limb stump lacking its nervous supply does not regenerate (Fig. 21–3). Nervous tissue is the source of regeneration-promoting factors, one of which has been identified as **glial growth factor (GGF).**

An intriguing and enigmatic phenomenon remains to be explained: why does the blastema form successively more distal structures, beginning at exactly the appropriate level so that nothing will be missed and nothing will be supernumerary? For example, a blastema on a stump containing a humerus does not form another humerus and does not form digits prematurely. Expressed in formal terms, new positional values are laid down in an orderly manner so that the sequence from 0 (shoulder) to 10 (tip of digits) is complete and correct.

A surprising experimental result shows that this sophisticated developmental program can be reset: when the blastema of an amputated forearm is treated with **retinoic acid** (vitamin A acid), the positional value of the blastema cells is set to 0. The stump ignores the existing humerus and residual radius and ulna, and forms a complete limb, beginning with the shoulder and ending with digits (Fig. 21–4).

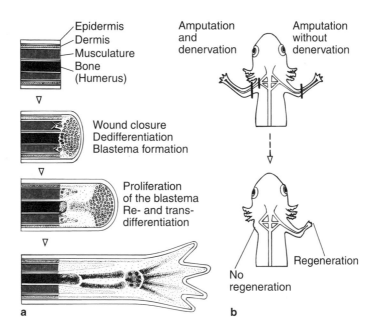

Figure 21–3 Regeneration of an extremity in urodeles (newts, salamanders, axolotls). (a) Wound healing; dedifferentiation of various cell types and formation of a blastema; proliferation, pattern formation, and differentiation result in a regenerated forelimb. Note: Regeneration restores those elements that are normally present distal to the cut surface. (b) Regeneration does not occur when the brachial nerves are removed completely (left side), whereas the innervated forelimb regenerates (right side).

Positional value is believed to be based in the expression pattern of homeobox-containg genes (*HOM/Hox* genes, Chapter 10) and is coded in the equipment of cell membranes with cell adhesion molecules (CAMs, Chapter 12). Retinoic acid influences both the expression pattern of *Hox* genes and that of the CAMs.

21.4.2 Transdifferentiation (Metaplasia)

Transformation of one differentiated cell type into another is observed even in vertebrates. A classic example is **wolffian lens regeneration** (Fig. 21–5). During embryo development of newts and salamanders, the lens is formed from the epidermis after the underlying eyecup has emitted an inducing signal. If the lens is removed later, a new lens forms from the dorsal iris rather

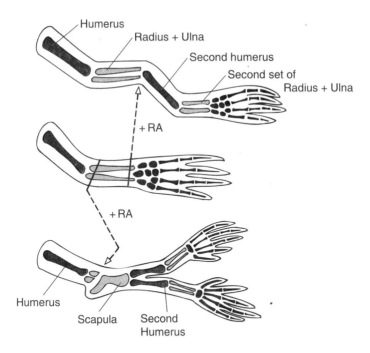

Figure 21–4 Regeneration of an extremity in urodeles under the influence of retinoic acid (RA). Middle: A normal forelimb is cut at one of the positions shown. Top: If the cut forelimb is bathed in a solution containing RA, the blastema makes a new start and begins with a humerus, after which the remainder of the forelimb is made. Bottom: With high concentrations of RA, positional values are reset to the lowest level. The blastema starts with the shoulder and sometimes splits up, in which case a mirror-image forelimb develops. (Experiments by Maden 1984.)

than from the epidermis. The iris is a derivative of the optic cup and thus of the brain, and consists of a type of smooth muscle cell that contains melanosomes. The transformation begins with dedifferentiation: melanosomes and muscular filaments are lost. Cell divisions are followed by differentiation into lens cells producing proteins called **lens crystallins.**

Frequently, but not in all instances, transdifferentiation assumes previous DNA replication. DNA replication would facilitate reprogramming and the acquisition of a new state of differentiation. Transdifferentiation of jellyfish striated muscle fibers into nerve cells depends on previous cell division but not transdifferentiation into smooth muscle cells.

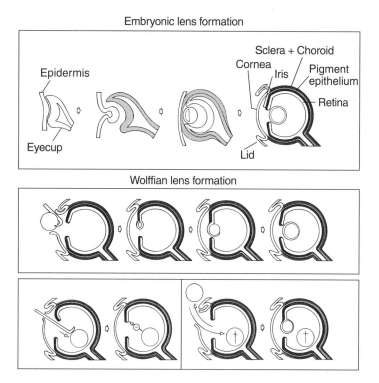

Figure 21–5   Transdifferentiation in the regeneration of a lens in the eye of a newt. Top: Normal development of the lens. As the optic cup touches the epidermis, it induces the formation of a lens placode. The placode invaginates, detaches from the epidermis, and differentiates to form the translucent lens. Middle: The lens is removed surgically. A new lens is formed not from the epidermis but from the dorsal margin of the iris. The iris is composed of smooth muscle cells; these undergo transdifferentiation. A lens pushed into the eyeball inhibits regenerative lens formation. If previously killed by heat treatment before being pushed into the eyeball, a lens no longer prevents regenerative lens formation.

21.5   The Limits of Regeneration Are the Seeds of Death

Apart from a few organisms, such as sponges and *Hydra*, multicellular animals—including, of course, humans—are not able to renew all of their cell types. For instance, aged nerve cells cannot be replaced by new ones. By the time we are 20 years old, our nerve cells begin to decay (not in all parts of our brain, as recent studies indicate). The successive loss of nerve cells leads inevitably to death. In some birds, however, recent research has discovered stem cells from which new nerve cells can be recruited when the behavior is adjusted to seasonal changes of the environment.

Stem cell biology, as it relates to the nervous system, is entering a new— exciting or terrifying—period. Mammalian embryonic nerve cells and glia cells are cultured as potential sources of neuronal tissue for clinical transplantation. However, to date, no efficient fountains of youth are known for human beings.

22

Life and Death:
What Is the Major Mystery?

22.1 "Immortal" Life Comes from the Capacity to Perpetuate Cell Divisions

Fundamental concepts linking programmed death to the evolution of multicellularity were advanced as early as 1881 by August Weismann, a zoologist and pioneer of genetic theories designed to explain development and cell differentiation. Weismann proposed that aging and decay are not inherent to life itself but are events that became integral to development only in the course of evolution of multicellular organisms. Only the multicellular organism inevitably would be doomed, through **senescence**—the process of aging.

A single-celled organism, such as an amoeba, increases in size when supplied with ample food and divides into two separate cells. Does division imply the end of individual life? We may debate whether the doubling of an individual means that the original individual's life has ended. After all, a corpse is not left behind. In a sense, organisms like bacteria or protists are immortal. The life of bacteria and protists is terminated by being devoured, or destroyed by parasites, or through environmental influences, but not through aging.

In protists, Weismann argued, there is no regular process (we would now say "program") of death in order to terminate the individual life, because every individual cell is simultaneously a **generative cell** that secures the continuation of the species. By contrast, in the multicellular organism a segregation of generative and somatic cells has evolved. **Somatic cells** can focus on a few functions and optimize them because they do not need to meet all requirements and accomplish all functions of life, including reproduction. Somatic cells can be fine-tuned to perform their particular task for the benefit of the entire cell community: some cells secrete enzymes and maximize the exploitation of food; other cells optimize sensory functions to exploit the environment and to recognize dangers or opportunities.

To focus on specialized functions to the detriment of cellular reproduction was possible because generative cells took over the task of reproducing the entire organism, standing proxy for the whole community. Gametes focus on the job of reproduction and propagation without being restrained by other occupations. Although both male and female gametes are highly specialized cells, they give rise to a totipotent zygote. The zygote is unable to perform any somatic functions by itself but is able to transmit to its daughter cells the ability to adopt every specialization and to differentiate into every cell type that is provided for by the genetic program of the species. Universal potency and perfection of particular somatic functions are mutually exclusive. A single-celled organism remains bound to compromises.

Death in metazoans is the regular fate of somatic cells. However, even in metazoans, there are, surprisingly, exceptions to this rule. Cells of metazoans maintained in culture for long periods of time, even human cells, are immortal just like protists. Because of their immortality, such cells can be kept alive and grown for years. Also, the freshwater polyp *Hydra* is, as a whole, potentially immortal.

However, cells grown in petri dishes have been subjected to selection: they represent stem cells with the programmed ability to divide, or are cells on the way to becoming cancer cells that have recovered immortality. *Hydra* is another peculiar case: all of its terminally differentiated cells, in fact, die, but every decaying cell is replaced by a substitute generated from immortal stem cells (Chapter 3, Section 3.3, and Chapter 21, Section 21.3). Only cells that have preserved the ability to divide are immortal. Apparently, perpetual cell divisions are a prerequisite to immortality, while loss of the capacity for cell division leads to death.

22.2 We Know Several Molecular, Cellular, and Organismal Causes of Aging

22.2.1 Even Proteins Age

Enzymes and other proteins are not in the energetically lowest conformational state when they are biologically active. Rather, they are in a metastable state, created during their synthesis, sometimes with the assistance of chaperones. Many influences, such as thermal energy, changing ionic strength and the pH of the solute, or collision with radicals, may throw a protein out of its metastable condition into a deeper energetic level: it **denatures.** The probability of spontaneous renaturation is minimal. Denatured proteins are useless and must be replaced by newly synthesized ones if the cell's metabolic or developmental programs still require their presence.

A terminally differentiated cell will have difficulties replacing all of its proteins. How could a heart muscle cell replace its contractile apparatus without interrupting its rhythmic beating? How could a nerve cell in the brain repair its numerous dendritic and axonal fibers and hundreds of synaptic connections while staying prepared to fire? Also, could such a nerve cell even procure all the needed building materials, crowded together with myriads of other competing nerve cells?

22.2.2 Limits Set by the Stability of Genetic Information

Above all, limits on the replacement of proteins comes from the condition of the DNA—the source of information needed for the synthesis of proteins. DNA itself suffers damage: thermal collision with water molecules, ionizing and UV irradiation, aggressive oxygen radicals—all cause mutations and irreversible loss of information. Every day a human cell loses about 5,000 purine bases (A or G) by irradiation-induced strand breaks, and 100 cytosine bases are converted into uracil. Cytosine spontaneously deaminates to uracil. After birth, when neuroblasts stop proliferating, the enzymes responsible for repair largely disappear from these cells.

A cell possesses multiple mechanisms of DNA repair. The mechanisms function as long as one of the two DNA strands remains intact and can serve as a template of the correct base sequence. Presumably, an extensive and complete correction of defects is only possible when the entire genome is replicated. This would explain why the only immortal cells are those that divide again and again. In mammals a correlation has been found between the capacity of the DNA repair system and the average life span of the species.

In addition, in unicellular organisms and cell communities alike, natural selection eliminates faulty cells. This may be the reason why so many cells undergo cell death in the germ line: only cells with (more or less) intact genetic information survive. Thus, sexual reproduction helps to perpetuate species. On the other hand, positive natural selection of cells with intact DNA presupposes continuing cell multiplication by mitosis. Therefore, the principle is invalid in tissues composed entirely of terminally differentiated cells, such as nervous tissue.

22.2.3 Limits Set by the Accumulation of Damage at the Cellular and Organismal Levels

Numerous age-associated irreversible changes and accumulating deficiencies have been identified. For example:

- Decline in the activity of antioxidant enzymes: damage to many biological molecules by oxygen-free radicals and oxygen reactive species accumulate. Due to oxidative impairment, the cytoskeleton is injured and many enzymes lose activity: **the efficiency of all physiological functions and regenerative processes decreases.**
- Deterioration of the thymus and disappearance of functional lymphocytes. As a result, **infectious diseases are less effectively resisted and overcome.**
- Decomposition of proteoglycans and structural proteins (in particular, hexauronates and elastic fibers) in the extracellular matrices of the cartilage, the dermis, and the blood vessels. Bond water is lost, skin and cartilage shrink, elasticity is lost, and **arteriosclerosis** (the hardening of the arteries) is favored. The resulting hypertension and mechanical stress favor **atherosclerosis,** the deposition of encrusting plaques in blood vessels.
- Accumulation of heart cells that are deficient in respiration: **the heart becomes enfeebled.**
- Renal blood flow and glomerular filtration decrease: **the efficacy of the kidneys decrease by 1% annually.**
- Nerve cells decay daily in our brain without being replaced, most abundantly in Alzheimer's disease but to a lesser extent in every human: the **performance and productivity of our brain diminishes.** The ability to cope with the tasks of everyday life decreases, and degenerative disorders eventually may result in dementia.

The last two causes alone restrict **the maximum longevity of humans to approximately 100 years.** Death is to be expected.

22.3   Even Senescence Still Harbors Mysteries: Are "Gerontogenes" Involved?

Although at the molecular, cellular, and organismal level many causes for irresistible aging can be found, we still do not understand why **longevity,** the maximum life span, is species specific. Medical science and public sanitation have increased our **life expectancy,** the average number of years an individual may expect to live, but they have scarcely increased longevity. Also, it is still enigmatic why in some organisms (humans, for example) senescence and aging occur slowly, while in others (mayflies and salmon) aging and death occur suddenly and dramatically.

The range of longevity among metazoans is considerable, of course: *Hydra* is immortal, *Caenorhabditis elegans* lives for 3 weeks, and *Homo sapiens* has a life span potential of at least 120 years, the record among mammals. Humans are among the longest-lived of all animals.

Apparently, the maximum life span in mammals is correlated with body size. Small mammals, whose consumption of energy and metabolic turnover are high and whose hearts beat fast, arrive at the end of their lives earlier than big animals with their slower way of life. Even among a taxonomically narrow clade, such as the primates, life spans range over an order of magnitude and generally follow body size (see Table 22–1).

TABLE 22–1   MAXIMUM LIFE SPAN POTENTIAL FOR SOME PRIMATES

Primate	Years
Tree shrews (*Tupaia, Tana*)	7
Marmorsets (*Callithrix, Leontides*)	15
Squirrel monkey (*Cercopithecus*)	21
Rhesus monkey (*Macaca mulatta*)	29
Gibbon	32
Baboon (*Papio*)	36
Gorilla	40
Chimpanzee	45
Orangtan	50
Human	120

Source: Gilbert 1991, extended.

However, this correlation is somewhat superficial and may hide more fundamental causes. Aging appears to have a genetic background. In organisms as varied as fungi, plants, mice, and even humans, mutations causing premature aging are known. In humans, several hereditary diseases, collectively termed **progeria,** have been described. These (fortunately infrequent) diseases result in premature senescence and death. In *progeria adultorum* (Werner syndrome) senility is diagnosed at the age of 39 years, on average, and death occurs at the age of 47 years, on average. In *progeria infantorum* (Hutchinson-Gilford syndrome) aging becomes apparent as early as in the child's first year. The children die when they are 12 to 18 years old, often from heart attacks, after having acquired wrinkled skin, white hair, and other attributes of the elderly.

22.4   Do Organisms Embody a Genetically Released Suicide Program?

Apoptosis, programmed cell death, affects certain cells as early as embryogenesis (Chapter 11, Section 11.4). Later in life many cells, for example, blood cells (Chapter 16, Section 16.4), die a few days or weeks after their "birth" (last cell division). However, because blood cells and other short-lived cells are replaced by substitutes generated from stem cells (Chapter 16), it is not so much the short-lived but the long-lived, irreplaceable cell that sets the limit to the life span.

Ever since it became possible to grow cells in culture, we learned that cell populations below a critical cell density die. Recent results have enabled a new interpretation of those observations: most cells survive only in social communities. Cells cut off from society commit suicide. The actual cause is the presence of **survival factors,** with which cells prevent each other from switching on their suicide program. Only in dense cultures do these factors accumulate sufficiently to exceed the threshold of critical concentration. Survival factors have been found, for instance, in cultures of oligodendrocytes, eye lens cells, and kidney cells.

However, all of the survival factors should be present in our body in sufficient quantities. Are we also under the regime of an internal clock that eventually starts a programmed death of the entire body?

Two hypotheses on genetically programmed **mechanisms of self-destruction** are recently discussed on the basis of preliminary experimental evidence.

1. Programmed decay of the mitochondrial DNA. Such a mechanism has been identified in *Podospora anserina*, a filamentous fungus belonging to the Ascomycetes and a favorite model system for genetically determined aging. The fungus gives rise to a filamentous network (mycelium) with a life span of 25 days in most wild-type strains. Senescence can be overcome by the synergistic action of two nuclear mutations. The mutants continue to grow without showing symptoms of aging, even after extended periods of vegetative growth. It must be emphasized that in these mutant fungi the acquired immortality is associated with an acquired ability of perpetual cell divisions. In the wild-type, senescent strain an **internal clock** initiates the expression of a few killing genes; these initiate a cascade of molecular events that destroy the mitochondrial DNA, and hence terminate the life of the organism.

2. **The telomere theory of cell senescence.** Cells in culture usually show a striking reluctance to continue proliferating forever, even when supplied with abundant food. After some lively rounds of division, they get tired, cease dividing, and die. The number of divisions they undergo may be 50 if the cells are taken from a human fetus; the number may be 40 if the cells are taken from a 40-year-old human; it may be 30 if the cells derive from a 80-year-old man. The diminishing readiness to divide is correlated with the gradual loss of nucleotides at the ends of the chromosomes.

Eukaryotic chromosomes have a particularly structured end region, called the **telomere.** This end region contains repetitive sequences. For example, in human cells it comprises hundreds of AGGGTT repeats. In each cell division a special enzyme, called **telomerase,** is responsible for adding the telomere repeat sequence to the daughter chromosomes in full length. However, this enzyme is always present and busily occupied with its job only in germ cells (and cancer cells). In ordinary somatic cells the amount of telomerase diminishes, telomeres become shorter and shorter, and eventually the chromosomes lose not only noncoding repeats but also coding genes. Once the chromosomes fall short of a critical length, the cells die. It is still a matter of discussion to what extent this phenomenon contributes to the termination of our life.

22.5   The Biological Meaning of Death

In multicellular organisms death has become an integral part of the life cycle. Through selection and evolution, mechanisms of programmed cell death have emerged and have been fixed in the genome. What then is the meaning of death?

Organisms have to make room for their offspring: old humans have to clear the way for their children. However, why has potentially immortal life not emerged with offspring generated sparingly (for instance, through asexual budding, as in *Hydra*) in order to compensate for losses due to accidents and to generate colonists for new living spaces?

Hydra has persisted over millions of years, but its achievements were modest in terms of the development of complexity and conquest of novel ecological niches. In the history of life sexual reproduction emerged as the dominant means by which eukaryotes perpetuate and evolve. Sexual reproduction makes possible the introduction of mutations into populations and their rapid spread; more importantly, it facilitates extensive recombination of allelic variants of genes. Sexual propagation enables organisms to endlessly offer new variants to natural selection. It favors adaptability in a changing world, and favors optimization of lifestyles, even in a constant environment.

Death makes way for new life, new not only in the sense of fresh and renewed but also in the sense of novel. An incessant sequence of ontogenies, an incessantly repeating sequence of being born, giving birth, and dying, has given rise to humans and to the inspiring diversity that the developmental patterns explored in this book have made possible.

Box 7

CONTEMPORARY TECHNIQUES IN DEVELOPMENTAL BIOLOGY

Cell lineages and **genealogical pedigrees** can be traced when the founder cells are labeled by microinjection of fluorescent dyes that are coupled to nondiffusible, high molecular weight dextran. Recently, **reporter genes** have been introduced (by retroviral vectors, DNA injection, or electroporation) to label founder cells. Well-designed reporter genes are replicated faithfully during mitosis and therefore are not diluted out in the course of cell divisions. A favorite reporter gene is the bacterial β-galactosidase gene *LacZ*. Before being injected, the *LacZ* gene is linked to a eukaryotic promoter that allows its expression in the recipient cell. The presence of transcribed and translated β-galactosidase (LacZ) can be detected by the β-galactosidase-catalyzed generation of a blue stain.

Two spectacular additions to the repertoire of reporter genes are the *luciferase* gene and the gene for the *green fluorescent protein (GFP)*. Both genes are derived from luminescing organisms.

Natural **luciferase** is produced by symbiotic bacteria in fireflies and many other luminescent species. Luciferase is an enzyme that catalyzes the oxidation of a substrate, generally termed "luciferin," that subsequently emits light. When luciferin and ATP are added to permeabilized cells or cell extracts, light flashes detected by photomultipliers indicate the presence of luciferase in recipient cells expressing the gene.

Even more convenient is the **green fluorescent protein. GFP** is responsible for the green bioluminescence of the jellyfish, *Aequorea victoria,* and other cnidarian species.

Box 7   Contemporary Techniques in Developmental Biology   331

GFP contains a chromophore composed of modified amino acids (Ser-Tyr-Gly). When the *GFP* gene is expressed, the chromophore within the GFP forms spontaneously. Due to this chromophore, the gene product is easily detected, even in living cells, with conventional fluorescence microscopes using standard long-wave UV sources. It is even possible to construct fusion genes by joining the *GFP* gene with a second gene and putting the fused gene under the control of a common promoter. The expression of the corresponding multifunctional protein encoded by the fused genes is manifested by intense fluorescence generated by the GFP moiety.

Spatial and temporal patterns of gene expression can be made visible at the following two levels: at the level of mRNA through in situ hybridization with labeled antisense RNA, and at the level of the translated protein by immunostaining with antibodies. **Fusion genes** are also useful tools to demonstrate expression patterns. The gene of interest is cloned, together with its promoter, to allow tissue- and position-specific expression in the recipient. Before its introduction into a recipient cell, the cloned gene is fused to a reporter gene such as *LacZ* or *GFP*, such that functional β-galactosidase or green fluorescent protein will be attached to the expressed gene product. The construct is introduced into the target cell, for example, a fertilized egg cell. If the gene of interest carries a homeobox, the gene product and its attached reporter is found in the nucleus. The presence of translated fusions is indicated by the fluorescence generated by the GFP moiety, or by the blue stain developed with the enzymatic assistance of LacZ.

Promoter trapping and **enhancer detection** are further methods to detect genes and to trace their region- and stage-dependent expression. Promoters and enhancers are regulatory DNA sequences located upstream of genes and controlling their transcription. Egg cells or their derivatives (e.g., blastomeres) are transfected with a foreign gene such as *LacZ,* whose transcripts are easily detected. The transgene lacks its own promoter, but it is flanked by sequences allowing its insertion into the genome of the host cell. Random insertion may integrate by chance the transgene between the regulatory sequences and the transcribed sequence of a native gene. If the promoter of the resident gene is activated by naturally occurring transcription factors, the transgene may be transcribed together with the neighboring resident gene. In the case of *LacZ,* a blue stain indicates when and where both adjacent genes—*LacZ* and the resident gene—are being transcribed. For example, genes transcribed in an imaginal disc of *Drosophila* can be detected by randomly inserting *LacZ* and selecting those transgenic lines that show staining specifically in the imaginal disc (Fig. 10–1, 10–2).

MUTAGENESIS AND "KNOCKOUT MICE"

To track down the function and significance of a particular gene, a powerful tool is the ability to completely knock out the gene (loss-of-function mutation), or to influence its function in a more subtle way by introducing planned mutations. Random mutations can easily be produced in certain model organisms introducing transposons (such as P elements in certain strains of *Drosophila*) or by introducing foreign DNA sequences that are flanked by insertion-mediating sequences taken from transposons or retroviruses (**insertional mutagenesis**). However, this method does not alter predictably a specific target gene, other genes are also likely affected. **Targeted mutagenesis** can be achieved when a DNA sequence is introduced that is a homolog of the gene in question but carries the mutation of choice. Sometimes, the resident gene is actually replaced by the introduced gene. Because such an event takes place only in rare cases (except in some haploid fungi, notably yeast, where it is routine), usually only one of two homologous chromosomes will carry the mutation. Nevertheless, researchers have developed techniques that allow the generation of mice having both alleles in the mutated version.

Standard "knockout mice" are made by specifically inactivating the gene of choice through targeted mutagenesis in cultured cells known as embryonic stem (ES) cells (Box 2). The ES cells are then injected into mouse embryos, where they have the potential to develop into all of the various cell types of a mouse, including germ cells. The resulting animals are then bred, and those whose germ cells—eggs or sperm—are derived from the ES cells will pass the mutated gene to their progeny. By subsequent conventional inbreeding, the mutation can be made homozygous.

TRANSGENIC ANIMALS

How can entire animals be provided with a foreign (e.g., human) gene? Experiments of this type are designed to identify the function of a gene and its contribution to inherited diseases. If medical interest is the context, most pioneer studies are made with mice. The method is similar to that used to introduce mutations. Mouse ES cells are loaded with an engineered gene construct by means of retroviral vectors or with physical methods such as microinjection. Frequently, the construct contains insertion-mediating sequences and an additional gene providing resistance to an antibiotic. This facilitates the identification of successfully transfected ES cells. Subsequently, the selected cells are introduced into blastocysts (Box 2). If such cells contribute to the production of germ cells, transgenic eggs or sperm are available, and by classical inbreeding of offspring, as described previously for mutated genes, animals that are heterozygous or homozygous for the transgene are obtained.

SWITCHING ON FOREIGN GENES

A promoter trapping procedure has been used to turn on the *eyeless* master gene introduced into imaginal discs (primordia of, for example, wings and legs) of *Drosophila*. Because the intact wild-type *eyeless+* gene was introduced into the egg cell, and was activated in the discs during metamorphosis, a fly hatched with eyes on its wings or legs (Fig. 10–2)! The method involved the previous introduction of two additional foreign elements that together led to the activation of the *eyeless+* gene in the imaginal disc: the construct inserted downstream of the disc-specific promoter contained a yeast gene coding for a yeast transcription factor. This factor was transcribed, together with the resident gene that is under the control of the same disc-specific promoter, and therefore was expressed regularly in the disc. The yeast transcription factor in turn switched on the *eyeless* gene because *eyeless* had been linked previously to the corresponding yeast promoter.

Related methods to artificially direct gene activity use transgenes linked to promoters containing a steroid hormone responsive element, or genes linked to a heat shock gene (*hsp-70*) promoter. The gene of interest then can be switched on by applying the steroid hormone or a heat shock. However, steroid hormones and heat shock transcription factors may also occur naturally, and not only when the researcher wants to turn on the gene. Moreover, the effectors (steroid hormones, heat shock) will turn on not only the transgene but also the resident hormone sensitive genes; they evoke pleiotropic effects.

Therefore, several new methods make use of prokaryotic promoters linked to eukaryotic genes. The promoters are selected carefully so that they can be turned on or off easily by applying nontoxic activators or suppressors. A prokaryotic control sys-

Box 7   Contemporary Techniques in Developmental Biology   333

tem used successfully is based on the resistance of *Escherichia coli* to the antibiotic **tetracycline.** Low levels of tetracycline switch on a tetracycline-resistance operon in the bacterium. When sequences from this operon are integrated into eukaryotic promoters at appropriate sites, the eukaryotic promoter can be switched on by adding nontoxic, cell membrane-permeating tetracycline to the host cell.

REVERSE GENETICS AND DIFFERENTIAL SCREENING TECHNIQUES

Meaningful cloning of genes and targeted manipulation of their activities presuppose the identification of developmentally relevant genes. "Developmentally relevant" genes are those that are involved in the control of development, or whose products (mRNA or protein) can be used as specific markers of advancing cell differentiation. However, there are only a few organisms in which a large variety of developmentally relevant genes could be identified using mutagenesis (random or targeted) and subsequent inbreeding schemes. Among the animals allowing a direct genetic approach are the fruit fly, *Drosophila;* the roundworm, *Caenorhabditis elegans;* the zebra fish, *Danio rerio;* and the mouse.

Of course, we also want to study gene expression in other animals, such as the sea urchin, *Hydra, Xenopus,* or the human, where a direct genetic approach is tedious or impossible. At present, two strategies are used to find developmentally relevant genes in such organisms.

The first strategy is known as **reverse genetics.** If the organism in question, say, *Xenopus,* contains genes that are identical or similar to known genes of, say, *Drosophila* (70% to 100% sequence identity), DNA fragments of the known *Drosophila* gene are used as hybridization probes to isolate ("fish") similar sequences from a library of *Xenopus* cDNA. For example, the homeobox of the *Antennapedia* gene is selected as a probe. When the probe is copied, usually by the Klenow fragment of the *Escherichia coli* DNA polymerase I, it is labeled by incorporation of a radioactive or chemically tagged nucleotide into the newly synthesized DNA or RNA strand. The labeled probe is used for nucleic acid hybridization reactions (described in numerous molecular biology laboratory manuals). By using a labeled *Antennapedia* sequence as a heterologous (foreign) probe, the presence of *Antennapedia*-like sequences in *Xenopus* can be detected in Northern or Southern hybridizations under reduced stringency (under conditions allowing the formation of stable hybrid molecules even in the presence of some mismatches between the probe and the target sequence). If a related sequence is present in the target library, it is isolated, multiplied by cloning, sequenced, and used itself as a probe to isolate adjacent sequences of the gene. By repeating the procedure, the entire gene can be reconstructed. Polymerase chain reaction (PCR) is a convenient method to speed up the whole procedure.

The second strategy uses **differential screening techniques.** Many such techniques, all using PCR, have been developed. The principle of differential screening starts with isolating mRNA from different developmental stages, or from different parts of the body. The vast majority of transcripts are identical in the two extracts and are not analyzed further. Of interest are the few mRNAs that are transcribed differentially. Using these mRNAs, or fragments thereof, genes can be identified that are unique to the developmental stage and body part under investigation. Even genes unique to the species can be identified. The biological significance of differentially expressed genes, however, is not evident generally merely from their identification. Targeted mutagenesis or targeted overexpression are two methods that can be used subsequently to trace the function of the genes.

To summarize, (fragments of) an identified gene, or oligonucleotides derived from the gene or its mRNA, are used as follows:

- As probes for in situ hybridization, to study the spatiotemporal expression pattern in embryo preparations

- As probes to find overlapping sequences in a genomic library, until repetition of the procedure has yielded the entire sequence of the gene (including its control region)

- To construct probes for targeted mutagenesis

- To construct fusions with reporter genes

- To introduce the gene, or selected fragments of it, into expression vectors. Such vectors are designed to synthesize the protein or peptide coded by the gene. The proteins or peptides can be used subsequently to raise specific antibodies. These antibodies can in turn be used to localize the proteins in the embryo by immunostaining.

The number of available methods is large and is growing daily. These methods have truly caused a revolution in developmental biology.

Bibliography

---◆---

Review Journals, General Biology

Bioessays
Current Opinion in Genetics & Development
Nature
Science
Scientific American
Trends in Biochemical Sciences
Trends in Cell Biology
Trends in Genetics
Trends in Neurosciences

Journals of Developmental Biology

Anatomy and Embryology
Cell
Development
Development, Genes and Evolution (Formerly: *Roux's Archives of Developmental Biology*)

Developmental Biology
Developmental Genetics
Differentiation, Ontogeny, Neoplasia, Differentiation Therapy
The International Journal of Developmental Biology
Mechanisms of Development

Monographs and Textbooks, General Developmental Biology

Alberts, B., et al. (1994): *Molecular Biology of the Cell*. 3rd ed. Garland, New York.

Balinsky, B.I. (1981): *An Introduction to Embryology*. 5th ed. Holt-Saunders, Philadelphia.

Gilbert, S.F. (1994): *Developmental Biology*. 4th ed. Sinauer Associates, Sunderland, MA.

Huettner, A.F. (1949): *Comparative Embryology of Vertebrates*. Macmillan, New York.

Kalthoff, K. (1996): *Analysis of Biological Development*. McGraw-Hill, New York.

Langman, J. (1981): *Medical Embryology*. 4th ed. Williams & Wilkins, Baltimore.

Morgan, T.H. (1927): *Experimental Embryology*. Columbia Univ. Press, New York.

Saunders, J.W., Jr. (1970): *Patterns and Principles of Animal Development*. Macmillan, New York.

Waddington, C.H. (1956): *Principles of Embryology*. Allen & Unwin, London.

Chapter 3. Model Organisms

3.1 Sea Urchin

Billet, F.S., and Wild, A.E. (1975): *Practical Studies of Animal Development, Echinoderms and Ascidians*. Chapman & Hall, London.

Czihak, G. (1975): *The Sea Urchin Embryo*. Springer-Verlag, Heidelberg.

Davidson, E.H. (1989): Lineage specific gene expression and the regulative capacities of the sea urchin embryo: A proposed mechanism. *Development* 105:421–445.

Hardin, J. (1994): The sea urchin. *In* Bard, J.B.L. (ed.) *Embryos, Color Atlas of Development*, pp. 37–53. Wolfe, London.

Hörstadius, S. (1973): *Experimental Embryology of Echinoderms*. Clarendon Press, Oxford.

Whitaker, M., and Swann, K. (1993): Lighting the fuse at fertilization. *Development* 117:1–12.

3.2 Dictyostelium

Bozzaro, S. (1992): *Dictyostelium:* From unicellularity to multicellularity. *In* Russo, V.E.A., et al. (eds.) *Development: The Molecular Genetic Approach*, pp. 137–149. Springer-Verlag, Heidelberg.

Brookman, J.J., Jermyn, K.A., and Kay, R.R. (1987): Nature and distribution of the morphogen DIF in the *Dictyostelium* slug. *Development* 100: 119–124.

Kay, R.R., Berks, M., and Traynor, D. (1989): Morphogen hunting in *Dictyostelium*. *Development* (Suppl.):81–90.

Kay, R., and Insall, R. (1994): *Dictyostelium discoideum*. *In* Bard, J.B.L. (ed.) *Embryos, Color Atlas of Development*, pp. 23–35. Wolfe, London.

Konijin, T.M., van der Meene, J.G.C., Bonner, J.T., and Barkley, D. (1967): The acrasin activity of adenosine-3′, 5′-cyclic phosphate. *Proc. Natl. Acad. Sci. USA* 58:1152–1154.

Loomis, W.F. (1975): *Dictyostelium discoideum. A Developmental System.* Academic Press, New York.

Ohmori, T., and Maeda, Y. (1987): The developmental fate of *Dictyostelium discoideum* cells depends greatly on the cell-cycle position at the onset of starvation. *Cell Differ.* 22:11–18.

Oohata, A.A. (1995): Factors controlling prespore cell differentiation in *Dictyostelium discoideum:* Minute amounts of differentiation-inducing factor promote prespore cell differentiation. *Differentiation* 59:283–288.

Schaap, P. (1991): Intercellular interactions during *Dictyostelium* development. *In* Dworkin, M. (ed.) *Microbial Cell-Cell Interactions*, pp. 147–178. American Society for Microbiology, Washington, D.C.

Takeuchi, I., Tosaka, M., Okamoto, K., and Maeda, Y. (1994): Regulation of cell differentiation and pattern formation in *Dictyostelium* development. *Int. J. Dev. Biol.* 38:311–319.

Weijer, C.J., Duschl, G., and David, C.N. (1984): Dependence of cell-type proportioning and sorting on cell cycle phase in *Dictyostelium discoideum*. *Exp. Cell Res.* 70:133–145.

3.3 Hydra and Other Hydrozoa

Berking, S. (1986): Transmethylation and control of pattern formation in hydrozoa. *Differentiation* 32:10–16.

Bode, P.M., and Bode, H.R. (1984): Patterning in *Hydra*. *In* Malacinski, G.M., and Bryant, S.V. (eds.) *Pattern Formation*, Vol. I, pp. 213–241, Macmillan, New York.

Gierer, A. (1977): Biological features and physical concepts of pattern formation exemplified by *Hydra*. *In* Moscona, A.A., and Monroy, A. (eds.) *Pattern Development. Curr Top. Dev. Biol.* 11:17–58.

Lange, R.G., Müller, W.A. (1991): SIF, a novel morphogenetic inducer in hydrozoa. *Dev. Biol.* 11:17–58.

Müller, W.A. (1975): *Hydractinia echinata*. Ablaichen, Embryonalentwicklung, Metamorphose. Film E2080 mit Begleittext. Institut für Wissenschaftlichen Film, Göttingen, Germany.

Müller, W.A. (1995): Competition for factors and cellular resources as a principle of pattern formation in *Hydra*. *Dev. Biol.* 167:159–174 (Part I); 175–189 (Part II).

Müller, W.A. (1996): Pattern formation in the immortal *Hydra*. *Trends Genet.* 11:91–96.

3.4 Caenorhabditis elegans

Bossinger, O., and Schierenberg, E. (1992): Cell-cell communication in the embryo of *Caenorhabditis elegans*. *Dev. Biol.* 151:401–409.

Boveri, T. (1904 and 1910): See Bibliography for Box 1.

Brenner, S. (1974): The genetics of *Caenorhabditis elegans*. *Genetics* 77: 71–94.

Edgar, L. (1992): Embryogenesis in *Caenorhabditis elegans*. *In* Russo, V.E.A., et al. (eds.) *Development: The Molecular Genetic Approach*, pp. 273–294. Springer-Verlag, Heidelberg.

Hope, I.A. (1994): 4. *Caenorhabditis elegans In* Bard, J.B.L. (ed.) *Embryos, Color Atlas of Development*, pp. 55–75. Wolfe, London.

Ruvkun, G. (1992): Generation of temporal and cell lineage asymmetry during *C. elegans* development. *In* Russo, V.E.A., et al. (eds.) *Development: The Molecular Genetic Approach*, pp. 295–307. Springer-Verlag, Berlin.

Schierenberg, E. (1982): Development of the nematode *Caenorhabditis elegans*. *In Developmental Biology of Freshwater Invertebrates*, pp. 249–281. Alan R. Liss, New York.

3.5 Spiralians

Anderson, D.T. (1973): *Embryology and Phylogeny in Annelids and Arthropods*. Pergamon Press, Oxford.

Atkinson, J.W. (1987): An atlas of light micrographs of normal and lobe-less larvae of the marine gastropood *Ilyanassa obsoleta*. *Int. J. Invert. Reprod. Dev.* 9:169–178.

Biggelaar, J.A.M., van den, Dictus, W.J.A.G., and Serras, F. (1994): Molluscs. *In* Bard, J.B.L. (ed.) *Embryos, Color Atlas of Development*, pp. 77–91. Wolfe, London.

de Laat, S.W., et al. (1980): Intercellular communication patterns are involved in cell determination in early molluscan development. *Nature (London)* 287:546–548.

Dorrestejin, A., et al. (1993): Molecular specification of cell lines in the embryo of *Platynereis* (Annelida). *Roux's Arch. Dev. Biol.* 202:260–269.

Freeman, G., and Lundelius, J.W. (1982): The developmental genetics of dextrality and sinistrality in the gastropod *Lymnea peregra*. *Roux's Arch. Dev. Biol.* 191:69–83.

Gourrier, P., et al. (1978): Significance of the polar lobe for the determination of dorsoventral polarity in *Dentalium vulgare* (da Costa). *Dev. Biol.* 53: 233–242.

Harrison, W., and Cowden, R.R. (1982): *Developmental Biology of Freshwater Invertebrates*. Alan R. Liss, New York.

Raven, C.P. (1966): *Morphogenesis: The Analysis of Molluscan Development*. Pergamon Press, Oxford.

Reverberi, G. (1971): Experimental embryology of marine and freshwater invertebrates. North-Holland Publishing Co., Amsterdam.

Weisblat, D.A. (1994): The leech. *In* Bard, J.B.L. (ed.) *Embryos, Color Atlas of Development*, pp. 93–112. Wolfe, London.

3.6 Drosophila

Beachy, P.A. (1990): A molecular view of the Ultrabithorax homeotic gene of *Drosophila*. *Trends Genet.* 6(2):46–51.

Campos-Ortega, J.A., and Hartenstein, V. (1996): *Embryonic Development of* Drosophila melanogaster. 2nd ed. Springer-Verlag, Heidelberg.

Campos-Ortega, J.A., and Knust, E. (1992): Genetic mechanisms in early neurogenesis of *Drosophila melanogaster*. *In* Russo, V., et al. (eds.) *Development: The Molecular Genetic Approach*, pp. 341–354. Springer-Verlag, Heidelberg.

Driever, W., and Nüsslein-Volhard, C. (1988): The *bicoid* protein determines position in the *Drosophila* embryo in a concentration-dependent manner. *Cell* 54:95–104.

Driever, W., Siegel, V., and Nüsslein-Volhard, C. (1990): Autonomous determination of anterior structures in the early *Drosophila* embryo by the bicoid morphogen. *Development* 109:811–820.

Govind, S., and Steward, R. (1991): Dorsoventral pattern formation in *Drosophila*. *Trends Genet.* 7:119–124.

Heemskerk, J., et al. (1994): *Drosophila hedgehog* acts as a morphogen in cellular patterning. *Cell* 76:449–460.

Lawrence, P.A. (1992). *The Making of a Fly. The Genetics of Animal Design.* Blackwell Scientific, Oxford.

Leptin, M. (1994): *Drosophila. In* Bard, J.B.L. (ed.) *Embryos, Color Atlas of Development*, pp. 113–134. Wolfe, London.

Micklem, D.R. (1995): mRNA localisation during development. *Dev. Biol.* 172:377–395.

Rongo, C., and Lehmann, R. (1996): Regulated synthesis, transport and assembly of the *Drosophila* germ plasm. *Trends Genet.* 12:102–109.

St. Johnston, D., and Nüsslein-Volhard, C. (1992): The origin of pattern and polarity in the *Drosophila* embryo. *Cell* 68:201–219.

Struhl, G. (1981): A homeotic mutation transforming leg to antenna in *Drosophila*. *Nature (London)* 292:635–638.

Struhl, G., Struhl, K., and Macdonald, P.M. (1989): The gradient morphogen bicoid is a concentration-dependent transcriptional activator. *Cell* 57:1259–1273.

Tautz, D. (1992): Genetic and molecular analysis of early pattern formation in *Drosophila. In* Russo, V.E.A., et al. (eds.) *Development: The Molecular Genetic Approach*, pp. 308–327. Springer-Verlag, Heidelberg.

3.7 Tunicates

Bates, W.R., and Jeffery, W.R. (1987): Localization of axial determinants in the vegetal pole region of ascidian eggs. *Dev. Biol.* 124:65–76.

Conklin, E.G. (1905): Mosaic development in ascidian eggs. *J. Exp. Zool.* 2:145–223.

Jeffery, W.R. (1990): An ultraviolet-sensitive maternal mRNA encoding a cytoskeletal protein may be involved in axis formation in the ascidian embryo. *Dev. Biol.* 141:141–148.

Meedel, T.H., Crowthier, R.J., and Wittaker, J.R. (1987): Determinative properties of muscle lineages in ascidian embryos. *Development* 100:245–260.

Nishida, H. (1990): Determinative mechanisms in secondary muscle lineages of ascidian embryos: Development of muscle-specific features in isolated muscle progenitor cells. *Development* 108:559–568.

Nishida, H. (1992): Developmental potential for tissue differentiation of fully dissociated cells of the ascidian embryo. *Roux's Arch. Dev. Biol.* 201: 81–87.

Sardet, C., et al. (1989): Fertilization and ooplasmic movements in the ascidian egg. *Development* 105:237–249.

Whittaker, J.R. (1979): Cytoplasmic determinants of tissue differentiation in the ascidian egg. *In* Subtelny, S., and Konigsberg, I.R. (eds.) *Determinants of Spatial Organization*, pp. 29–51. Academic Press, New York.

Whittaker, J.R. (1980): Acetylcholinesterase development in extra cells caused by changing the distribution of myoplasm in ascidian embryos. *J. Embryol. Exp. Morphol.* 55:343–354.

Whittaker, J.R. (1987): Cell lineages and determinants of cell fate in development. *Am. Zool.* 27:607–622.

3.8 Xenopus, *Amphibians*

Billet, F.S., and Wild, A.E. (1975): *Practical Studies of Animal Development, Amphibians.* Chapman & Hall, London.

Bolce, M.E., et al. (1992): Ventral ectoderm of *Xenopus* forms neural tissue, including hindbrain, in response to activin. *Development* 115:681–688.

Cho, K.W.Y., et al. (1991): Molecular nature of Spemann's organizer: The role of the *Xenopus* Homeobox gene *goosecoid*. *Cell* 67:1111–1120.

Chui, Y., et al. (1995): *Xwnt*-8b: A maternally expressed *Xenopus Wnt* gene with a potential role in establishing the dorsoventral axis. *Development* 121: 2177–2186.

Dohrmann, C.E., et al. (1993): Expression of activin mRNA during early development in *Xenopus laevis*. *Development* 157:474–483.

Gerhart, J., et al. (1981): A reinvestigation of the role of the grey crescent in axis formation in *Xenopus laevis*. *Nature* 292:511–516.

Gerhart, J., et al. (1986): Amphibian early development. *BioScience* 36: 541–549.

Gilbert, S.F., Saxén. L. (1993): Spemann's organizer: Models and molecules. *Mech. Dev.* 41:73–89.

Grunz, H. (1993): The dorsalization of Spemann's organizer takes place during gastrulation in *Xenopus laevis* embryos. *Dev. Growth Differ.* 35(1):25–32.

Grunz, H., Schüren C., and Richter, K. (1995): The role of vertical and planar signals during the early steps of neural induction. *Int. J. Dev. Biol.* 39: 539–543.

Gurdon, J.B. (1987): Embryonic induction—molecular prospects. *Development* 99:285–306.

Gurdon, J.B., et al. (1994): Activin signalling and response to a morphogen gradient. *Nature* 371:487–492.

Hausen, P., and Riebesoll, M. (1991): *The Early Development of* Xenopus laevis. Springer-Verlag, Heidelberg.

Hemmati-Brivaniou, A., Kelly, O.G., and Melton, D.A. (1994): Follistatin, an antagonist of activin, is expressed in the Spemann organizer and displays direct neutralizing activity. *Cell* 77:283–295.

Hemmati-Brivaniou, A., and Melton, D.A. (1994): Inhibition of activin receptor signaling promotes neuralization in *Xenopus*. *Cell* 77:273–281.

Henry, J.J., and Grainger, R.M. (1990): Early tissue interactions leading to embryonic lens formation in *Xenopus laevis*. *Dev. Biol.* 141:149–163.

Jacobson, A.G. (1994): Normal neurulation in amphibia. *In* Bock, G., and Marsh, J. (eds.) *Neural Tube Defects*, pp. 49–65. John Wiley & Sons, New York.

Keller, R.E. (1986): The cellular basis of amphibian gastrulation. *In* Browder, L. (ed.) *Developmental Biology: A Comprehensive Synthesis*. Vol. 2, pp. 241–327. Plenum, New York.

Kessler, D.S., and Melton, D.A. (1995): Induction of dorsal mesoderm by soluble, mature Vg1 protein. *Development* 121:2155–2164.

Kimelman, D., et al. (1992): Synergistic principles of development: Overlapping patterning systems in *Xenopus* mesoderm induction. *Development* 116: 1–9.

Micklem, D.R. (1995): mRNA localisation during development. *Dev. Biol.* 172:377–395.

Niehrs, C., et al. (1993): The homeobox gene *goosecoid* controls cell migration in *Xenopus* embryos. *Cell* 72:491–503.

Niehrs, C., Steinbeisser, H., and De Robertis, E.M. (1994): Mesodermal patterning by a gradient of the vertebrate homeobox gene *goosecoid*. *Science* 263:817–820.

Nieuwkoop, P.D. (1977): Origin and establishment of embryonic polar axes in amphibian development. *In* Moscona, A.A., and Monroy, A. (eds.) *Pattern Development. Curr. Top. Dev. Biol.* 11:115–132.

Nieuwkoop, P.D., and Faber, J. (1975): *Normal Table of* Xenopus laevis (*Daudin*), 2nd ed., North-Holland Publishing Co., Amsterdam.

Otte, A.P., et al. (1988): Protein kinase C mediates neural induction in *Xenopus laevis*. *Nature* 334:618–620.

Pieler, T. (1992): *Xenopus* embryogenesis. *In* Russo, V.E.A., et al. (eds.) *Development: The Molecular Genetic Approach*, pp. 355–369. Springer-Verlag, Heidelberg.

Rugh, R. (1962): *Experimental Embryology*. Burgess, Minneapolis.

Sasai, Y., et al. (1994): *Xenopus chordin:* A novel dorsalizing factor activated by organizer-specific homeobox genes. *Cell* 79:779–790.

Sasai, Y., et al. (1995): Regulation of neural induction by the *Chd* and *Bmp-4* antagonistic patterning signal in *Xenopus. Nature* 376:333.

Sive, H.L. (1993): The frog prince-ss: A molecular formula for dorsoventral patterning in *Xenopus. Genes Dev.* 7:1–12.

Slack, J.M.W. (1993): Embryonic induction. *Mech. Dev.* 41:91–107.

Slack, J.M.W. (1994): *Xenopus. In* Bard, J.B.L. (ed.) *Embryos, Color Atlas of Development,* pp. 149–166. Wolfe, London.

Smith, W.C., and Harland, R.M. (1992): Expression cloning of noggin, a new dorsalizing factor localized to the Spemann organizer in *Xenopus* embryos. *Cell* 70:829–840.

Smith, W.C., et al. (1993): Secreted *noggin* protein mimics the Spemann organizer in dorsalizing *Xenopus* mesoderm. *Nature* 361:547–549.

Sokol, S.Y., and Melton, D.A. (1992): Interaction of wnt and activin in dorsal mesoderm induction in *Xenopus. Dev. Biol.* 154:348–355.

Sosoi, Y., et al. (1994): *Xenopus chordin:* A novel dorsalizing factor activated by organizer-specific homeobox genes. *Cell* 79:779–790.

Steinbeisser, H., et al. (1993): *Xenopus* axis formation: Induction of *goosecoid* by injected *WXwnt-8* and activin mRNAs. *Development* 118:499–507.

Vincent, J.P., and Gerhart, J.C. (1987): Subcortical rotation in *Xenopus* eggs: An early step in embryonic axis formation. *Dev. Biol.* 123:526–529.

Wischnitzer, S. (1975): *Atlas and Laboratory Guide for Vertebrate Embryology.* McGraw-Hill, New York.

3.9 Danio (formerly *Brachydanio,* Zebra Fish)

Hisaoka, K.K., and Battle, H.I. (1985): The normal developmental stages of the zebrafish *Brachydanio rerio* (Hamilton-Buchanan). *J. Morphol.* 102: 311–328.

Laale, H.W. (1977): The biology and use of zebrafish, *Brachydanio rerio* in fisheries research. *J. Fish Biol.* 10:121–173.

Metcalfe, W.K. (1994): The zebrafish. *In* Bard, J.B.L. (ed.) *Embryos, Color Atlas of Development,* pp. 135–147. Wolfe, London.

Rugh, R. (1962): *Experimental Embryology.* Burgess, Minneapolis.

Strähle, U., and Ingham, P.W. (1992): Zebrafish development: Flight of fancy or a major new school? *Curr. Biol.* 2:135–139.

Warga, R.M., and Kimmel, C.B. (1990): Cell movements during epiboly and gastrulation in zebrafish. *Development* 108:569–580.

Woo, K., and Fraser, S.E. (1995): Order and coherence in the fate map of the zebrafish nervous system. *Development* 121:2595–2609.

3.10 Bird

Billet, F.S., and Wild, A.E. (1975): *Practical Studies of Animal Development, Birds.* Chapman & Hall, London.

Patten, B.M. (1951): *Early Embryology of the Chick.* 5th ed. McGraw-Hill, New York.

Romanoff, A.L. (1960): *The Avian Embryo.* Macmillan, New York.

Rugh, R. (1962): *Experimental Embryology.* Burgess, Minneapolis.

Stern, C.D. (1994): The chick. *In* Bard, J.B.L. (ed.) *Embryos, Color Atlas of Development*, pp. 167–182. Wolfe, London.

Stern, C.D., and Canning, D.R. (1990): Origin of cells giving rise to mesoderm and endoderm in chick embryo. *Nature* 343:273–275.

Wischnitzer, S. (1975): Atlas and laboratory guide for vertebrate embryology. McGraw-Hill, New York.

3.11 Mouse

Bard, J.B.L., and Kaufmann, M.H. (1994): The mouse. *In* Bard, J.B.L. (ed.) *Embryos, Color Atlas of Development*, pp. 183–206. Wolfe, London.

Barlow, D.P. (1992): Cloning developmental mutants from the mouse t complex. *In* Russo, V.E.A., et al. (eds.) *Development: The Molecular Genetic Approach*, pp. 394–408. Springer-Verlag, Heidelberg.

Billet, F.S., and Wild, A.E. (1975): *Practical Studies of Animal Development, Mammals.* Chapman & Hall, London.

Bürki, K. (1986): *Experimental Embryology of the Mouse.* S. Karger, Basel.

Hogan, B., Constantini, F., and Lacy, E. (1986): *Manipulating the Mouse Embryo.* Cold Spring Harbor Laboratory, Cold Spring Harbor, NY.

Lobe, C.G., and Gruss, P. (1992): From *Drosophila* to mouse. *In* Russo, V.E.A., et al. (eds.) *Development: The Molecular Genetic Approach*, pp. 382–393. Springer-Verlag, Heidelberg.

Rugh, R. (1967): *Experimental Embryology.* Burgess, Minneapolis.

Surani, M.A.H., Barton, S.C., and Norris, M.L. (1986): Nuclear transplantation in the mouse: Hereditable differences between parental genomes after activation of the embryonic genome. *Cell* 45:127–136.

Theiler, K. (1989): *The House Mouse. Atlas of Embryonic Development.* 2nd printing. Springer-Verlag, Heidelberg.

Thomson, J.A., and Solter, D. (1989): The developmental fate of andro-genetic, parthenogenetic, and gynogenetic cells in chimeric gastrulating mouse embryos. *Genes Dev.* 2:1344–1351.

Wagner, E.F., and Keller, G. (1992): The introduction of genes into mouse embryos and stem cells. *In* Russo, V.E.A., et al. (eds.) *Development: The Molecular Genetic Approach*, pp. 440–458. Springer-Verlag, Heidelberg.

Wischnitzer, S. (1975): *Atlas and Laboratory Guide for Vertebrate Embryology.* McGraw-Hill, New York.

3.12 Human

England, M.A. (1994): The human (by M. England). *In* Bard, J.B.L. (ed) *Embryos, Color Atlas of Development*, pp. 207–220. Wolfe, London.

Hinrichsen, K.V. (1990): *Human Embryology.* Springer-Verlag, Heidelberg.

Langman, J. (1989): *Medical Embryology.* The Williams & Wilkins Co., Baltimore.

Moore, K.L. (1990): Grundlagen der medizinischen Embryologie. Enke, Stuttgart, Germany.

Tuchmann-Duplessis, H., David, G., and Haegel, P. (1972): *Illustrated Human Embryology.* Vol. 1 and 2. Springer-Verlag, New York.

Chapter 4. Comparative Review:
The Phylotypic Stage of Vertebrates, Aspects of Evolution

Davidson, E.H., Peterson, K.J., and Cameron, R.A. (1995): Origin of bilaterian body plans: Evolution of developmental regulatory mechanisms. *Science* 270:1319–1325.

De Robertis, E.M., and Sasai, Y. (1996): A common plan for dorsoventral patterning in bilateria. *Cell* 380:37–40.

Gilbert, S.F., Opitz, J.M., and Raff, R.A. (1996): Resynthesizing evolutionary and developmental biology. *Dev. Biol.* 173:357–372.

Haeckel, E. (1892): The history of creation. Translation of *Natürliche Schöpfungsgeschichte.* Kegan Paul, Trench, Trubner; London.

Hogan, B.L.M. (1995): Upside-down ideas vindicated. *Nature* 376: 210–211.

Holland, P.W.H., and Carcia-Fernandez, J.G. (1996): *Hox* genes and chordate evolution. *Dev. Biol.* 173:382–395.

Hunt, P., et al. (1991): Homeobox genes and models for patterning the hindbrain and brachial arches. *Development* (Suppl. 1):187–169.

Nübler-Jung, K., and Arendt, D. (1994): Is ventral in insect dorsal in vertebrates? *Roux's Arch. Dev. Biol.* 203:357–366.

Richardson, M.K. (1995): Heterochrony and the phylotypic period. *Dev. Biol.* 172:412–421.

Sander, K. (1983): The evolution of patterning mechanisms: Gleanings from insect embryogenesis and spermatogenesis. *In* Goodwin, B.C., et al. (eds.) *Developmental and Evolution*, pp. 123–159. Cambridge Univ. Press, New York.

Valentine, J.W., Erwin, D.H., and Jablonski, D. (1996): Developmental evolution of metazoan body plans: The fossil evidence. *Dev. Biol.* 173: 373–381.

von Baer, K.E. (1828): *Über Entwickelungsgeschichte der Thiere.* Königsberg, Germany.

Chapter 5. Gametogenesis, Genomic Imprinting

Eddy, E.M., et al. (1981): Origin and migration of primordial germ cells in mammals. *Gamete* 4:333–362.

Hill, R.S., and MacGregor, H.C. (1980): The development of lampbrush chromosome-type transcription in the early diplotene oocytes of *Xenopus laevis:* An electron microscope analysis. *J. Cell Sci.* 44:87–101.

Latham, K.E. (1995): Stage-specific and cell type-specific aspects of genomic imprinting effects in mammals. *Differentiation* 59:269–282.

Monk, M. (1987): Genomic imprinting. Memories of mother and father. *Nature* 328:203–204.

Rongo, Chr., and Lehmann, R. (1996): Regulated synthesis, transport and assembly of the *Drosophila* germ plasm. *Trends Genet.* 12:102–109.

Sanford, J.P., et al. (1987): Differences in DNA methylation during oogenesis and spermatogenesis and their persistence during early embryogenesis in the mouse. *Genes Dev.* 1:1039–1046.

Schatten, G., et al. (1991): Maternal inheritance of centrosomes in mammals? Studies on parthenogenesis and polyspermy in mice. *Proc. Natl. Acad. Sci. USA* 88:6785–6789.

Schuetz, A.W. (1971): Induction of oocyte maturation in starfish by 1-methyladenosine. *Exp. Cell Res.* 66:5–10.

Silva, A.L., and White, R. (1988): Inheritance of allelic blueprints for methylation patterns. *Cell* 54:145–152.

Surani, M.A.H., Barton, S.C., and Norris, M.L. (1984): Development of reconstituted mouse eggs suggests imprinting of the genome during gametogenesis. *Nature* 308:548–550.

Surani, M.A.H., Barton, S.C., and Norris, M.L. (1986): Nuclear transplantation in the mouse: Hereditable differences between parental genomes after activation of the embryonic genome. *Cell* 45:127–136.

Chapter 6. Fertilization, Egg Activation

Austin, C.R. (1952): The "capacitation" of the mammalian spermatozoa. *Nature* 170:326.

Ciapa, B., et al. (1992): Phosphoinositide metabolism during the fertilization wave in sea urchin eggs. *Development* 115:187–195.

Eddy, E.M. (1988): The spermatozoon. *In* Knobil, E., and Neill, D.J. (eds.) *Physiology of Reproduction*, pp. 27–68, Raven, New York.

Eisenbach, M., and Ralt, D. (1992): Precontact mammalian sperm-egg communication and role in fertilization. *Am. J. Physiol.* 262 (*Cell Physiol.* 31): C1095–C1101.

Foltz, K.R., and Lennarz, W.J. (1993): The molecular basis of sea urchin gamete interactions at the egg plasma membrane. *Dev. Biol.* 158:46–61.

Gallicano, G.I., et al. (1993): Protein kinase C, a pivotal regulator of hamster egg activation, functions after elevation of intracellular free calcium. *Dev. Biol.* 156:94–106.

Giles, R.E., et al. (1980): Maternal inheritance of human mitochondrial DNA. *Proc. Natl. Acad. Sci. USA* 77:6715–6719.

Grainger, J.L., and Winkler, M. (1987): Fertilization triggers unmasking of maternal mRNAs in sea urchin eggs. *Mol. Cell Biol.* 7:3947–3954.

Larabell, C., and Nuccitelli, R. (1992): Inositol lipid hydrolysis contributes to the Ca^{2+} wave in the activating egg of *Xenopus laevis*. *Dev. Biol.* 153:347–355.

Miller, R.L. (1985): Sperm chemo-orientation in the metazoa. *In* Metz, C.B., and Monroy, A. (eds.) *Biology of Fertilization*. Vol. 2., pp. 275–337, Academic Press, New York.

Nuccitelli, R., Cherr, G.N., and Clark, W.H. (eds.) (1989): *Mechanisms of Egg Activation*. Plenum Press, New York.

Rosier, T.K., and Wassarman, P.M. (1992): Identification of a region of mouse zona pellucida glycoprotein mZP3 that possesses sperm receptor activity. *Dev. Biol.* 154:309–317.

Ward, C.R., and Kopf, G.S. (1993): Molecular events mediating sperm activation. *Dev. Biol.* 158:9–34.

Wassarman, P.M. (1989): Fertilization in mammals. *Sci. Am.* 256:78–84.

Whitaker, M., and Swann, K. (1993): Lighting the fuse at fertilization. *Development* 117:1–12.

Chapter 7. (Embryonic) Cell Cycle

Draetta, G. (1990): Cell cycle control in eukaryotes: Molecular mechanisms of cdc2 activation. *Trends Biochem. Sci.* 15:378–383.

Hunt, T. (1989): Embryology: Under arrest in the cell cycle. *Nature* 342:483–484.

Hunter T., and Pines, J. (1994): Cyclins and cancer II: Cyclin D and CDK inhibitors come of age. *Cell* 79:573–582.

Kimelman, D., Kirschner, M., and Scherson, T. (1987): The events of the midblastula transition in *Xenopus* are regulated by changes in the cycle. *Cell* 48:399–407.

Kirschner, M. (1992): The cell cycle then and now. *Trends Biochem. Sci.* 17:281–285.

Lock, L.F., and Wickramasinghe, D. (1994): Cycling with CDKs. *Trends Cell Biol.* 4:404–405.

Murray, A.W. (1992): Creative blocks: Cell-cycle check points and feedback controls. *Nature* 359:599–604.

Murray, A.W., and Kirschner, M.W. (1989): Cyclin synthesis drives the early embryonic cell cycle. *Nature* 339:275–280.

Murray, A.W., and Kirschner, M.W. (1991): What controls the cell cycle. *Sci. Am.* 3/1991:34–41.

Murray, A.W., Solomon, M.J., and Kirschner, M.W. (1989): The role of cyclin synthesis and degradation in the control of maturation promoting factor activity. *Nature* 339:280–286.

Nurse, P. (1990): Universal control mechanism regulating onset of M-phase. *Nature* 344:503–508.

Sherr, C.J. (1993): Mammalian G1 cyclins. *Cell* 73:1059–1065.

Whitaker, M., and Patel, R. (1990): Calcium and cell cycle control. *Development* 108:525–542.

Chapter 8. Determination (See also Chapter 9)

Briggs, R., and King, T.J. (1952): Transplantation of living nuclei from blastula cells into enucleated frog's eggs. *Proc. Natl. Acad. Sci. USA* 38:455.

Davidson, E.H. (1990): How embryos work: A comparative review of diverse models of cell fate specification. *Development* 108:365–389.

DiBernadino, M.A. (1988): Genomic multipotentiality of differentiated somatic cells. *In* G. Eguchi, et al. (eds.) *Regulatory Mechanisms in Developmental Processes*, pp. 129–136. Elsevier, New York.

Freeman, G. (1988): The role of egg organization in the generation of cleavage patterns. *In* Jeffery, W.R., and Raff, R.A. (eds.) *Time, Space and Pattern in Embryonic Development*, pp. 176–169. Alan R. Liss, New York.

Gurdon, J.B. (1962): The developmental capacity of nuclei taken from intestinal epithelium cells of feeding tadpoles. *J. Embryol. Exp. Morphol.* 10: 622–640.

Horvitz, H.R., and Herskowitz, I. (1992): Mechanisms of asymmetric cell division: Two BS or not two BS, that is the question. *Cell* 68:237–255.

Rhyu, M.S., and Knoblich, J.A. (1995): Spindle orientation and asymmetric cell fate. *Cell* 82:523–526.

Chapter 9. Pattern Formation, Positional Information, Inductors, Morphogens (See also Bibliography for model organisms, in particular, *Drosophila* and *Xenopus*, and Bibliography for Box 4)

9.2 External Cues Guide Determination of Spatial Coordinates

Freeman, G. (1988): The role of egg organization in the generation of cleavage patterns. *In* Jeffery, W.R., and Raff, R.A. (eds.) *Time, Space and Pattern in Embryonic Development*, pp. 176–169. Alan R. Liss, New York.

Fujisue, M., Kobayakawa, Y., and Yamana, K. (1993): Occurrence of dorsal axis-inducing activity around the vegetal pole of an uncleaved *Xenopus* egg and displacement to the equatorial region by cortical rotation. *Development* 118:163–170.

Gerhart, J., et al. (1989): Cortical rotation of the *Xenopus* egg: Consequences for the anterioposterior pattern of embryonic dorsal development. *Development* 1989 (Suppl.):37–51.

Govind, S., and Steward, R. (1991): Dorsoventral pattern formation in *Drosophila*. *Trends Genet.* 7:119–124.

Houliston, E. (1994): Microtubuli translocation and polymerization during cortical rotation in *Xenopus* eggs. *Development* 120:1213–1220.

9.4 Pattern Formation by the Exchange of Signals between Adjacent Cells

Artavanis-Tsakonas, S., Matsuno, K., and Fortini, M.E. (1995): Notch signaling. *Science* 268:225.

Basler, K., and Hafen, E. (1989): Ubiquitous expression of *sevenless*: Position-dependent specification of cell fate. *Science* 243: 931–934.

Campos-Ortega, J.A., and Knust, E. (1992): Genetic mechanisms in early neurogenesis of *Drosophila melanogaster*. *In* Russo, V.E.A., et al. (eds.) *Development: The Molecular Genetic Approach*, pp. 343–354. Springer-Verlag, Heidelberg.

Hafen, E., et al. (1987): Sevenless, a cell-specific homeotic gene of *Drosophila*, encodes a putative transmembrane receptor with a tyrosine kinase domain. *Science* 236:55–63.

Hassan, B., and Vaessin, H. (1996): Regulatory interactions during early neurogenesis in *Drosophila. Dev. Genet.* 18:18–27.

Hoppe, P.E., and Greenspan, R.J. (1986): Local function of the Notch gene for embryonic ectodermal pathway choice in *Drosophila. Cell* 46: 773–783.

9.5 The Principle of Embryonic Induction

Chui, Y., et al. (1995): *Xwnt*-8b: A maternally expressed *Xenopus Wnt* gene with a potential role in establishing the dorsoventral axis. *Development* 121: 2177–2186.

Cooke, J., and Wong, A. (1991): Growth-factor-related proteins that are inducers in early amphibian development may mediate similar steps in amniote (bird) embryogenesis. *Development* 111:197–212.

Cornell, R.A., Musci, T.J., and Kimelman, D. (1995): FGF is a prospective competence factor for early activin-type signals in *Xenopus* mesoderm induction. *Development* 121:2429–2437.

Doniach, T. (1995): Basic FGF as an inducer of anteroposterior neural pattern. *Cell* 83:1067–1070.

Echelard, Y., et al. (1993). *Sonic hedgehog*, a member of a family of putative signaling molecules, is implicated in the regulation of CNS polarity. *Cell* 75: 1414–1430.

Grainger, R.M., Henry, J.J., and Henderson, R.A. (1988): Reinvestigation of the role of optic vesicle in embryonic lens induction. *Development* 102: 517–526.

Green, J.B.A. (1994): Roads to neuralness: Embryonic neural induction as derepression of a default state. *Cell* 77:317–330.

Grimes, G.W., and Aufderheide, K.J. (1991): Cellular aspects of pattern formation. The problem of assembly. S. Karger, Basel.

Grunz, H., Schüren, C., and Richter, K. (1995): The role of vertical and planar signals during the early steps of neural induction. *Int. J. Dev. Biol.* 39: 539–543.

Gurdon, J.B. (1987): Embryonic induction—molecular prospects. *Development* 99:285–306.

Gurdon, J.B., et al. (1994): Activin signalling and response to a morphogen gradient. *Nature* 371:487–492.

Harrison, R.G. (1920): Experiments on the lens in *Ambystoma. Proc. Soc. Exp. Biol. Med.* 17:413–461.

Heemskerk, J., et al., (1994): *Drosophila hedgehog* acts as a morphogen in cellular patterning. *Cell* 76:449–460.

Hemmati-Brivaniou, A., Kelly, O.G., and Melton, D.A. (1994): Follistatin, an antagonist of activin, is expressed in the Spemann organizer and displays direct neuralizing activity. *Cell* 77:283–295.

Hemmati-Brivaniou, A., and Melton, D.A. (1994): Inhibition of activin receptor signaling promotes neuralization in *Xenopus. Cell* 77:273–281.

Henry, J.J., and Grainger, R.M. (1990): Early tissue interactions leading to embryonic lens formation in *Xenopus laevis. Dev. Biol.* 141:149–163.

Hoppe, P.E., and Greenspan, R.J. (1986): Local function of the Notch gene for embryonic ectodermal pathway choice in *Drosophila. Cell* 46: 773–783.

Jacobson, A.G., and Sater, A.K. (1988): Features of embryonic induction. *Development* 104:341–359.

Jessel, T.M., and Melton, D.A. (1992): Diffusible factors in vertebrate embryonic induction. *Cell* 68:257–270.

John, M., Born, J., Tiedemann, He., and Tiedemann, Hi. (1984): Activation of a neuralizing factor in amphibian ectoderm. *Roux's Arch. Dev. Biol.* 193: 13–18.

Kelly, O.G., and Melton, D.A. (1995): Induction and patterning of the vertebrate nervous system. *Trends Genet.* 11(7):273–278.

Kessler, D.S., and Melton, D.A. (1995): Induction of dorsal mesoderm by soluble, mature Vg1 protein. *Development* 121:2155–2164.

Malacinski, G.M., and Bryant, S.V. (eds.) (1984): *Pattern Formation. A Primer in Development Biology.* Macmillan, New York.

Mavilio, F. (1993): Regulation of vertebrate homeobox-containing genes by morphogens. *Eur. J. Biochem.* 212:273–288.

Mitrani, E., et al. (1990): Activin can induce the formation of axial structures and is expressed in the hypoblast of the chick. *Cell* 63:495–501.

Niehrs, C., et al. (1993): The homeobox gene goosecoid controls cell migration in *Xenopus* embryos. *Cell* 72:491–503.

Nüsslein-Volhard, C. (1991): From egg to organism—Studies on embryonic pattern formation. *JAMA* 266:1848–1849.

Perrimon N. (1995): Hedgehog and beyond. *Cell* 80:517–520.

Robinson, M.L., et al. (1995): Extracellular FGF-1 acts as a lens differentiation factor in transgenic mice. *Development* 121:505–514.

Ruiz i Altaba, A., and Melton, D.A. (1989): Interaction between peptide growth factors and homeobox genes in the establishment of antero-posterior polarity in frog embryos. *Nature* 341:33–38.

Sasai, Y., et al. (1994): *Xenopus chordin:* A novel dorsalizing factor activated by organizer-specific homeobox genes. *Cell* 79:779–790.

Sasai, Y., et al. (1995): Regulation of neural induction by the Chd and Bmp-4 antagonistic patterning signal in *Xenopus. Nature* 376:333.

Slack, J.M.W. (1987): Morphogenetic gradients—past and present. *Trends Biochem. Sci.* 12:201–204.

Slack, J.M.W. (1991): *From Egg to Embryo: Determinative Events in Early Development,* 2nd ed., Cambridge Univ. Press, Cambridge.

Slack, J.M.W. (1993): Embryonic induction. *Mech. Dev.* 41:91–107.

Sosoi, Y., et al. (1994): *Xenopus chordin:* A novel dorsalizing factor activated by organizer-specific homeobox genes. *Cell* 79:779–790.

Spemann, H. (1938): *Embryonic Development and Induction.* Yale Univ. Press, New Haven, CT (Reprinted by Hafner, New York, 1962).

Spemann, H. (1968): *Experimentelle Beiträge zu einer Theorie der Entwicklung.* Nachdruck. Springer-Verlag, Heidelberg.

Wolpert, L., and Brown, N.A. (1995): Hedgehog keeps to the left. *Nature* 377:103–104.

9.7 How to Create a Field, Subdivide It, and Define a Point within It

Campbell, G., and Tomlinson, A. (1995): Initiation of the proximodistal axis in insect legs. *Development* 121:619–628.

9.8 The Avian Wing as a Model Limb

Bryant, S.V., and Gadiner, D.M. (1992): Retinoic acid, local cell-cell interactions, and pattern formation in vertebrate limbs. *Dev. Biol.* 152:1–25.

Chen, Y.P., Huang, L., and Solursh, M. (1994): A concentration gradient of retinoids in the early *Xenopus* embryo. *Dev. Biol.* 161:70–76.

Chen, Y.P., et al. (1996): Hensen's node from vitamin A-deficient quail embryo induces chick limb bud duplication and retains its normal asymmetric expression of *Sonic hedgehog (shh). Dev. Biol.* 173:256–264.

Eichele, G. (1989): Retinoic acid induces a pattern of digits in anterior half wing buds that lack the zone of polarizing activity. *Development* 107:863–867.

Maden, M., Ong, D.E., Summerbell, D., Chytil, F., and Hirst, E.A. (1989): Cellular retinoic acid-binding protein and the role of retinoic acid in the development of the chick embryo. *Dev. Biol.* 135:124–132.

Nelson, C.E., et al. (1996): Analysis of *Hox* gene expression in the chick limb bud. *Development* 122:1449–1466.

Ogura, T., et al. (1996): Evidence that *Shh* cooperates with a retinoic acid inducible co-factor to establish ZPA-like activity. *Development* 122:537–542.

Riddle, R.D., Johnson, R.L., Laufer, E., and Tabin, C. (1993): *Sonic hedgehog* mediates the polarizing activity of the ZPA. *Cell* 75:1401–1416.

Selleck, M.A., et al. (1996): Digit induction by Hensen's node and notochord involves the expression of *shh* but not RAR-β2. *Dev. Biol.* 173:318–326.

Tabin, C.J. (1995): The initiation of the limb bud: Growth factors, *hox* genes, and retinoids. *Cell* 80:671–674.

Thaller, C., and Eichele, G. (1987): Identification and spatial distribution of retinoids in the developing chick limb bud. *Nature* 327:625–628.

Wolpert, L. (1978): Pattern formation in biological development. *Sci. Am.* 10:124–137.

9.9 *Positional Information and Positional Memory in* Hydra

Müller, W.A. (1996): Pattern formation in the immortal *Hydra*. *Trends Genet.* 11:91–96.

9.10 *Intercalation in Insect Appendages*

Bohn, H. (1976): Regeneration of proximal tissues from a more distal amputation level in the insect leg *(Blaberus craniifer, Blattaria)*. *Dev. Biol.* 53: 285–293.

Chapter 10. Differentiation and Its Molecular Basis Development Genetics
(See also Bibliography to Chapter 3, Section 3.6, *Drosophila*)

10.2 *Chromosome Puffing in Flies: Genes Activated but Equivalence Lost*

Andres, A.J., and Thummel, C.S. (1992): Hormones, puffs and flies: The molecular control of metamorphosis by ecdysone. *Trends Genet.* 8: 132–138.

Ashburner, M. (1990): Puffs, genes, and hormones revisited. *Cell* 61:1–3.

Ashburner, M., and Berondes, H.D. (1978): Patterns of puffing activity in the salivary glands of *Drosophila*. *In Genetics and Biology of* Drosophila, Vol. 2B, pp. 316–395. Academic Press, New York.

Segraves, W.A., and Hogness, D.S. (1990): The E75 ecdysone-inducible gene responsible for the 75B early Puff in *Drosophila* encodes two new members of the steroid receptor superfamily. *Genes Dev.* 4:204–209.

10.3 MyoD/myogenin, Muscle Cell Development

Braun, T., Rudnicki, M.A., Arnold, H.-H., and Jaenisch, R. (1992): Targeted inactivation of the muscle regulatory gene Myf-5 results in abnormal rib development and perinatal death. *Cell* 71:369–382.

Cossu, G., et al. (1996): Activation of different myogenetic pathways: myf-5 is induced by the neural tube and MyoD by the dorsal ectoderm in mouse paraxial mesoderm. *Development* 122:429–437.

Hasty P., et al. (1993): Muscle deficiency and neonatal death in mice with targeted mutation in the *myogenin* gene. *Nature* 364:501–506.

Hopwood, N.D., and Gurdon, J.B. (1990): Activation of muscle genes without myogenesis by ectopic expression of MyoD in frog embryo cells. *Nature* 347:197–199.

Pinney, D.F., and Emerson, C.P. (1992): Skeletal muscle differentiation. *In* Russo, V.E.A., et al. (eds.) *Development: The Molecular Genetic Approach*, pp. 459–478. Springer-Verlag, Heidelberg.

Thayer, M.J., et al. (1989): Positive autoregulation of the myogenic determination gene MyoD1. *Cell* 58:241–248.

Weintraub, H. (1993): The MyoD family and myogenesis: Redundancy, networks, and thresholds. *Cell* 75:1241–1244.

10.4 to 10.9 Master Genes, Homeoboxes, Homeotic Genes

Burke, A.C., et al. (1995): *Hox* genes and the evolution of vertebrate axial morphology. *Development* 121:333–346.

Carroll, S.B. (1995): Homeotic genes and the evolution of arthropods and chordates. *Nature* 376:479–485.

Cho, K.W.Y., et al. (1991): Molecular nature of Spemann's organizer: The role of the *Xenopus* Homeobox gene *goosecoid*. *Cell* 67:1111–1120.

Davidson, E.H. (1991): Spatial mechanisms of gene regulation in metazoan embryos. *Development* 113:1–26.

De Robertis, E.M., Morita, E.A., and Cho, K.W.Y. (1991): Gradient fields and homeobox genes. *Development* 112:669–678.

De Robertis, E.M., Oliver, G., and Wright, C.V.E. (1990): Homeobox genes and the vertebrate body plan. *Sci. Am.* 7/1990:26–32.

Ekker, S.C., et al. (1995): Distinct expression and shared activities of members of the *hedgehog* gene family of *Xenopus laevis. Development* 121: 2337–2347.

Gehring, W. (1992): The homeobox in perspective. *Trends Biochem. Sci.* 8: 277–280.

Graham, A., Papalopulu, N., and Krumlauf, R. (1989): The murine and *Drosophila* homeobox gene complexes have common features of organization and expression. *Cell* 57:367–378.

Gruss, P., and Walther, C. (1992): Pax in development. *Cell* 69:719–722.

Haack, H., and Gruss, P. (1993): The establishment of murine Hox-1 expression domains during patterning of the limb. *Dev. Biol.* 157:410–422.

Halder, G., Callaerts, P., and Gehring, W.J. (1995): Induction of ectopic eyes by targeted expression of the *eyeless* gene in *Drosophila*. *Science* 267:1788–1792.

Holland, P.W.H., and Carcia-Fernandez, J.G. (1996): *Hox* genes and chordate evolution. *Dev. Biol.* 173:382–395.

Izpisua-Belmonte, J.C., et al. (1991): Expression of the homeobox *Hox-4* genes and the specification of position in chick wing development. *Nature* 350:585–589.

Joyner, A.L. (1996): *Engrailed, Wnt* and *pax* genes regulate midbrain-hindbrain development. *Trends Genet.* 12:15–20.

Lewis, J., and Martin, P. (1989): Limbs: A pattern emerges. *Nature* 342:734–735.

Marti, E., et al. (1995): Distribution of Sonic hedgehog peptides in the developing chick and mouse embryo. *Development* 121:2537–2547.

Mavilio, F. (1993): Regulation of vertebrate homeobox-containing genes by morphogens. *Eur. J. Biochem.* 212:273–288.

McGinnis, W., and Krumlauf, R. (1992): Homeobox genes and axial patterning. *Cell* 68:283–302.

Nelson, C.E., et al. (1996): Analysis of *Hox* gene expression in the chick limb bud. *Development* 122:1449–1466.

Niehrs, C., Steinbeisser, H., and De Robertis, E.M. (1994): Mesodermal patterning by a gradient of the vertebrate homeobox gene *goosecoid*. *Science* 263:817–820.

Nusse, R., and Varmus, H.E. (1992): *Wnt* Genes. *Cell* 69:1073–1078.

Patel, N.H., et al., (1989): Expression of *engrailed* proteins in arthropods, annelids, and chordates. *Cell* 58:955–968.

Tabin, C.J. (1992): Why we have (only) five fingers per hand: Hox genes and the evolution of paired limbs. *Development* 116:289–296.

Tabin, C.J. (1995): The initiation of the limb bud: Growth factors, *hox* genes, and retinoids. *Cell* 80:671–674.

Zecca, M., Basler, K., and Struhl, G. (1995): Sequential organizing activities of engrailed, hedgehog and decapentaplegic in the *Drosophila* wing. *Development* 121:2265–2278.

10.10 to 10.11 Gene Silencing, Cell Heredity

Gartler, S.M., and Riggs, A.D. (1983): Mammalian X-chromosome inactivation. *Annu. Rev. Genet.* 17:155–190.

Monk, M., and Harper, M.I. (1979): Sequential X chromosome inactivation coupled with cellular differentiation in early mouse embryos. *Nature* 281:311–313.

Sanford, J.P., et al. (1987): Differences in DNA methylation during oogenesis and spermatogenesis and their persistence during early embryogenesis in the mouse. *Genes Dev.* 1:1039–1046.

Chapter 11. Reversibility-Irreversibility of Differentiation, Programmed Cell Death

Collins, M.K.L., Rivas, A.L. (1993): The control of apoptosis in mammalian cells. *Trends Biochem. Sci.* 18:307–309.

Hannun, Y.A., and Obeid, L.M. (1995): Ceramide: An intracellular signal for apoptosis. *Trends Biochem. Sci.* 20:73–77.

Oppenheim, R.W., Prevette, D., Tytell, M., and Homma, S. (1990): Naturally occurring and induced neuronal cell death in the chick embryo in vivo requires protein and RNA synthesis: Evidence for the role of cell death genes. *Dev. Biol.* 138:104–113.

Raff, M.C., et al. (1993): Programmed cell death and the control of cell survival: Lessons from the nervous system. *Science* 262: 695–700.

Ruoslahti, E., and Reed, J.C. (1994): Anchorage dependence, integrins and apoptosis. *Cell* 77:477–478.

Spradling, A.C. (1981): The organization and amplification of two chromosomal domains containing *Drosophila* chorion genes. *Cell* 27:193–201.

Williams, G.T., and Smith, C.A. (1993): Molecular regulation of apoptosis. Genetic controls on cell death. *Cell* 74:777–780.

Chapter 12. Cell Migration; Cell Adhesion Molecules

Adams, J.C., and Watt, F.M. (1993): Regulation of development and differentiation by the extracellular matrix. *Development* 117:1183–1198.

Armstrong, P.B. (1989): Cell sorting out: The self-assembly of tissues in vitro. *Crit. Rev. Biochem. Mol. Biol.* 24:119–149.

Begovac, P.C., and Shur, B.D. (1990): Cell surface galactosyltransferase mediates the initiation of neurite outgrowth from PC12 cells of laminin. *J. Cell Biol.* 110:461–470.

Brown, N.H. (1993) Integrins hold *Drosophila* together. *Bioessays* 15: 383–390.

Cunningham, B.A. (1991) Cell adhesion molecules and the regulation of development. *Am. J. Obstet. Gynecol.* 164:939–948.

Drubin, D.G., and Nelson, W.J. (1996): Origins of cell polarity. *Cell* 84: 335–344.

Edelman, G.M. (1983): Cell adhesion molecules. *Science* 219:450–457.

Edelman, G.M. (1986): Cell adhesion molecules in the regulation of animal form and tissue pattern. *Annu. Rev. Cell Biol.* 2:81–116.

Edelman, G.M. (1988): *Topobiology: An Introduction to Molecular Embryology*. Basic Books, New York.

Edelstein-Keshet, L., and Ermentrout, B.G. (1990): Contact response of cells can mediate morphogenetic pattern formation. *Differentiation* 45: 147–159.

Foty, R.A., et al., (1966): Surface tensions of embryonic tissues predict their mutual envelopmental behavior. *Development* 122:1611–1620.

Grenningloh, G., et al. (1990). Molecular genetics of neuronal recognition in *Drosophila:* Evolution and function of immunoglobulin superfamily cell adhesion molecules. *Cold Spring Harbor Symp. Quant. Biol.* 55:327–340.

Gullberg, D., and Ekblom, P. (1995): Extracellular matrix and its receptors during development. *Int. J. Dev. Biol.* 39:845–854.

Gumbiner, B.M. (1996): Cell adhesion: The molecular basis of tissue architecture and morphogenesis. *Cell* 84:345–357.

Hardin, J., and Keller, R. (1988): Behavior and function of bottle cell during gastrulation of *Xenopus laevis. Development* 103:211–230.

Holtfreter, J. (1946): Structure, motility and locomotion in isolated amphibian cells. *J. Morphol.* 79:27–62.

Hynes, R.O. (1992): Integrins: Versatility, modulation, and signaling in cell adhesion. *Cell* 69:11–25.

Hynes, R.O., and Lander, A.D. (1992): Contact and adhesive specificities in the associations, migrations, and targeting of cells and axons. *Cell* 68: 303–322.

Jacobson, A.G. (1994): Normal neurulation in amphibia. *In* Bock, G., and Marsh, J. (eds.) *Neural Tube Defects*, pp. 6–24. John Wiley & Sons, New York.

Keller, R.E. (1986): The cellular basis of amphibian gastrulation. *In* Browder, L. (ed.) *Developmental Biology: A Comprehensive Synthesis*, Vol. 2, pp. 241–327. Plenum Press, New York.

Phillips, D.R., Charo, I.F., and Scarborough, R.M. (1991): GPIIb-IIIA: The responsive integrin. *Cell* 65:359–362.

Ruoslahti, E., and Reed, J.C. (1994): Anchorage dependence, integrins and apoptosis. *Cell* 77:477–478.

Shur, B.D. (1991): Cell surface β1,4galactosyltransferase. Twenty years later. *Glycobiology* 1:563–575.

Steinberg, M.S. (1970): Does differential adhesion govern self-assembly processes in histogenesis? Equilibrium configurations and the emergence of a hierarchy among populations and animal morphogenesis. *J. Exp. Zool.* 173: 395–434.

Steinberg, M.S., and Takeichi, M. (1994): Experimental specification of cell sorting, tissue spreading, and specific spatial patterning by quantitative differences in cadherin expression. *Proc. Natl. Acad. Sci. USA* 91:206–209.

Takeichi, M. (1988): The cadherins: Cell-cell adhesion molecules controlling animal morphogenesis. *Development* 102:639–656.

Townes, P.L., and Holtfreter, J. (1955): Directed movements and selective adhesion of embryonic amphibian cells. *J. Exp. Zool.* 128:53–120.

Chapter 13. Cells on Travels, Neural Crest Derivatives, and Primordial Germ Cells

Bronner-Fraser, M. (1993): Environmental influences on neural crest cell migration. *J. Neurobiol.* 24:233–247.

Bronner-Fraser, M., and Fraser, S.E. (1988): Cell lineage analysis reveals multipotency of some avian neural crest cells. *Nature* 335: 161–164.

Collazo, A., Bronner-Fraser, M., and Fraser, S.E. (1993): Vital dye labelling of *Xenopus laevis* trunk neural crest reveals multipotency and novel pathways of migration. *Development* 118:363–376.

Eddy, E.M. (1975): Germ plasm and the differentiation of the germ line. *Int. Rev. Cytol.* 43:229–280.

Erickson, C.A., and Goins, T.L. (1995): Avian neural crest cells can migrate in the dorsolateral path if they are specified as melanocytes. *Development* 121:915–924.

Groves, A.K., and Anderson, D.J. (1996): Role of environmental signals and transcriptional regulators in neural crest development. *Dev. Genet.* 18: 64–72.

Perris, R., von Boxburg, A., and Löfberg, J. (1988): Local embryonic matrices determine region-specific phenotypes in neural crest cells. *Science* 241:86–89.

Richardson, M.K., and Sieber-Blum, M. (1993): Pluripotent neural crest cells in the developing skin of the quail embryo. *Dev. Biol.* 157:348–358.

Chapter 14. Nervous System; Nerve Growth Factor

Ang, S.-L. (1996): The brain organization. *Nature* 380:25–27.

Campos-Ortega, J.A., and Knust, E. (1992): Genetic mechanisms in early neurogenesis of *Drosophila melanogaster. In* Russo, V.E.A., et al. (eds.) *Development: The Molecular Genetic Approach*, pp. 343–354. Springer-Verlag, Heidelberg.

Coulombe, J.N., and Bronner-Fraser, M. (1987): Cholinergic neurones aquire adrenergic neurotransmitters when transplanted into an embryo. *Nature* 324:569–572.

Doniach, T. (1995): Basic FGF as an inducer of anteroposterior neural pattern. *Cell* 83:1067–1070.

Henderson, C., et al. (1993): Neurotrophins promote motor neuron survival and are present in embryonic limb bud. *Nature* 363:266–363.

Hubel, D.H., and Wiesel, T.N. (1962): Receptive fields, binocular interaction and functional architecture in the cat's visual cortex. *J. Physiol.* 160: 106–154.

Hubel, D.H., and Wiesel, T.N. (1963): Receptive fields of cells in striate cortex of very young, visually inexperienced kittens. *J. Neurophysiol.* 26: 944–1002.

Hubel, D.H., Wiesel, T.N., and LeVay, S. (1977): Plasticity of ocular dominance columns in monkey striate cortex. *Philos. Trans. R. Soc. London B* 278:377–409.

Hunt, P., et al. (1991): Homeobox genes and models for patterning the hindbrain and brachial arches. *Development* (Suppl. 1):187–169.

Hynes, R.O., and Lander, A.D. (1992): Contact and adhesive specifities in the associations, migrations, and targeting of cells and axons. *Cell* 68:303–322.

Jacobson, A.G. (1994): Normal neurulation in amphibia. *In* Bock, G., and Marsh, J. (eds.) *Neural Tube Defects*, pp. 6–24. John Wiley & Sons, New York.

Jacobson, M. (1967): Retinal ganglion cells: Specification of central connections in larval *Xenopus laevis. Science* 155:1106–1108.

Johnson, F., and Bottjer, S.W. (1994): Afferent influences on cell death and birth during development of a cortical nucleus necessary for learned vocal behavior in zebra finches. *Development* 120:13–24.

Joyner, A.L. (1996): *Engrailed, Wnt* and *pax* genes regulate midbrain-hindbrain development. *Trends Genet.* 12:15–20.

Kelly, O.G., and Melton, D.A. (1995): Induction and patterning of the vertebrate nervous system. *Trends Genet.* 11:273–278.

Klacheim, C., et al. (1987): *In vivo* effect of brain-derived neurotrophic factor on the survival of neural crest precursor cells of the dorsal root ganglia. *EMBO J.* 6:2871–2873.

LeDourain, N., and Smith, J. (1988): Development of the peripheral nervous system from the neural crest. *Annu. Rev. Cell Biol.* 4:375–404.

Levi-Montalcini, R. (1987): The nerve growth factor 35 years later. *Science* 237:1154–1161.

Snider, W.D. (1994): Functions of the neurotrophins during nervous system development: What the knockouts are teaching us. *Cell* 77:627–638.

Yamada, K.M., Spooner, B.S., and Wessells, N.K. (1971): Ultrastructure and function of growth cones and axons of cultured nerve cells. *J. Cell Biol.* 49:614–635.

Chapter 15. Heart and Blood Vessels

Flame, I., and Risau, W. (1992): Induction of vasculogenesis and hematopoiesis in vitro. *Development* 116:435–439.

Folkman, J., and Klagsbrun, M. (1987): Angiogenic factors. *Science* 235:442–447.

Linask, K.K., and Lash, J.W. (1986): Precardiac cell migration: Fibronectin localization at mesoderm-endoderm interface during directional movement. *Dev. Biol.* 114:87–101.

Liotta, L.A. (1991): Cancer metastasis and angiogenesis: An imbalance of positive and negative regulation. *Cell* 84:327–336.

Parnanaud, L., Yassine, F., and Dieterlen-Lièvre, F. (1989): Relationship between vasculogenesis, angiogenesis, and hematopoiesis during avian ontogeny. *Development* 105:473–485.

Pepper, M.S., and Montesano, R. (1990): Proteolytic balance and capillary morphogenesis. *Cell Differ. Dev.* 32:319–328.

Chapter 16. Stem Cells, Blood Cells

Golde, D.W. (1991): The stem cell. *Sci. Am.* 265:86–93.

Hall, P.A., and Watt, F.M. (1989): Stem cells: The generation and maintenance of cellular diversity. *Development* 106:619–633.

Introna, M., and Golay, J. (1992): The role of oncogenes in myeloid differentiation. *In* Russo, V.E.A., et al. (eds.) *Development: The Molecular Genetic Approach*, pp. 504–518. Springer-Verlag, Heidelberg.

Potten, C.S., and Loeffler, M. (1990): Stem cells, attributes, cycles, spirals, pitfalls and uncertainties. Lessons for and from the crypt. *Development* 110:1001–1020.

Spangrude, G.J., Heimfeld, S., and Weissman, I. (1988): Purification and characterization of mouse hematopoietic stem cells. *Science* 241:58–62.

Svendsen, C.N., and Rosser, A.E. (1995): Neurones from stem cells? *Trends Neurosci.* 18:465–467.

Whetton, A.D., and Dexter, T.M. (1986): Hemopoietic growth factors. *Trends Biochem. Sci.* 11:207–211.

Chapters 17, 18, 19, 20. Signal Substances: Morphogens, Growth Factors, Inductors, and Hormones

Beato, M. (1989): Gene regulation by steroid hormones. *Cell* 56:335–344.

Beato, M., Herrlich, P., and Schütz, G. (1995): Steroid hormone receptors: Many actors in search of a plot. *Cell* 83:851–857.

Bownes, M., et al. (1988): Evidence that insect embryogenesis is regulated by ecdysteroids released from yolk proteins. *Proc. Natl. Acad. Sci. USA* 85: 1554–1557.

Bückmann, D. (1983): The phylogeny and polytropy of hormones. *Horm. Metab. Res.* 15:211–217.

Cross, M., and Dexter, T.M. (1991): Growth factors in development, transformation, and tumorigenesis. *Cell* 64:271–280.

Harvey, M.B., and Kaye, P.L. (1990): Insulin increases the cell number of the inner cell mass and stimulates morphological development of mouse blastocysts *in vitro. Development* 110:963–967.

Introna, M., Golay, J., and Ottolenghi, S. (1992): Cellular differentiation in the hematopoietic system. *In* Russo, V.E.A., et al. (eds.) *Development: The Molecular Genetic Approach*, pp. 499–503. Springer-Verlag, Heidelberg.

Jones, N. (1990): Transcriptional regulation by dimerization: Two sides to an incestuous relationship. *Cell* 61:9–11.

Kaye, P.L. (1993): Insulin-like growth factors: Growth in the family. *Cell* 73:1059–1065.

Koolman, J., and Spindler, K.-D. (1983): Mechanisms of action of ecdysteroids. *In Endocrinology of Insects*, pp. 179–201. Alan R. Liss, New York.

Lee, J.E., et al. (1990): Pattern of the insulin-like growth factor II gene expression during early mouse embryogenesis. *Development* 110:151–159.

Schwabe, J.W.R., and Rhodes, D. (1991): Beyond zinc fingers: Steroid hormone receptors have a novel structural motif for DNA recognition. *Trends Biochem. Sci.* 16:291–296.

Sporn, M.B., and Roberts, A.B. (1988): Peptide growth factors are multifunctional. *Nature* 332:217–219.

Umesono, K., and Evans, R.M. (1989): Determinants of target gene specificity for steroid/thyroid hormone receptors. *Cell* 57:1139–1146.

Chapter 18. Growth Control, Cancer

Adamson, E.L. (1987): Oncogenes in development. *Development* 99: 449–471.

Bishop, J.M. (1982): Oncogenes. *Sci. Am.* 246:81–92.

Bishop, J.M. (1989): Viruses, genes, and cancer. *Am. Zool.* 29:653–666.

Bishop, J.M. (1991): Molecular themes in oncogenesis. *Cell* 64:235–248.

Cantley, L.C., et al. (1991): Oncogenes and signal transduction. *Cell* 64: 281–302.

Crook, T., et al. (1994): Transcriptional activation by p53 correlates with suppression of growth but not transformation. *Cell* 79:817–827.

Cross, M., and Dexter, T.M. (1991): Growth factors in development, transformation, and tumorigenesis. *Cell* 64:271–280.

DeCaprio, J.A., et al. (1989): The product of the retinoblastoma susceptibility gene has properties of a cell cycle regulatory element. *Cell* 58: 1085–1095.

Gutman, D.H. (1995): Tumor suppressor genes as negative growth regulators in development and differentiation. *Int. J. Dev. Biol.* 39:895–907.

Hartwell, L.H., and Kastan, M.B. (1994): Cell cycle control and cancer. *Science* 266:1821–1828.

Hunter, T. (1986): Cell growth control mechanismus. *Nature* 322: 14–16.

Hunter, T. (1991): Cooperation between oncogenes. *Cell* 64:249–270.

Introna, M., and Golay, J. (1992): The role of oncogenes in myeloid differentiation. *In* Russo, V.E.A., et al. (eds.) *Development: The Molecular Genetic Approach*, pp. 504–518. Springer-Verlag, Heidelberg.

Lewin, B. (1991): Oncogenic conversion by regulatory changes in transcription factors. *Cell* 64:303–312.

Linzer, D.I.H. (1988): The marriage of oncogenes and anti–oncogenes. *Trends Genet.* 4(9):245–247.

Liotta, L.A. (1991): Cancer metastasis and angiogenesis: An imbalance of positive and negative regulation. *Cell* 84:327–336.

Lowe, S.W., et al. (1993): p53 is required for radiation-induced apoptosis in mouse thymocytes. *Nature* 362:847–849.

Marshall, C.J. (1991): Tumor suppressor genes. *Cell* 64:313–326.

Matzuk, M.M., et al. (1992): α-Inhibin is a tumor-suppressor gene with gonadal specifity. *Nature* 360:313–319.

Olson, E.L. (1992): Interplay between proliferation and differentiation within the myogenic lineage. *Dev. Biol.* 154:261–272.

Philipson, L., and Sorrentino, V. (1992): Growth control in animals. *In* Russo, V.E.A., et al. (eds.) *Development: The Molecular Genetic Approach*, pp. 537–553. Springer-Verlag, Heidelberg.

Ruoslahti, E., and Reed, J.C. (1994): Anchorage dependence, integrins and apoptosis. *Cell* 77:477–478.

Slack, R.S., and Miller, F.D. (1996): Retinoblastoma gene in mouse neural development. *Dev. Genet.* 18:81–91.

Weinberg, R.A. (1983). A molecular basis of cancer. *Sci. Am.* 249: 126–142.

Weinberg, R.A. (1988): Finding the anti-oncogene. *Sci. Am.* 259:44–51.

Ziegler, A., et al. (1994): Sunburn and p53 in the onset of skin cancer. *Nature* 372:773–775.

Chapter 19. Metamorphosis

Etkin, W. (1968). Hormonal control of amphibian metamorphosis. *In* Etkin, W., and Gilbert, I.I. (eds.) *Metamorphosis.* Appleton-Century-Crofts, New York.

Etkin, W., and Gilbert, I.I. (1968): Metamorphosis. Appleton-Century-Crofts, New York.

Frieden, E. (1981): The dual role of thyroid hormones in vertebrate development and calorigenesis. *In* L.I. Gilbert, and Frieden, E. (eds.) *Metamorphosis: A Problem in Developmental Biology*, pp. 545–564. Plenum Press, New York.

Gilbert, L.I., and Goodman, W. (1981): Chemistry, metabolism, and transport of hormones controlling insect metamorphosis. *In* Gilbert, L.I., and Frieden, E. (eds.) *Metamorphosis: A Problem in Developmental Biology*, pp. 139–176. Plenum Press, New York.

Karim, F.D., and Thummel, C.S. (1992): Temporal coordination of regulatory gene expression by the steroid hormone ecdysone. *EMBO J.* 11: 4083–4093.

Koch, P.B., and Bückmann, D. (1987): Hormonal control of seasonal morphs by the timing of ecdysteroid release in *Araschnia levana*. *J. Insect Physiol.* 33:823–829.

Neumann, D., and Spindler, K.-D. (1991): Circaseminular control of imaginal disc development in *Clunio marinus*. *J. Insect Physiol.* 37:101–109.

Pratt, G.E., et al. (1980): Lethal metabolism of precocene-I to a reactive epoxide by locust corpora allata. *Nature* 284:320–323.

Schmutterer, H. (ed.) (1995): *The Neem Tree*. VCH, Weinheim.

Spindler, K.-D. (1991): Roles of morphogenetic hormones in the metamorphosis of arthropods other than insects. *In* Gupta, A.P. (ed.) *Morphogenetic Hormones of Arthropods*, pp. 131–149. Rutgers Univ. Press, New Brunswick.

Thummel, C.S. (1995): From embryogenesis to metamorphosis: The regulation and function of *Drosophila* nuclear receptor superfamily members. *Cell* 83:871–877.

Yaoita, Y., and Brown, D.D. (1990): A correlation of thyroid hormone receptor gene expression with amphibian metamorphosis. *Genes Dev.* 4: 1917–1924.

Chapter 20. Sexual Development

Arnold, A.P. (1980): Sexual differences in the brain. *Am. Sci.* 68: 165–173.

Baker, B.S. (1989): Sex in flies: The splice of life. *Nature* 340: 521–524.

Bell, L.R., Main, E.M., Schedl, P., and Cline, T.W. (1988): Sex-lethal, a *Drosophila* sex determination switch gene, exhibits sex-specific RNA splicing and sequence similary to RNA binding proteins. *Cell* 55:1037–1046.

Bogan, J.S., and Page, D.C. (1994): Ovary? Testis?—A mammalian dilemma. *Cell* 76:603–607.

Bottjer, S.W. (1991): Neural and hormonal substrates for song learning in zebra finches. *Neurosciences* 3:481–488.

Breedlove, S.M. (1994): Sexual differentiation of the human nervous system. *Annu. Rev. Psychol.* 45:389–418.

Bull, J.J. (1990): Sex determination in reptiles. *Q. Rev. Biol.* 55:3–21.

Cline, T.W. (1993): The *Drosophila* sex determination signal: How do flies count to two? *Trends Genet.* 9:385–390.

Crews, D. (1994): Animal sexuality. *Sci. Am.* January:97–103.

Crow, J.F. (1994): Advantages of sexual reproduction. *Dev. Genet.* 15:205–213.

Goodfellow, P.N., and Lovell-Badge, R. (1993): *SRY* and sex determination in mammals. *Annu. Rev. Genet.* 27:71–92.

Gustafson, M.L., and Donahoe, P.K. (1994): Male sex determination: Current concepts of male sex differentiation. *Annu. Rev. Medicine* 45:505–524.

Hampson, E., and Kimura, D. (1992): Sex differences and hormonal influences on cognitive function in humans: *In* Becker, J.B., Breedlove, S.M., and Crews, D. (eds.) *Behavioral Endocrinology*, pp. 347–400. MIT Press, Cambridge, MA.

Haqq, C.M., et al. (1994): Molecular basis of mammalian sexual determination: Activation of Müllerian inhibiting substance gene expression by SRY. *Science* 266:1494–1497.

Hodgkin, J. (1992): Genetic sex determination mechanisms and evolution. *Bioessays* 14:253–261.

Kelly, D.B., and Brenowitz, E. (1992): Hormonal influences on courtship behaviors. *In* Becker, J.B., Breedlove, S.M., and Crews, D. (eds.) *Behavioral Endocrinology*, pp. 187–218. MIT Press, Cambridge, MA.

Keyes, L.N., Cline, T.W., and Schedl, P. (1992): The primary sex determination signal of *Drosophila* acts at the level of transcription. *Cell* 68: 933–943.

Kimura, D. (1992): Sex differences in the brain. *Sci. Am.* 267:119–125.

Koopman, P., Gubbay, J., Goodfellow, P., and Lovell-Badge, R. (1991): Male development of chromosomally female mice transgenic for *Sry*. *Nature* 351:117–121.

Lee, M.M., and Donahoe, P.K. (1993): Mullerian inibiting substance: A gonadal hormone with multiple functions. *Endocrine Rev.* 14:152–156.

LeVay, S. (1991): A difference in hypothalamic structure between heterosexual and homosexual men. *Science* 253:1034–1037.

MacKeown, M. (1994): Sex determination and differentiation. *Dev. Genet.* 15:201–204.

MacLaren, A. (1988): Sex determination in mammals. *Trends Genet.* 4: 153–156.

Marler, P., et al. (1988): The role of sex steroids in the acquisition and production of birdsong. *Nature* 336:770–772.

Oliver, B., Kim, Y.J., and Baker, B.S. (1993): Sex-lethal master and slave: A hierarchy of germ-line sex determination in *Drosophila*. *Development* 119: 103–114.

Packhurst, S.M., and Meneely, P.M. (1994): Sex determination and dosage compensation: Lessons from flies and worms. *Science* 264: 924–932.

Steinmann-Zwicky, M. (1994): *Sxl* in the germline of *Drosophila:* A target for somatic late induction. *Dev. Genet.* 15:275–276.

Zarkower, D., and Hodgkin, J. (1992): Molecular analysis of the *C. elegans* sex-determining gene *tra-1:* A gene encoding two zinc finger proteins. *Cell* 70:237–249.

Zhou, J.-N., et al. (1995): A sex difference in the human brain and its relation to transsexuality. *Nature* 378:68–69.

Chapter 21. Regeneration, Transdifferentiation

Baguñà, J., Saló, E., and Auladell, C. (1989): Regeneration and pattern formation in planarians. III. Evidence that neoblasts are totipotent stem cells and the source of blastema cells. *Development* 107:77–86.

Chandebois, R. (1984): Intercalary regeneration and level interactions in the fresh-water planarian *Dugesia lugubris*. I. The anteroposterior system. *Roux's Arch. Dev. Biol.* 193:149–157.

Goss, R.J. (1969): *Principles of Regeneration*. Academic Press, New York.

Hadorn, E. (1968): Transdetermination in cells. *Sci. Am.* 219(5):110–120.

Maden, M. (1984): Retinoids as probes for investigating the molecular basis of pattern formation. *In* G.M. Malacinsky (ed.) *Pattern Formation*. pp. 539–579. Macmillan, New York.

Morgan, T.H. (1901): *Regeneration*. Macmillan, New York.

Saló, E., and Baguñà, J. (1989): Regeneration and pattern formation in planarians. II. Local origin and role of cell movements in blastema formation. *Development* 107:69–76.

Schmid, V., Alder, H., Plickert, G., and Weber, C. (1988): Transdifferentiation from striated muscle of medusae in vitro. *In* Eguche, G., et al. (eds.) *Regulatory Mechanisms in Developmental Processes*. pp. 137–146. Elsevier, Ireland.

Steward, F.C. (1970): From cultured cells to whole plants: The induction and control of their growth and morphogenesis. *Proc. Roy. Soc. London B* 175: 1–30.

Stocum, D.L. (1991): Limb regeneration: A call to arms (and legs). *Cell* 76:5–8

Chapter 22. Aging and Death

Blackburn, E. (1991): Structure and function of telomeres. *Nature* 350:569–573.

Harley, C.B., et al. (1992): The telomere hypothesis of cellular aging. *Exp. Gerontol.* 27:375–382.

Hayflick, L. (1980): The cell biology of human aging. *Sci. Am.* 242(1): 58–65.

Johnson, F., and Bottjer, S.W. (1994): Afferent influences on cell death and birth during development of a cortical nucleus necessary for learned vocal behavior in zebra finches. *Development* 120:13–24.

Oppenheim, R.W., Prevette, D., Tytell, M., and Homma, S. (1990): Naturally occurring and induced neuronal cell death in the chick embryo in vivo requires protein and RNA synthesis: Evidence for the role of cell death genes. *Dev. Biol.* 138:104–113.

Raff, M.C., et al. (1993): Programmed cell death and the control of cell survival: Lessons from the nervous system: *Science* 262: 695–700.

Svendsen, C.N., and Rosser, A.E. (1995): Neurones from stem cells? *Trends Neurosci.* 18:465–467.

Wallace, D.C. (1992): Mitochondrial genetics: A paradigm for aging and degenerative diseases. *Science* 256:628–632.

Wickelgren, I. (1996): Is hippocampal cell death a myth? *Science* 271: 1229–1230.

Williams, G.T., and Smith, C.A. (1993): Molecular regulation of apoptosis. Genetic controls on cell death. *Cell* 74:777–780.

Zakian, V.A., et al. (1990): How does the end begin. *Trends Genet.* 6: 12–16.

Zwilling, R., and Balduini, C. (1992): *Biology of Aging.* Springer-Verlag, Heidelberg.

Box 1. History of the Developmental Biology

Aristotle. *De Anima. In* Hicks, R.D. (1965) *Aristotle's De Anima*, Adolf M. Hakkert Publ., Amsterdam.

Aristotele. *Historia Animalium. In* Smith, J.A., and Ross, W.D. (eds.) (1949): The works of Aristotle translated into English, Vol. IV, Historia animalium, translated by Thomson, D.W.; at the Clarendon Press, Oxford.

Bonner, J.T. (1962): *Ideas in Biology.* Harper & Row, New York.

Boveri, T. (1904): Ergebnisse über die Konstitution der chromatischen Substanz. Gustav Fischer, Jena, Germany.

Boveri, T. (1910): Die Potenzen der Ascaris-Blastomeren bei abgeänderter Furchung. Festschrift für Richard Hertwig, vol3. Gustav Fischer, Jena, Germany.

Dampier, W.C. (1948): *A History of Science and Its Relations with Philosophy and Religion.* 4th ed., The Claredon Press, Cambridge.

Driesch, H. (1892): The potency of the first two cleavage cells in echinoderm development. Experimental production of partial and double formations. *In* Willer, B.H., and Oppenheimer, J.M. (eds.) *Foundations of Experimental Embryology*, pp. 38–50. Hafner, New York.

Driesch, H. (1908): *The Science and Philosophy of the Organism. I. Gilford Lectures 1907; II. Gilford Lectures 1908.* A & C. Black, London.

Gardner, E.J. (1965): *History of Biology.* Burgess, Minneapolis.

Gould, S.J. (1977): *Ontogeny and Phylogeny.* Belkamp, Harvard Univ. Press, Cambridge, MA.

Haeckel, E. (1892): The history of creation. Translation of *Natürliche Schöpfungsgeschichte*. Kegan Paul, Trench, Trubner; London.

Hamburger, V. (1988): *The Heritage of Experimental Embryology: Hans Spemann and the Organizer*. Oxford Univ. Press, New York.

Harvey, W. (1651): *De generatione animalium*. English translation by R. Willis. Encyclopedia Brittanica, Inc., Great Books of the Western World, Chicago; reprinted 1952.

Müller, W.A. (1996): From the Aristotelian soul to genetic and epigenetic information. *Int. J. Dev. Biol.* 40:21–26.

Sander, K. (1991a): Landmarks in developmental biology: Wilhelm Roux and his programme for developmental biology. *Roux's Arch. Dev. Biol.* 200: 1–3.

Sander, K. (1991b): Wilhelm Roux's treatise on "qualitative" mitoses—a "classic" by either definition. *Roux's Arch. Dev. Biol.* 200:61–63.

Sander, K. (1991c): Wilhelm Roux on embryonic axes, sperm entry and the grey crescent. *Roux's Arch. Dev. Biol.* 200:117–119.

Sander, K. (1991d): When seeing is believing: Wilhelm Roux's misconceived fate map. *Roux's Arch. Dev. Biol.* 200:177–179.

Sander, K. (1991e): "Mosaic work" and "assimilating effects" in embryogenesis: Wilhelm Roux's conclusions after disabling frog blastomeres. *Roux's Arch. Dev. Biol.* 200:237–239.

Sander, K. (1991f): Wilhelm Roux and the rest: Developmental theories 1885–1895. *Roux's Arch. Dev. Biol.* 200:331–333.

Sander, K. (1992a): Shaking a concept: Hans Driesch and the varied fates of sea urchin blastomeres. *Roux's Arch. Dev. Biol.* 201:265–267.

Sander, K. (1992b): Hans Driesch's "philosophy really ab ovo" or why to be a vitalist. *Roux's Arch. Dev. Biol.* 202:1–3.

Sander, K. (1993): Entelechy and the ontogenetic machine-work and views of Hans Driesch from 1895 to 1910. *Roux's Arch. Dev. Biol.* 202:70–76.

Spemann, H. (1938): *Embryonic Development and Induction*. Yale Univ. Press, New Haven, CT. (Reprinted by Hafner, New York, 1962).

von Baer, K.E. (1828): *Über Entwickelungsgeschichte der Thiere*. Königsberg, Germany.

Box 2. Experiments with Eggs and Early Embryos, Cloning, Chimeras, and Transgenic Animals

Briggs, R., and King, T.J. (1952): Transplantation of living nuclei from blastula cells into enucleated frog eggs. *Proc. Natl. Acad. Sci. USA* 38:455–463.

Brun, R.B. (1978): Developmental capacities of *Xenopus* eggs, provided with erythrocyte or erythroblast nuclei from adults. *Dev. Biol.* 65:271–284.

Campbell, K.H.S., McWhir, J., Ritchie, W.A., and Wilmut, I. (1996): Sheep cloned by nuclear transfer from a cultured cell line. *Nature* 380:64–66.

Cappecchi, M.R. (1980): High efficiency transformation by direct microinjection of DNA into cultured mammalian cells. *Cell* 22:479–488.

DiBernadino, M.A. (1988): Genomic multipotentiality of differentiated somatic cells. *In* G. Eguchi, et al. (eds.) *Regulatory Mechanisms in Developmental Processes*, pp. 129–136. Elsevier, Amsterdam, New York.

First, N.L., and Prather, R.S. (1991): Genomic potentials in mammals. *Differentiation* 48:1–8.

Gordon, J.W. (1988): Transgenic mice. *In* Malacinsky G.M. (ed.) *Developmental Genetics of Higher Organisms: A Primer in Developmental Biology*, pp. 477–498. Macmillan, New York.

Gossler, A., et al. (1986): Transgenesis by means of blastocyst-derived stem cell lines. *Proc. Natl. Acad. Sci. USA* 83:9065–9069.

Gurdon, J.B. (1962): The developmental capacity of nuclei taken from intestinal epithelium cells of feeding tadpoles. *J. Embryol. Exp. Morphol.* 10: 622–640.

Gurdon, J.B. (1968): Transplanted nuclei and cell differentiation. *Sci. Am.* 219(6):24–35.

Hanahan, D. (1989): Transgenic mice as probes into complex systems. *Science* 245:1265–1274.

McGrath, J., and Solter, D. (1984): Inability of mouse blastomere nuclei transferred to enucleated zygotes to support development in vitro. *Science* 226:1317–1319.

Mintz, B. (1957): Does embryological development of primordial germ cells affect its development? *Symp. Br. Soc. Dev. Biol.* 7:225–227.

Solter, D. (1996): Lambing by nuclear transfer. *Nature* 380:24–25.

Thompson, S., et al. (1989): Germ line transmission and expression of a corrected HPRT gene produced by gene targeting in embryonic stem cells. *Cell* 56:313–321.

Wagner, E.F. (1990): Mouse genetics meet molecular biology at Cold Spring Harbor. *New Biologist* 2:1971–1074.

Wagner, E.F., and Keller, G. (1992): The introduction of genes into mouse embryos and stem cells. *In* Russo, V.E.A., et al. (eds.) *Development: The Molecular Genetic Approach*, pp. 440–458. Springer-Verlag, Heidelberg.

Wakimoto, T., and Karpen, G.H. (1988): Transposable elements and germ-line transformation in *Drosophila*. *In* Malacinsky, G.M. (ed.) *Developmental Genetics of Higher Organisms: A Primer in Developmental Biology*, pp. 275–303. Macmillan, New York.

Box 3. The PI Signal Transduction System

Ciapa, B., et al. (1992): Phosphoinositide metabolism during the fertilization wave in sea urchin eggs. *Development* 115:187–195.

Gallicano, G.I., et al. (1993): Protein kinase C, a pivotal regulator of hamster egg activation, functions after elevation of intracellular free calcium. *Dev. Biol.* 156:94–106.

Larabell, C., and Nuccitelli, R. (1992): Inositol lipid hydrolysis contributes to the Ca^{2+} wave in the activating egg of *Xenopus laevis*. *Dev. Biol.* 153:347–355.

Otte, A.P., et al. (1988): Protein kinase C mediates neural induction in *Xenopus laevis*. *Nature* 334:618–620.

Whitaker, M., and Swann, K. (1993): Lighting the fuse at fertilization. *Development* 117:1–12.

Box 4. Models of Biological Pattern Formation

Edelstein-Keshet, L., and Ermentrout, B.G. (1990): Contact response of cells can mediate morphogenetic pattern formation. *Differentiation* 45:147–159.

Haken, H. (1978): *Synergetics*. Springer-Verlag, Heidelberg.

Meinhardt, H. (1982): *Models of Biological Pattern Formation*. Academic Press, New York.

Meinhardt, H. (1995): *The Algorithmic Beauty of Seashells*. Springer-Verlag, Heidelberg.

Murray, J.D. (1989): *Mathematical Biology*. Springer-Verlag, Heidelberg.

Prigogine, I., and Nicolis, G. (1967): On symmetry-breaking instabilities in dissipative systems. *J. Chem. Phys.* 46:3542–3550.

Turing, A.M. (1952): The chemical basis of morphogenesis. *Philos. Trans. Roy. Soc. London B* 237:37–72.

Steinberg, M.S., and Takeichi, M. (1994): Experimental specification of cell sorting, tissue spreading, and specific spatial patterning by quantitative differences in cadherin expression. *Proc. Natl. Acad. Sci. USA* 91:206–209.

Wolpert, L. (1969): Positional information and the spatial pattern of cellular formation. *J. Theor. Biol.* 25:1–47.

Wolpert, L. (1978): Pattern formation in biological development. *Sci. Am.* 239(4):154–164.

Wolpert, L. (1989): Positional information revisited. *Development* 1989 (Suppl.):3–12.

Box 5: Signal Molecules Acting through Nuclear Receptors

(See also Chapter 17)

Beato, M., Herrlich, P., and Schütz, G. (1995): Steroid hormone receptors: Many actors in search of a plot. *Cell* 83:851–857.

Chen, Y.P., Huang, L., and Solursh, M. (1994): A concentration gradient of retinoids in the early *Xenopus* embryo. *Dev. Biol.* 161:70–76.

Conlon, R.A. (1995): Retinoic acid and pattern formation in vertebrates. *Trends Genet.* 11(8):314–319.

Eichele, G. (1989): Retinoic acid induces a pattern of digits in anterior half wing buds that lack the zone of polarizing activity. *Development* 107:863–867.

Kastner, P., Mark., M., and Chambon, P. (1995): Nonsteroid nuclear receptors: What are genetic studies telling us about their role in real life? *Cell* 83:859–869.

Maden, M., et al. (1989): Cellular retinoic acid-binding protein and the role of retinoic acid in the development of the chick embryo. *Dev. Biol.* 135:124–132.

Mangelsdorf, D.J., et al. (1995): The nuclear receptor superfamily: The second decade. *Cell* 83:835–839.

Mangelsdorf, D.J., and Evans, R.M. (1995): The RXR heterodimers and orphan receptors. *Cell* 83:841–850.

Umesono, K., and Evans, R.M. (1989): Determinants of target gene specificity for steroid/thyroid hormone receptors. *Cell* 57:1139–1146.

Box 7. Some Cellular and Molecular Methods of Recent Developmental Biology (See also Bibliography for Box 2)

Barinaga, M. (1994): Knockout mice: round two. *Science* 265:26–28.

Chisaka, O., and Capecchi, M.R. (1991): Regionally restricted developmental defects resulting from targeted disruption of the mouse homeobox gene *Hox-1.5*. *Nature* 350:473–479.

Cubitt, A.B., et al. (1995): Understanding, improving and using green fluorescent protein. *Trends Biochem. Sci.* 20:448–455.

Friedrich, G., and Soriano, P. (1991): Promoter traps in amphibian stem cells: A genetic screen to identify and mutate developmental genes in mice. *Genes Dev.* 5:1513–1523.

Gossen, M., Bonin, A.L., and Bujard, H. (1993): Control of gene activity in higher eukaryotic cells by prokaryotic regulatory elements. *Trends Biochem. Sci.* 18:471–475.

Prasher, D.C. (1995): Using GFP to see the light. *Trends Genet.* 11: 320–323.

Wagner, E.F. (1990): Mouse genetics meet molecular biology at Cold Spring Harbor. *New Biologist* 2:1971–1074.

Wakimoto, T., and Karpen, G.H. (1988): Transposable elements and germ-line transformation in *Drosophila. In* Malacinsky, G.M. (ed.) *Developmental Genetics of Higher Organisms: A Primer in Developmental Biology*, pp. 275–303. Macmillan, New York.

Index